P9-CBT-476

Career Opportunities in Television and Video

Career Opportunities in Television and Video

By Maxine K. and Robert M. Reed

Facts On File® Publications
460 Park Avenue South
New York, N.Y. 10016

CAREER OPPORTUNITIES IN TELEVISION & VIDEO

Published by Facts On File, Inc., 460 Park Avenue South, New York, New York 10016

This book was produced for Facts On File by Gramercy Press, Inc. of New York City. It was developed and edited by Lawrence and Shirley Sarris.

Library of Congress Cataloging in Publication Data

Maxine K. and Robert M. Reed
Career opportunities in television and video.

1. Television broadcasting—Vocational guidance.
2. Television industry—Vocational guidance.
3. Community antenna television—Vocational guidance.
4. Video tapes—Vocational guidance. I. Title.
HE8700.4.E35 791.45′023′73 81-12600
ISBN 0-87196-613-1 AACR2
ISBN 0-87196-616-6 (pbk.)

Printed and bound in the United States of America
9 8 7 6 5 4 3 2 1

To Bob, Rick and Deri, whom we have counseled and from whom we have learned—and who make it all worthwhile.

CONTENTS

PART II—VIDEO AND TELECOMMUNICATIONS

PREFACE

Premise and Purpose

The 95 jobs described in this book come out of one of the most exciting developments in our modern world—television and video. Some observers call today's environment the "Communications Age," or the "Electronic Era." To others, the 1980s are a "Decade of Video."

Under whatever label, the incredible number of electronic devices that affect and influence our lives represent a new entertainment and information society, which is vital and dynamic. In less than 40 years, broadcast television has become a ubiquitous presence in almost every home in America. To add to the proliferation of TV, along come cable and pay television, videocassettes, and videodiscs. Satellites beam programming to an increasing number of outlets, and the family television set has become a display mechanism for video games, teletext information, and home computers. Education, health, corporate, and government agencies also use the new technologies with increasing frequency.

Underlying all of these new entertainment and information services are people. Television, with all of its show business allure, is often seen as a high-paying, glamorous field. The talented Performers who appear in our living rooms every day become a part of our lives. But for every individual we see on the air, there are hundreds of workers behind the scene who made that appearance possible.

The exploding television and video industries need such back stage individuals as trained engineers, talented programming and production professionals, skilled crafts people, diligent clerical workers, and aggressive managers. News and advertising specialists are in demand, as are persuasive sales persons. Health, private industry, education, and government media centers need individuals who can harness all of the new technologies to satisfy specific training and information objectives.

This book describes the exciting employment opportunities in these dynamic and changing fields. No radio positions are listed, and jobs in public relations, film, journalism, and theater are included only as they relate to television broadcasting and video. The job descriptions are at once more complete and more concise than those that are often found in similar career opportunity sources. An attempt has been made to present hard facts in the context of this expanding job market.

The purpose of the book is to serve as an introductory guide to some of the most common occupations in these new fields. High school or college students who are interested in a career in these areas should be able to learn who does what at a television or video operation. Those who are currently employed in these professions should also find the information useful as they plan their career paths.

Methods

The information for this book was gathered from printed job descriptions, employment documents, research studies, salary surveys, and tables of organization from more than 100 sources in the communications field. All of these data were carefully analyzed and re-analyzed by the authors. Information from that material and from discussions with more than 60 professional colleagues were used in developing this guide.

The career descriptions are not theoretical; they represent current practice and reflect the actual structure of jobs in the industries covered.

General Organization

The book is divided into two main parts, the first dealing with broadcasting and the second with video and telecommunications.

Part I, Television Broadcasting, begins with an Introduction that gives a brief historic perspective of the television industry, and provides a general overview of commercial and public TV. The Introduction also contains relevant statistical information outlining the growth patterns of the industry, its employment opportunities, and trends for the 80s. Typical organizational structures of a network, a commercial station, and a public TV station are explained. In addition, there is discussion of the use of television in advertising with a brief examination of how an ad agency develops commercials. Following the Introduction to Part I, there are 66 job descriptions organized in 8 sections, each representing a function within television or an advertising agency. The sections are: Management and Administration; Programming; Production; News; Engineering; Advertising Sales; Advertising Agency; and Technical/Craft.

Part II, Video and Telecommunications, opens with an Introduction that provides a general survey of each of the industries or telecommunications environments discussed. It contains relevant statistical information outlining the growth patterns of the various areas and discusses employment opportunities and trends for the 80s. Following the Introduction, there are 29 job descriptions, organized in 6 sections, each of which represents the major video or telecommunications area discussed in the Introduction: Consumer Electronics and Home Video; Cable Television, Subscription Television, and Multipoint Distribution Service; Education; Private Industry; Government; and Health.

The careers discussed in this book have been included because they are most frequently found within a TV- or video-related industry or department. Titles, however, are not universally consistent and their definitions often overlap from setting to setting. Those included here are most commonly used in today's professional environment.

The positions in Part I describe functions in broadcasting but many are also found in other television-related operations. For example, a Producer or a Camera Operator could work in programs produced for cable television or health care communications as well as for commercial or public broadcasting. The entry notes when there are opportunities for a position in other areas.

Whenever a title that is discussed in the book is mentioned, it appears with an initial capital letter. This will enable the reader to easily identify and refer to a related career. If a position is referred to but is not discussed in a separate description, it is printed in lower case only.

Following Part II's job descriptions, there are three useful appendixes that provide additional sources of career information. Appendix I, Degree and Non-Degree Programs, lists Colleges and Universities; Workshops and Seminars; Internships, Apprenticeships, and Training Programs; and Fellowships, Scholarships, and Grants. Appendix II, Unions and Associations, includes all groups mentioned in the book as well as others that might be important to particular careers. Appendix III, Bibliography, lists essential books and periodicals in categories that parallel the industry and department sections used in the book. In addition, an Index helps the reader to locate all positions discussed and cross references alternate job titles.

Job Descriptions

These represent the heart of the book and warrant a detailed statement of the elements and criteria used. Each description has three major elements: Career Profile, Career Ladder, and a narrative text.

1. Career Profile. This is a quick reference to help the reader judge whether the position is of interest and warrants further reading. It presents a concise summary of the major facts contained in the full text and includes:

Duties: A brief synopsis of the major responsibility(ies) of the job;

Alternate Title(s): Commonly used additional titles for the position;

Salary Range: High and low extremes that are discussed in the Salaries section of the text;

Employment Prospects and Advancement Prospects: Rates the overall opportunities for moving into and out of the job. The ratings—Excellent, Good, Fair, or Poor—indicate the chances for employment or advancement in terms of competition for the job, supply and demand, relative turnover in the position, and the number of jobs available. They may be generally interpreted as follows—

Excellent: demand greater than supply and/or high turnover;

Good: demand somewhat greater than supply; some competition and less turnover;

Fair: demand and supply balance out; quite competitive; little turnover;

Poor: supply exceeds demand; very specialized skills needed; highly competitive; little turnover.

The ratings represent the subjective evaluation of the authors, based on the statistical evidence available, discussions with professionals in the occupation, and the viewpoints of industry observers.

Prerequisites: Highlights three sets of criteria for obtaining the job, including—

Education: minimum schooling required and preferred;

Experience: the type of work and length of service usually needed; and

Special Skills: the major aptitudes and attributes that are useful in obtaining the job.

2. Career Ladder. Each Career Ladder shows the job(s) to which the position often leads, and the job(s) or schooling that usually lead to the position. Career employment and advancement, however, depend entirely upon the individual and his or her skills, initiative, ambition, and luck. The jobs listed in the Career Ladder indicate the common or most direct routes to employment or advancement, but they are certainly not the only ones. The variables, and the opportunities for career branching are many. Some of the positions in the Career Ladder are not fully discussed in the book because they are not specifically concerned with television or video, or they occur too infrequently to deserve full descriptions. The job(s) in the lower "rung" of the Ladder are usually referred to in the Employment Prospects or Experience/Skills portions of the text. The middle rung is always the position under discussion in the job description. The upper rung position(s) is usually mentioned in the paragraph on Advancement Prospects.

3. Narrative Text. Each job description contains the following eight categories.

Position Description: This is the core of each profile and describes the nature of the work. It presents an overview of what the job entails, including the environment, the reporting lines, whether it is supervisory in nature, and the main responsibilities of the position.

Salaries: Whenever available, ranges, average annual earnings in particular settings, and starting salaries are included. High and low extremes for the position are provided to show the broad range of salary possibilities and do not necessarily represent industry standards. It is noted if payment is for hourly, project-by-project, weekly, or per diem work. When salaries are dependent on specific education and seniority, that is also mentioned. Fringe benefits, commissions, and perquisites are included when the information is helpful. Most salary information is based on 1980 surveys, and is usually the most recent data available.

Employment Prospects: The rating in the Career Profile is elaborated upon here. The opportunities for employment are discussed in terms of the number of positions available nationally, the estimated amount of turnover, and the demand for special skills or talents. This section usually states whether individuals are promoted to the position from within or recruited from school, college, or other industry positions. As a rule, the routes to and probabilities of getting the job are discussed. Often, immediate prior experience is mentioned here (or in the Experience/Skills section), and the competition is noted.

Advancement Prospects: The Career Profile rating is discussed in this section. The possibilities of advancement from the position are explained in terms of the turnover at the next level, the special skills or experience required, and other career tracks that an individual might pursue. When career advancement is tied to an individual's geographic mobility, the text contains that information.

Education: The minimum educational requirements and any specific educational background or training preferred by employers are noted. Required courses, certifications, or training programs are listed, and recommended courses or degrees are indicated. To complement the Education section, Appendix I contains a list of institutions and organizations offering relevant degree and non-degree programs, training courses, seminars, workshops, internships, and apprenticeships.

Experience/Skills: This section indicates the nature and the amount of experience and vocational skills usually needed to obtain the job. There is also an assessment of the specific talents or intangible abilities that are generally expected by employers.

Minority/Women's Opportunities: The percentage of women and members of minority groups holding the position has been included when the information is available from national, government, or industry sources. Other information and estimates supplied by industry observers is also provided. In addition to current statistics, trends are supplied whenever possible.

Unions/Associations: The required craft, technical, or performing union membership is listed. Since union requirements vary considerably from job to job and among geographic locations, individuals interested in particular jobs are urged to contact local chapters or national headquarters for information. Many people advance their careers and enhance their professional growth through membership in associations and the groups most commonly associated with the job are included. The full address and phone number for each union and organization mentioned are listed in Appendix II.

General Comments

We have sorted through mounds of data and talked with many people in the field to develop the latest and best material and to provide a realistic picture of job responsibilities, opportunities, and environments. It is assumed that the typical reader will not be interested in every position described, but will scan the book selectively for specific information. For this reason, there is a deliberate repetition of terms, names of organizations and agencies, and acronyms in the job descriptions.

Conclusion

This book is the result of a lifetime of counseling and advising young people, formally and informally, about career opportunities in the communications fields, particularly in television, video, and telecommunications. Specifically, it is the result of rising to the challenge of a brilliant (but hard-nosed) editor, Larry Eidelberg, and three children who dared us to put down on paper all those dinner table conversations.

Exciting opportunities for employment are available in the rapidly changing and vigorous fields of information and communications. Our world is changing and television, video, and telecommunications are vital parts of that change. Nearly 15 years ago, in the movie, *The Graduate*, a family friend whispered "Plastics" to Dustin Hoffman. Today, he might more accurately shout "Television/Video!"

MKR and RMR
East Northport, NY
January 1982

ACKNOWLEDGEMENTS

The authors gratefully acknowledge the following individuals who contributed research materials, job descriptions, tables of organization—and advice and counsel—in the preparation of this book. Without their contributions in time, effort, and patience, the project could not have been completed.

Robert L. Allen, Director, Audio-Visual Services, Pennsylvania State U; Robert K. Avery, Associate Professor, U of Utah; Joe Bellon, Vice President, CBS News; Reba Benschoter, Director of Biomedical Communications, U of Nebraska; Pamela Blumenthal, Chief, Equal Employment Opportunity Unit, Federal Communications Commission; Douglas F. Bodwell, Director, Education Services, Corporation for Public Broadcasting; Frederick Breitenfeld, Jr., Executive Director, Maryland Center for Public Broadcasting; Bruce Brennan, Media Manager, American Hospital Association; Henry Brief, Executive Director, International Tape/Disc Association; Patricia J. Bryan, Director, Western Region, American Advertising Federation; Henry D. Caruthers, President, Caruthers Consulting, Inc.; Renee Cherow-O'Leary, Professor of Communications, City College of New York; Nazaret Cherkezian, Director, Office of Telecommunications, Smithsonian Institution; Fred L. Christen, Director, Biomedical Communications, U of Texas; S. Michael Collins, President, WNED-TV; Joseph Day, Director of Educational Media, Georgetown U; James A. Fellows, President, National Association of Educational Broadcasters; Avra Fliegelman, Executive Vice President, Broadcast Information Bureau; David Garloff, Health Sciences Learning Resources, U of Minnesota; Shel Goldstein, Director of Media Resources, U of Minnesota; Barton L. Griffith, Professor, U of Missouri; George Hall, Director, Telecommunications, Commonwealth of Virginia; Robert N. Hall, Staff Coordinator, Speech Communication Association; Stan Hankin, Director, Media Resource Center, U.S. Department of Labor; George Heinemann, Vice President, NBC; Michael Hobbs, Esq., Senior Vice President, PBS; Ron W. Irion, Vice President, Station Services, National Association of Broadcasters; Ralph W. Jones, Director of Communications, Electronic Industries Association; Erling S. Jorgensen, Assistant Dean, Michigan State U; Barbara F. Kurka, Director, Programs and Services, International Radio and Television Society; Joan Lemnah, Consumer Assistance Office, Federal Communications Commission; Alan Lewis, Director, Public Television Archive; Jean McCauley, Chief, Information Branch, National AudioVisual Center; John H. McLean, Director, National AudioVisual Center; Harry R. McGee, Executive Vice President, National Audio-Visual Association; James S. Neal, Survey Branch, Equal Employment Opportunity Commission; Richard Nibeck, Deputy Executive Director, Association for Educational Communications and Technology; Dovie W. Nichols, Director, Corporate Relations, NBC; Frank Norwood, Executive Director, Joint Council on Educational Telecommunications; Carolyn Palmer, U.S. Office of Personnel Management; James Poteat, Manager, Research Services, Television Information Office; Donald R. Quayle, Vice President, WETA-TV; Rick Reed, Associate Producer, Major League Baseball Productions; Robert G. Reed, MDS Manager, Digital Paging Systems, Inc.; E. L. Richardson, Executive Director, Audio-Visual Center, Indiana U; Tom Richter, President, Richter Associates; Neil Rosenthal, Chief, Occupational Outlook, U.S. Department of Labor; Barbara Schiltges, Information Services Specialist, Corporation for Public Broadcasting; Ernie Schultz, Executive Vice President, Radio-Television News Directors Association; Bill Scott, Program Manager, WETV; Harvey Seslowsky, President, National Video Clearinghouse; Mary Anna Seversen, Human Resources Development, Corporation for Public Broadcasting; Jerry Storch, Bureau of Labor Statistics, U.S. Department of Labor; Alfred L. Sweeney, Director, Office of Public Affairs, Equal Employment Opportunity Commission; Bob Temple, Senior Vice President, KUTV; Charles S. Tepfer, Publisher/Editor, C. S. Tepfer Publishing Co.; James Thompson, Vice President, Member Services, National Audio-Visual Association; I. Marlene Thorn, Director, Training and Development, Corporation for Public Broadcasting; Charles W. Vaughan, President, WCET-TV; Ken Warren, Coordinator for TV and Media, U of Wisconsin; Elmon Webb, Secretary, United Scenic Artists; Ken Winslow, President, Winslow Associates; Charles M. Woodliff, Director, Division of Instructional Communications, Western Michigan U; S. Anders Yokum, Jr., Vice President, WTTW

In addition to the individuals cited above, the following associations and organizations graciously provided research materials, salary information, employment surveys, and study documents for the preparation of this book. Their contributions are gratefully acknowledged.

American Association of Advertising Agencies; American Federation of Musicians; American Federation of State, County and Municipal Employees; American Federation of Television and Radio Artists; American Guild of Authors and Composers; American Women in Radio and Television; Directors Guild of America; Institute of Electrical and Electronics Engineers; International Alliance of Theatrical Stage Employees; International Television Association; Music Teachers National Association; National Association of Broadcast Employees and Technicians; National Association of Television and Electronic Servicers of America; National Cable Television Association; National Electronic Service Dealers Association; Paul Kagan Associates; Screen Actors Guild; Society of Broadcast Engineers; Society of Cable Television Engineers; Society of Motion Picture and Television Engineers; Women in Communications; Writers Guild of America, West

PART I
TELEVISION BROADCASTING
Introduction

Experiments in sending pictures through the air were conducted as early as 1884, but the first real television broadcast occurred in 1927. The first major telecast is generally considered to be the transmission of President Franklin D. Roosevelt opening the 1939 World's Fair in New York City.

The number of over-the-air electronic transmissions that can be made without interfering with each other are effectively limited by the "physics of scarcity." The federal government, therefore, in accordance with the Communications Act of 1934, established the Federal Communications Commission (FCC)* to regulate radio and television stations. The FCC licenses stations to operate "in the public interest, convenience, and necessity," and assigns them to specific frequencies and power. The FCC also limits stations to particular communities, assigns their call letters, allocates channels, processes requests to transfer control of a station to another party, and monitors all stations to ensure that they are in compliance with its rules and technical regulations. At license renewal time, the FCC reviews a station's record to verify that it is operating in the public interest.

By the end of World War II, the available television channels were becoming crowded and, in 1948, the FCC imposed a ban on all new license applications. Finally, in 1952, it adopted a plan that allocated about 2,000 TV channels to 1,300 communities, and the age of television was born. By 1955, there were 448 stations on the air. By 1981, 1,035 were licensed and operating throughout the United States (see Chart I-1).

Television stations broadcast on assigned channels 2 through 13 in the VHF (very high frequency) band and on channels 14 through 83 in the UHF (ultra high frequency) band. The majority operate in the VHF band.

Public (educational) non-commercial TV stations account for more than one-third of the licensees, and are operated by universities, schools, state agencies, and community groups. These stations were originally designed and licensed to serve the educational needs of local communities. The first such station began broadcasting in 1953. With passage of the Public Broadcasting Act by Congress in 1967, the Corporation for Public Broadcasting (CPB)* was formed as an administrative funding agency, and the Public Broadcasting Service (PBS) was started as the programming agency. Today, most public television (PTV) stations transmit instructional television (ITV) programs during the day, and offer cultural programs, documentaries, art, and drama series during the evening as an alternative to

*The Federal Communications Commission is an independent federal agency headed by seven commissioners who are appointed by the President with the advice and consent of the Senate. No more than four commissioners can belong to the same political party.

*The CPB is a non-profit corporation responsible for administering federal funds for public broadcasting stations and for research and development, interconnection, and training programs.

Chart I-1 — Television Stations on the Air as of July 30, 1981

	Commercial TV	Public TV	Total
VHF	522	108	630
UHF	243	162	405
Total	765	270	1,035

Source: FCC

Chart I-2 — Typical Commercial Television Network Table of Organization

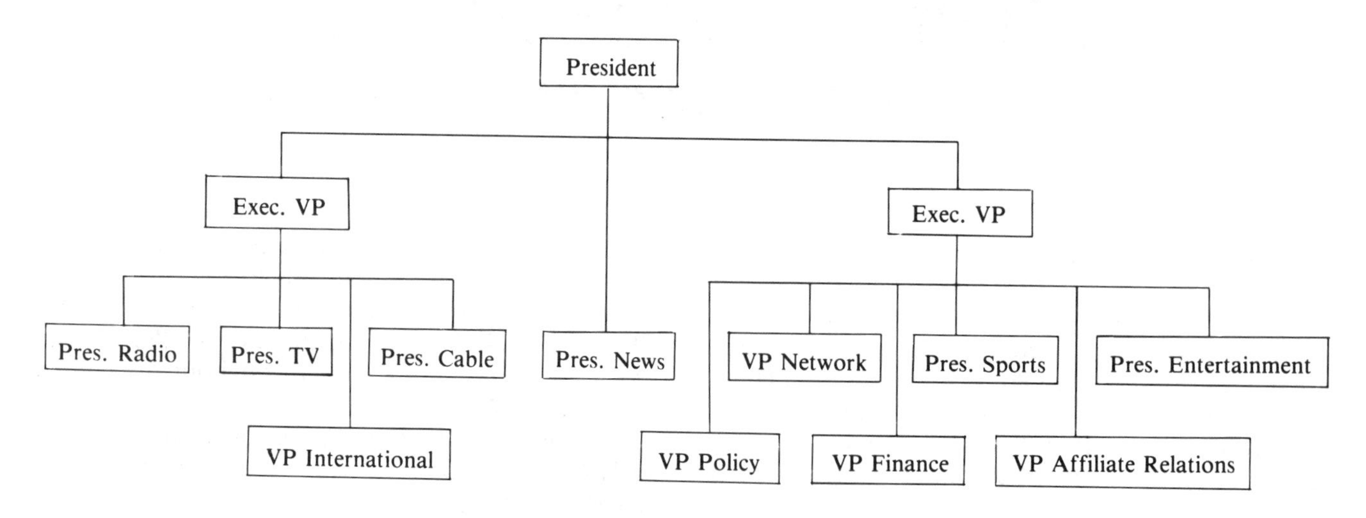

commercial TV fare. All public stations are members of PBS, which serves as a national source of programming to the interconnected PTV stations.

Commercial television stations and networks make their money primarily by selling portions of their air time to advertisers. Public TV stations, on the other hand, sell no commercial time. Their incomes are derived from federal, state, and local governments, viewer donations, and grants from corporations and foundations. The gross revenues for all commercial TV operations in 1980 was $8.1 billion. Public television had an income of $581.4 million in 1980, of which 26.1 percent was received from the federal government.

Local television programming is created by station staffs, by independent producers, and by production companies who conceive and develop programs that are sold to commercial and some public TV stations. Many of the larger independent production firms produce shows and sell the broadcast rights to national syndication companies, which, in turn, make the programs available to individual stations. Some national independent production companies produce programs for the networks, which then sell time on them to sponsors through advertising agencies.

In 1981, over 76 million homes in the United States—more than 98 percent of the total—had a television set; and nearly one half had more than one. The average household watched TV for over six hours a day, according to survey statistics compiled by the A.C. Nielsen Company.

Typical Organizational Structures

Commercial Networks

The three major commercial networks are ABC (the American Broadcasting Company), CBS (formerly called the Columbia Broadcasting System), and NBC (the National Broadcasting Company). All three maintain corporate headquarters in New York City and have studio facilities there and in Hollywood. The networks supply programming to their affiliated stations via satellite and coaxial cable or by microwave relay. According to FCC statistics, the three networks employed 16,590 individuals in 1980, or nearly 22 percent of all positions in commercial television.

The networks are large corporate entities with staffs the equal of any major company. A typical Table of Organization (TO) is complicated and undergoes constant alterations to reflect changes in business emphases as well as the company's involvement in both media-related and non-media-related industries. For a recent example, see Chart I-2. Reporting to the executives shown in the chart are additional vice presidents in charge of advertising sales, programs and talent, and production, as well as many program managers and executives in charge of comedy, daytime programs, drama, children's programs, variety shows, and specials. Their efforts are supported by staffs that often number 50 or more.

Commercial Television Stations

Each of the three networks also owns and operates a few local television stations. About 15 percent of all commercial stations are non-affiliated "independents." Most local stations, although affiliated with one of the three commercial networks, are independently owned. The station signs an affiliation agreement with a specific network, and agrees to carry programs in certain time periods. The network can then obtain advertising sponsors for commercials within those programs by promising that they will be seen by a large national audience comprised of their local affiliates' viewers.

Regardless of its affiliation status, each station operates as a separate business and is usually organized into five major departments: Advertising Sales, Engineering, Business, News, and Programming. Every station, however, has its own specific structure and practices. The TOs in this introduction reflect some of the more common organizational patterns in the industry.

Chart I-3 shows a typical TO for a commercial TV station. Reporting lines are clearly shown, although not all stations follow these lines of authority. We have attempted to show positions in terms of their relative importance by drawing lines of varying lengths. This is not precise and the reader should consult each job description to get a sense of the significance of each position, salary levels, and reporting lines.

Many small commercial stations employ 25 people or less who double and triple in various jobs. Major market stations may employ staffs of 150 or more. The average number of full-time employees in 1979 was 79, according to industry sources. The number of employees often depends on the amount of local production originated by the station or on its emphasis on news programming. In most stations, the engineering and news departments have the largest number of employees.

Public Television Stations

In general, public stations have fewer employees and a somewhat simpler organizational structure. Most PTV stations operate a separate department known as development (fund-raising). The heavy emphasis on local production at public TV stations has led many of them to consider the production area as a separate department, reporting directly to the General Manager.

Few public stations employ in-house Attorneys. Normally there is no news department. Most do not have Community Relations Directors either, inasmuch as the entire station staff is actively engaged in these duties since all of PTV programming is designed to serve community needs.

In commercial stations, the Traffic/Continuity Supervisor normally reports to the General Sales Manager, and schedules commercials as well as station breaks and programming. Since public TV stations do not usually deal with advertising sales, the Traffic/Continuity Supervisor handles only programming and station breaks and reports directly to the Program Manager. At most commercial stations, engineering and technical employees are represented by a union and, as a part of an engineering/technical contract, that union often represents Camera Operators. Engineering and technical staff members in public television are usually not represented by a union. In addition, the heavier emphasis on production at most PTV stations requires that Camera Operators report to, and be part of, the production department.

As a rule, the production and engineering departments at a public television station have the largest staffs. The average number of employees at a public station was 25 in 1980, but some community-owned outlets in major markets employed more than 100 individuals.

A public TV station is likely to have the following people reporting directly to the General Manager: Chief Engineer, Program Manager, Business Manager, Production Manager, and Director of Development. Chart I-4 depicts a typical Table of Organization for a public TV station. As with the TO for a commercial station, reporting lines are clear but the reader should look at the job descriptions to get a more accurate picture of the relative importance of each position, salaries, and lines of authority.

Advertising Agencies

Television networks and stations seldom produce the commercials seen on the air. Most are created by advertising agencies, under contract from a specific sponsor. The agencies research the product and its potential customers, write the advertising copy, and produce the commercial on tape or film, using their own or independent production facilities. On occasion, they rent production equipment and facilities from local stations.

The ad agency also recommends the frequency and periods when the commercial should be broadcast, contracts with a network or station for air time, and researches the effectiveness of the advertising campaign. For their services, ad agencies usually receive 15 percent of the amount spent by the sponsor in the advertising campaign.

Employment Statistics

As of June 1981, 203,200 people were working in

Chart I-3 — Typical Commercial Television Station Table of Organization

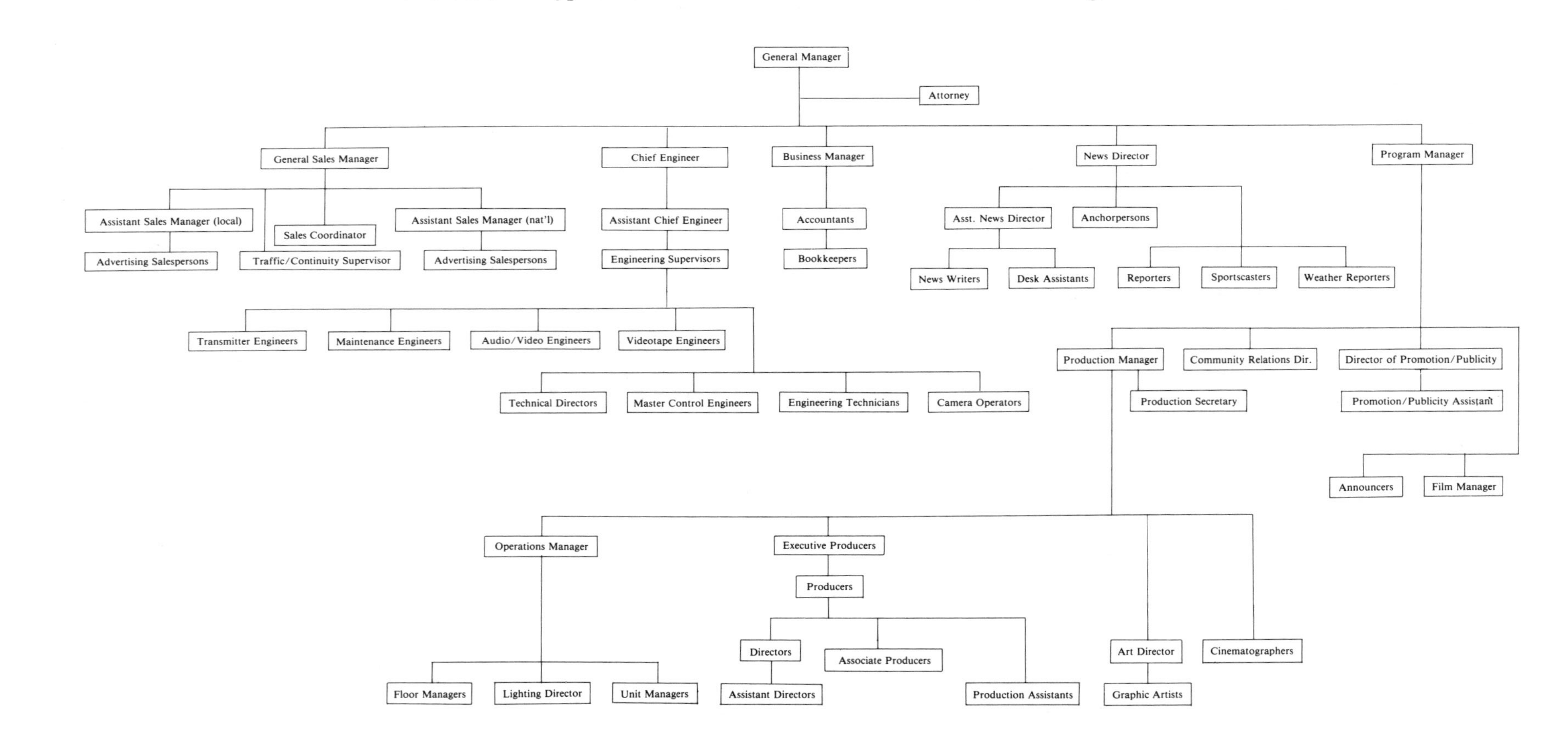

Chart I-4 — Typical Public Television Station Table of Organization

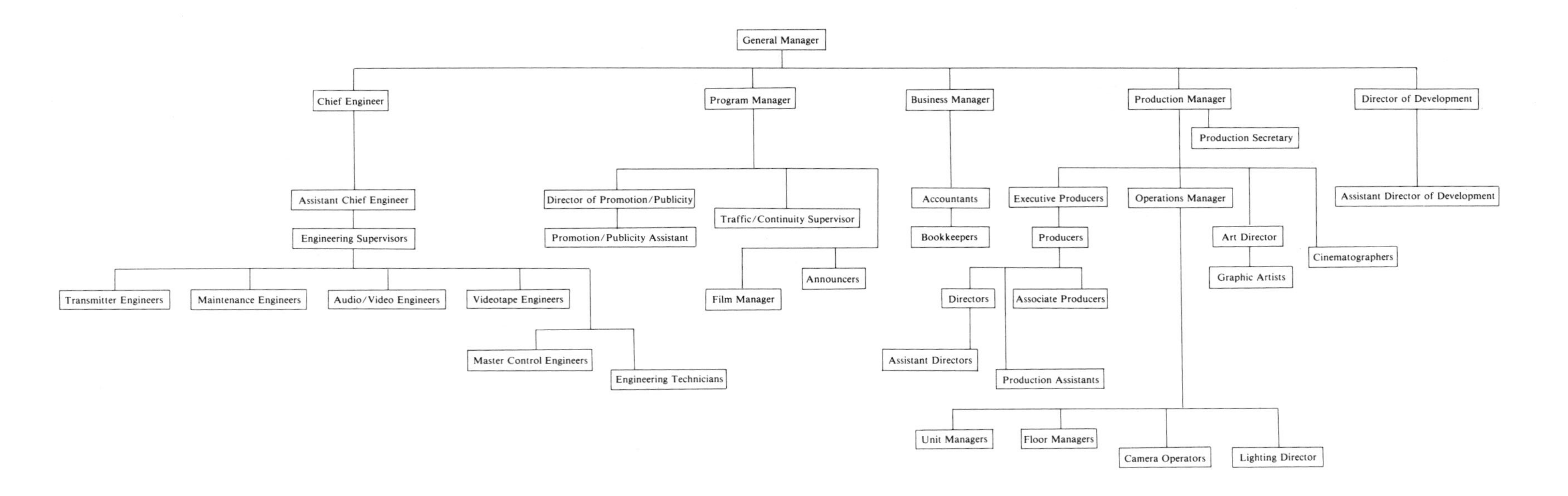

Chart I-5 — Commercial Television Network and Station Employment: 1980

	Full Time	Part Time	Total
Industry Total	70,822	7,449	78,271
Networks	14,838	1,752	16,590
Stations	55,984	5,697	61,681

Source: FCC, August 10, 1981

radio and television broadcasting in the U.S., according to the Bureau of Labor Statistics of the U.S. Department of Labor. Commercial TV employed 38 percent of the total, nearly 10 percent part-time, according to the FCC.

The payroll for 1980 in commercial television totalled $1.8 billion. The majority of workers were employed at local network-affiliated VHF stations. (See Chart I-5.) Nearly 50 percent held professional or technical jobs, 25 percent were in clerical and sales jobs, and 25 percent had managerial jobs, according to data from the Bureau of Labor Statistics and the FCC.

In 1980, almost 9,200 individuals were employed in public television, according to the Corporation for Public Broadcasting. More than 60 percent held professional and technical jobs, 20 percent held clerical and support positions, and the remaining 20 percent were in management.

Women and Minorities

The employment of women and members of minority groups has increased in both commercial and public television in the last ten years. In 1980, 31.4 percent of all commercial TV jobs, and 37 percent of all public television jobs were held by women, according to FCC statistics. Minority-group members held 16.2 percent of all jobs in commercial TV and 16.9 percent of all positions in public TV in 1980, also according to the FCC.

Employment Trends

Employment in the U.S. broadcast industry is expected to grow faster than the average for all U.S. industries through the 1980s, according to estimates from the Department of Labor. A considerable number of opportunities will occur through normal attrition. Within the decade, a few new independent television stations are expected to go on the air, and it is anticipated that the FCC will authorize low-power television (LPT) stations,* thereby opening up many new positions. For the most part, though, the number of jobs at stations, networks, and independent production companies will not experience great growth. The real potential for jobs will probably occur in the newer technologies of cable TV (CATV), subscription television (STV), and multipoint distribution service (MDS) (see Introduction to Part II).

Even though opportunities are expected to increase, the competition is predicted to be intense. More people are in the job market than ever before. The communications industries have traditionally attracted a large number of applicants. Most observers predict a continuation of this trend throughout the decade. Severe competition will probably raise the requirements for jobs so that those with training and education will have a better chance of getting positions.

Education

Nearly all the jobs described in this book require a high school education and, for at least 50 percent, a college degree is preferred or required. Even if an entry-level position does not require one, the advancement opportunities are often brighter with a degree. The more training and education an individual acquires, the easier it is to obtain a position or advance in this challenging field.

Although on-the-job training is very important, a general liberal arts education can be helpful for most positions. Targeted courses for specific jobs can also be useful. Courses in television production are beneficial for anyone entering the programming or production fields. Those involved in advertising sales benefit from courses in advertising and mar-

*LPT stations are designed to provide the opportunity for rural and small communities to establish their own stations. Operating with limited power and covering a distance of less than 20 miles, they will offer specialized programming designed for specific target audiences.

keting. Individuals working in the news department should take courses in political science and government. An ambitious person with an interest in engineering should have course work in the sciences and technology. Courses and levels of education are recommended in the appropriate job descriptions in this book.

Historically, the FCC has issued broadcasting licenses for First, Second, or Third Class Radiotelephone engineers. An examination was required for each level of competence. In late 1981, however, the FCC eliminated these distinctions, replacing them with a General Radiotelephone Operator designation. The new rules will take some time to implement and, in the meantime, most television stations will continue to use the old designations which are referred to in this book in the relevant job descriptions.

The Job Descriptions

The 66 positions discussed in Part I of this book are divided among eight sections. Six of the sections reflect major divisions indicated in the Table of Organization charts for typical commercial and public television stations. These six are labeled: Management and Administration, Programming, Production, News, Engineering, and Advertising Sales. A seventh section contains discussions of jobs at an Advertising Agency. Section 8, Technical/Craft, comprises descriptions of those specialty positions that are usually found only at networks, independent production companies, and some major market stations.

GENERAL MANAGER

CAREER PROFILE

Duties: Overall management of a television station

Alternate Title(s): Station Manager; Vice President/General Manager

Salary Range: $15,000 to $87,000

Employment Prospects: Poor

Advancement Prospects: Poor

Prerequisites:

Education—Undergraduate degree in communications, advertising, or journalism; master's degree preferable

Experience—Minimum of 10 years in television management

Special Skills—Leadership qualities; business judgment; extensive broadcasting knowledge

CAREER LADDER

General Manager (Larger Station); Network Executive; Program Syndication Executive

↑

General Manager

↑

Program Manager; Business Manager; General Sales Manager

Position Description

The General Manager is in charge of the overall management and operation of a television station. He or she is responsible for all business and financial matters, including income, expenses, short- and long-range planning, budgeting, forecasting, and profitability. He or she runs the station in accordance with all federal, state, and local laws, including Federal Communications Commission (FCC) regulations. The General Manager is also responsible for establishing and maintaining the station's image.

If the station is affiliated with a network, the General Manager has an obligation to uphold the policies of the affiliation agreement and to maintain contact with network officials. He or she also supports relationships with community leaders, advertisers, contributors, advertising agencies, program suppliers, and other outside station contacts.

The responsibilities of the General Manager are similar in both commercial and public broadcasting, except for their roles in generating income. In commercial TV, the General Manager is in overall charge of producing advertising revenues, and must constantly evaluate the effectiveness of the sales department. At a public station, the General Manager prepares and defends budgets submitted to legislatures or other public entities, and solicits funds from the viewing public, corporations, and foundations.

The General Manager hires the major department heads and establishes their goals, monitors their performance, and approves their budgets. Although they have day-to-day responsibility, the General Manager oversees each department's activities. In Programming, he or she approves what is produced locally, which syndicated programs are bought, and what should be broadcast in the future. He or she evaluates the schedule in response to competition and audience ratings. For Production, the General Manager monitors program budgets. In News, he or she approves newscast formats and editorial policies. In Engineering, the General Manager approves of the investment in new equipment and facilities. For Advertising Sales, he or she determines advertising policies, tracks income, evaluates the rate structure of the station and its competition, and helps identify potential advertisers.

General Managers are selected and employed by boards of directors at public TV stations, and, in commercial broadcasting, by the chief executive officer of the company that owns the station.

Additionally, the General Manager:

- delivers editorials on air;
- keeps up with changes in communications technology;
- represents the station at industry conferences.

Salaries

In keeping with the demands of the position, General Managers are paid well.

The most comprehensive salary surveys of this elite group of broadcasters have been undertaken for the past seven years by the Broadcast Information Bureau, Inc. (BIB). In the 1980 nationwide survey, the average salary reported by General Managers at commercial stations was $60,500 a year. The range was from $15,000 at very small stations to $70,000 or more at some of the top 100 market outlets. Most individuals are paid by a salary-plus-bonus arrangement and, according to the BIB survey, many General Managers had an expense account, a company profit-sharing plan, a fully paid pension plan, and company-paid life insurance.

General Managers at public TV stations are generally less well paid than their commercial counterparts. According to a 1980 Corporation for Public Broadcasting (CPB) study, the average annual salary was $36,400, which was an increase of 8.2 percent over the previous year. The range was from $16,000 at smaller school-owned stations, to $87,000 at the larger community-owned companies. Most General Managers of non-commercial stations contribute to their medical, life insurance, and pension plans, and, of course, do not participate in profit sharing.

Employment Prospects

Opportunities in both commercial and public broadcasting are extremely limited. There are fewer than 1,000 stations that employ General Managers and the number is not expected to rise significantly in the future.

Advancement Prospects

The majority of General Managers have reached the peak of their careers. Most are between 40 and 55 years of age. Some younger people take more responsible positions as General Managers at larger stations, as network executives involved in sales or relations with affiliated stations, or as managers in program syndication companies. Some establish their own advertising agencies or consulting firms. Still others purchase their own broadcasting stations. The overall prospects for advancement, however, are poor.

Education

As a rule, the minimum requirement is an undergraduate degree, especially in communications, advertising, or journalism. Marketing or business degrees are also useful, and many General Managers have received a master of business administration (MBA), or have taken graduate level business courses. A few have law degrees.

Experience/Skills

At least 10 years of supervisory work in various television positions are necessary. In commercial TV, the General Manager frequently comes from the sales or business areas. In public broadcasting, he or she often has middle management background in programming or fund raising and development. His or her experience should include a successful track record in the specific position of Program Manager, Business Manager, or General Sales Manager.

Leadership, sound business judgment, a deep knowledge of broadcasting, a sense of responsibility, honesty, and an ability to motivate people are essential qualities for a General Manager.

Minority/Women's Opportunities

The majority of General Managers in both public and commercial broadcasting are white males. In 1980, only 23.1 percent of employees in the category of officials and managers (which includes General Managers) were women, and 8.9 percent were members of minorities, according to the FCC. In public TV, in 1980, 31.1 percent were women, and 7.8 percent were minority-group members.

Unions/Associations

There are no unions that serve as bargaining agents or representatives for General Managers. Some belong to the National Association of Television Program Executives (NATPE). Most are active in various network-affiliate committees and boards, and some are members of the International Radio and Television Society (IRTS).

ATTORNEY

CAREER PROFILE

Duties: Advising and representing a communications company or government regulatory agency on legal matters

Alternate Title(s): Lawyer

Salary Range: $9,000 to $70,000+

Employment Prospects: Poor

Advancement Prospects: Poor

Prerequisites:

Education—Law degree

Experience—Two to three years in communications law

Special Skills—Interpersonal skills; logical reasoning; sense of responsibility; writing and verbal abilities

CAREER LADDER

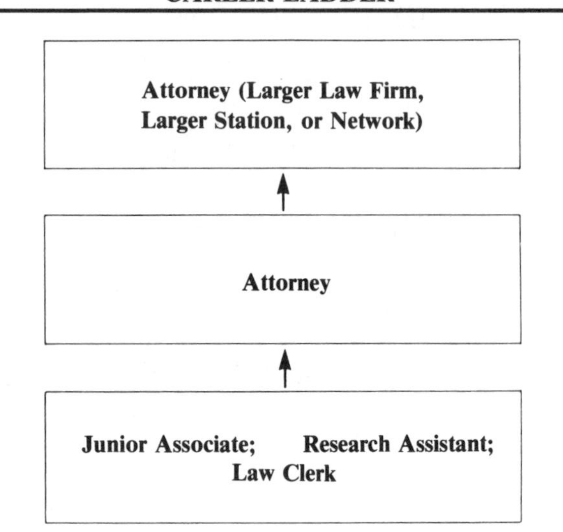

Position Description

An Attorney provides legal counsel to and represents an outside company, employer, or public agency active in the communications field. He or she minimizes the legal risks clients face and, generally, protects their interests in all legal matters. Attorneys are technically officers of the courts and must balance the needs of their clients with the requirements of the law. An Attorney specializing in communications addresses issues relating to contracts, copyright, libel, as well as regulations of the Federal Communications Commission (FCC) and other government agencies. He or she works either for a communications company or represents a regulatory arm of the government.

Attorneys are usually on staff only at networks, major market stations, or at multiple system operators (MSOs), companies that own a number of cable TV systems. The majority of television stations do not have a staff Attorney but are represented by outside law firms. Usually, these firms assign one to three attorneys to handle the legal needs of a station. Most public and commercial stations also retain a law firm in Washington, DC that specializes in communications law, and that represents the client in all federal regulatory matters, particularly with the FCC.

An Attorney representing a station must advise his or her client on how to protect and renew its FCC license. He or she is often called upon to handle copyright issues to protect the station from copyright infringement. An Attorney is usually in charge of, prepares, and negotiates contracts. These often include employment, Performer, equipment or program purchase, and advertising agreements. An Attorney is also frequently asked to review materials to be aired to make certain they are not libelous.

Many Attorneys work in federal, state, and local government regulatory agencies that oversee broadcasting or telecommunications. They assure that stations and cable TV systems are in compliance with all regulations. Attorneys working for these agencies help to write the rules, oversee compliance with federal, state, and local laws, and pass on various broadcast license or cable TV franchise applications. In this current deregulation environment, FCC Attorneys (and those with some state agencies) are involved in trying to lessen the regulations governing communications.

Attorneys give clients or employers advice about their legal obligations, interpret the law, analyze problems and proposals, and suggest possible actions. When necessary, an Attorney develops strategies for and prepares and argues cases in court for a communications company or a regulatory agency.

Additionally, the Attorney:

- handles real estate transactions, tax matters, and other non-communications legal issues;
- researches and investigates cases, precedents, and rulings;
- writes reports and briefs regarding the legal aspects of business activities.

Salaries

An Attorney's earnings depend somewhat on his or her academic record, and the size and location of the law firm. The starting salaries for law school graduates range from $9,000 at very small firms to $30,000 and more at large corporations, and average about $18,000, according to government estimates. The average earnings for experienced communications Attorneys are more than $50,000, and some heads of legal departments at the networks or major market stations earn more than $70,000.

In the federal government, starting salaries in 1980 at the FCC and other government agencies ranged from $15,900 to $19,300, depending on the Attorney's academic and personal qualifications. Experienced individuals in the federal government averaged $30,400.

Employment Prospects

The outlook for getting work as an Attorney in communications law is poor. Even though there has been employment growth, the sizeable number of law school graduates entering the employment market each year creates intense competition for jobs. In addition, employers are extremely selective and graduates of prestigious law schools who rank high in their classes are chosen first.

Most beginning communications Attorneys begin working as junior associates, research assistants, or clerks in communications law firms.

Advancement Prospects

There is relatively little turnover in communications law, and the chances for advancement are poor. With the increasing competition for positions, actual communications law experience is of great importance. It is often necessary to relocate to advance a legal communications career. Most of the large private firms specializing in communications law are in Washington, DC, New York City, and Los Angeles.

Some Attorneys with communications law experience at private law firms in small or middle market communities, obtain positions at specialized communications law firms in major market cities. Others join in-house legal staffs at networks or major market stations, and some obtain communications law positions with the federal government. A few become partners in communications law firms. Still fewer establish their own firms. A tiny number become General Managers at a commercial TV station.

Education

To practice law in the courts of any state, an individual must be admitted to its bar. To qualify for the bar examination, an applicant must complete at least three years of college and graduate from a law school approved by the American Bar Association (ABA) or the proper state authorities. Law school graduates receive a degree of either juris doctor (JD) or bachelor of law (LLB). An undergraduate degree in mass communications or electrical engineering is helpful, as are studies in business, economics, government, and public speaking.

Experience/Skills

Most communications law firms or network legal departments require that candidates have two to three years of work in communications law, usually as a junior associate, research assistant, or clerk. Actual experience in dealing with the FCC and its regulations is also necessary.

Attorneys should possess good interpersonal skills to deal with clients, and excellent writing and verbal abilities. They must be able to reason logically, handle complex issues, and be responsible individuals with integrity and good judgment.

Minority/Women's Opportunities

The majority of Attorneys specializing in communications law are white males. The number of law school graduates who are women and members of minority groups increased greatly in the past ten years, but less than 10 percent of the Attorneys in the communications field are women or minority members, according to professional estimates.

Unions/Associations

There are no unions that represent communications Attorneys. Some lawyers themselves serve as legal counsel to communications unions.

Many belong to the ABA or to their state and county bar associations to share industry concerns and make business contacts. Those in the Washington, DC area often belong to the Federal Communications Bar Association for the same purposes.

BUSINESS MANAGER

CAREER PROFILE

Duties: Managing all financial activities and planning for a television station

Alternate Title(s): Comptroller; Vice President of Business Affairs; Treasurer

Salary Range: $12,000 to $67,000

Employment Prospects: Poor

Advancement Prospects: Fair

Prerequisites:

Education—Undergraduate degree in accounting, business administration, or management

Experience—Minimum of five years in business and accounting, preferably in TV or related field

Special Skills—Understanding of business and finance; knowledge of computer operations; management ability; detail-orientation

CAREER LADDER

General Manager

↑

Business Manager

↑

Accountant

Position Description

The Business Manager of a television station is responsible for the management of all financial transactions, including accounts receivable and payable, general ledgers, journals, and vouchers, and is immediately responsible for all financial planning. He or she supervises the preparation of all billings and develops and analyzes the financial statements and records of various station operations.

As a primary aide to the General Manager, the Business Manager develops short- and long-range plans, and modifies and interprets financial data on a weekly, quarterly, and yearly basis. These plans enable the General Manager and the station's ownership to make financial projections.

The Business Manager develops and supervises all accounting policies and procedures, including the valuation of plant, equipment, programming, and other assets and prepares reports for the Internal Revenue Service and for corporate balance sheets. The responsibilities of the position include the development of profit and loss and cash flow statements and the monitoring of all departmental expenditures.

In performing these duties, the Business Manager supervises all members of the business department including Accountants, Bookkeepers, billing clerks, and benefits personnel. In addition, he or she oversees the station's personnel policies and is often directly in charge of hiring clerical and support employees (messengers, switchboard operators, file clerks, etc.).

Additionally, the Business Manager:

- assists the Sales Manager and the General Manager in establishing realistic advertising rates for the station;
- provides accurate data and records for union negotiations concerning arbitration and contracts;
- oversees the management of the station's physical plant, including furniture and equipment;
- establishes credit policies that achieve maximum sales results while reducing the number of late payments or defaults.

Salaries

Annual salaries for Business Managers at commercial stations are reasonably high, in keeping with their responsibilities. They ranged from $15,000 at a small market station to more than $50,000 for a Vice President of Business Affairs at a major market station in 1980. The average salary was $23,200 in 1980, according to industry sources. Most Business Managers are paid a salary and a yearly bonus.

Salaries for business or financial officers are similar in public television. The average in 1980 was $25,000, according to Corporation for Public

Broadcasting (CPB) surveys. Figures ranged from $12,000 in small school or university stations to $67,000 in major market community stations. According to the CPB, the 1980 salary average was 14.1 percent higher than that reported in the previous year.

Employment Prospects

The opportunity to obtain a position as a Business Manager of a television station is limited.

In small market TV stations, both commercial and public, the Business Manager is usually someone who has been promoted from within the station's business department and who has an intimate knowledge of the station's particular financial operations.

In major market commercial and public stations, business executives are sometimes recruited from other stations or associated media companies, or are brought in by corporate headquarters to improve a station's financial condition. The opportunities for individuals from outside a station, however, are generally poor, and the competition is heavy.

Advancement Prospects

Some Business Managers are promoted to the position of General Manager at their own stations, or join a smaller market station in that capacity. The specific skills and experience acquired as Business Manager are readily transferable to other TV-related fields, including subscription TV (STV) and multipoint distribution service (MDS) companies and cable TV systems, as well as to commercial production companies and advertising agencies.

Few Business Managers at television stations initially trained in communications and many view the position as broad-based and applicable to many fields. As a result, they often seek non-media business opportunities to advance their careers.

Education

Almost all Business Managers in commercial or public television have an undergraduate degree in accounting, business administration, or management. In larger market stations, a master's degree in business administration (MBA) is often a prerequisite for being hired or promoted.

Many are Certified Public Accountants (CPAs) and some have law degrees in addition to their undergraduate business education. A substantial number have had extensive education in computer science, and a few have degrees in this field. In general, however, an education in accounting is the primary academic background required for Business Managers.

Experience/Skills

Most Business Managers are detail-oriented fact-finders. Today, the majority have some experience and skills in computer science. Interpersonal skills in personnel management and the ability to deal fairly and effectively with people are also considered to be major advantages. Some employers require candidates to have experience in labor negotiations and some knowledge of law.

At most stations, prospective Business Managers are required to have at least five years of prior experience in business and accounting, preferably in television broadcasting or a related media field. Personal integrity, reliability, accuracy, and a realistic view of television as a business are also considered important qualifications.

Minority/Women's Opportunities

The majority of Business Managers in commercial television are white males. The percentage of women and minority-group members serving as "officials/managers" (which includes Business Managers) in commercial broadcasting has increased from 17.8 percent to 23.1 percent during the past four years, according to the Federal Communications Commission (FCC).

In public television, 39.3 percent of the top financial positions were held by women in 1980, according to CPB surveys. Members of minorities held 7.9 percent of the posts in public TV stations in 1980, also according to the CPB.

Unions/Associations

There are no unions that serve as representatives or bargaining agents. Business Managers in commercial television, however, often belong to the Broadcast Financial Management Association (BFMA), an organization devoted to developing new concepts of financial management. In public TV, the Public Telecommunications Financial Management Association (PTFMA) provides a similar forum.

ACCOUNTANT

CAREER PROFILE

Duties: Assisting the Business Manager in billings, payroll, and accounts receivable and payable

Alternate Title(s): Assistant Business Manager

Salary Range: $13,000 to $25,000

Employment Prospects: Fair

Advancement Prospects: Fair

Prerequisites:

Education—Minimum of two years of business school; undergraduate degree in business administration or accounting preferable

Experience—Minimum of two years of accounting work

Special Skills—Mathematical aptitude; accuracy; reliability; detail orientation; analytical mind

CAREER LADDER

Business Manager

↑

Accountant

↑

Bookkeeper; School

Position Description

An Accountant in a commercial or public television station is in charge of specific financial transactions. In performing these activities, the Accountant works closely with the General Sales Manager and with Advertising Salespersons. An Accountant usually reports directly to the Business Manager.

As a rule, he or she directly supervises some of the accounting and business staff, including Bookkeepers and billing, accounts receivable/payable, and payroll clerks. An Accountant often functions as a troubleshooter, working to solve specific day-to-day problems in the financial section of the station.

The role of an Accountant varies from station to station. Some large market stations have three or more Accountants, each with a specific responsibility and authority. At top market stations, he or she is sometimes assigned primary responsibilities for billing and accounts receivable.

At smaller stations, an Accountant's responsibilities are broader and encompass almost all activities in the business department, often including the direct supervision of clerical and support personnel and the management of office supplies, furniture, and equipment.

One of the major duties of an Accountant is to develop improved methods of cost accounting in specific financial areas so that the Business Manager can analyze specific problem areas more effectively. An Accountant systematically gathers and maintains financial data, using a variety of techniques, that provide a basis on which management can itemize income and expenses.

Many Accountants are skilled in computer science and word processing techniques, which improve financial data-gathering and increase the speed with which such data can be analyzed. Still others have strengths in such accounting areas as inventory assessment, assets and liabilities, and profit and loss statements, and are assigned specific responsibilities in those areas.

Additionally, the Accountant:

- double-checks vouchers, purchase orders, payroll timesheets, and invoices for accuracy, prior to the approval of the Business Manager;
- reviews network accounts and agency and client payments and accounts;
- prepares data and summaries of financial reports for the Business Manager;
- maintains records for Federal Communications Commission (FCC) reports, income tax returns, and insurance claims;
- continually monitors department budgets to balance expenses with allocated resources.

Salaries

The salary scale for Accountants in commercial TV is relatively low, ranging from $13,000 at small market stations to $25,000 in the major markets, according to industry estimates.

In public television, the average salary was $15,000 in 1980, an 8.9 percent increase over the previous year. As in commercial television, the salaries for Accountants in public broadcasting were higher in the top 50 market stations.

Employment Prospects

Because competition is heavy, the opportunity to obtain or be promoted to the position of Accountant at a commercial or public television station is only fair. The majority of Accountants are normally not recruited from outside the station staff unless specific skills, such as computer science training or tax expertise, are required. For recent college graduates with accounting or related degrees, employment opportunities are brighter at subscription television (STV) and multipoint distribution service (MDS) operations and at cable TV systems.

Advancement Prospects

The top promotion for an Accountant at a television station is to the position of Business Manager. The opportunities for this move, however, are somewhat limited because of the general stability of such employees, and the severity of the competition. Some Accountants assume the role of Business Manager at a smaller station, or join a commercial production firm or media-related organization in a similar capacity. Still others use their experience to move into a top financial spot in a small, unrelated business.

Education

A minimum of two years of business school training in accounting is usually required, and an undergraduate degree in business administration or accounting is preferable. Increasingly, Business Managers seek individuals who have training in computer science.

Courses in bookkeeping, personnel management, or human resources are also valuable. Some Accountants have undergraduate degrees in business administration. Others have certification as a Certified Public Accountant (CPA).

Experience/Skills

Most Accountants are young—between the ages of 25 and 30—and the promotion to Accountant is their first career advancement. In many instances, they have worked as a Bookkeeper in the business department of the station for at least two years, and have taken additional course work and training in accounting.

In large network-affiliated stations, promising young college graduates are employed in the position. In general, however, the job is most often filled by promotion internally.

An Accountant should be a bright, alert individual with a particular specialty in business, accounting, law, or computer science. Accuracy, reliability, and aggressiveness are prime requisites for the position. A successful Accountant is a fact-finder and problem-solver who can bring new ideas and techniques to the financial operation of a station.

Minority/Women's Opportunities

The employment opportunities for women and minority-group members are quite good. According to FCC statistics, 89 percent of the employees in the "office/clerical" job category (which includes Accountants) in 1980 were women. Minority-member employment in this category represented 25.3 percent of all persons employed, and is increasing yearly.

In public television, 55.4 percent of all Accountants were women in 1980, and 15.4 percent were members of minorities, according to Corporation for Public Broadcasting (CPB) surveys.

Unions/Associations

There are no unions that serve as bargaining agents for Accountants. Many individuals, however, belong to the Broadcast Financial Management Association (BFMA) which is devoted to developing new concepts of financial management. The Public Telecommunications Financial Management Association (PTFMA) provides a similar function for Accountants in public television.

BOOKKEEPER

CAREER PROFILE

Duties: Responsibility for financial record-keeping at a television station

Alternate Title(s): None

Salary Range: $7,500 to $16,800

Employment Prospects: Good

Advancement Prospects: Fair

Prerequisites:

Education—Minimum of one year of business school training in bookkeeping and accounting

Experience—Minimum of one year of bookkeeping experience preferable

Special Skills—Mathematical aptitude; accuracy; reliability; ability to work rapidly under pressure

CAREER LADDER

Accountant

↑

Bookkeeper

↑

Business School/College; Bookkeeper (Another Station or Business)

Position Description

The basic duties of a Bookkeeper are to maintain journals, ledgers, and financial books for use by a television station's management in day-to-day operations and budget forecasting. TV stations and non-broadcast organizations employ from two to ten Bookkeepers, depending upon the size of the operation and the market. Bookkeepers report to the Business Manager or, at larger stations, to an Accountant.

At larger stations, the Business Manager or Accountant assigns Bookkeepers to specific areas of responsibility. An important specialty for a Bookkeeper can be maintaining the station's advertising billing accounts and files. Other individuals may be responsible for the preparation of all payroll elements, including time sheets, vacation and sick leave records, payroll and Social Security deductions, and overtime compensation. Some Bookkeepers may be assigned to accounts payable, where they post bills received at the station, including telephone bills, subscriptions, power and lighting bills, invoices for videotape, and other normal operational expenses. The Bookkeeper records all bills, matches them with purchase orders or vouchers, and checks their accuracy with the department head who purchased the items or services.

Some Bookkeepers are assigned to maintain the day-to-day accounts of a specific department, such as engineering or programming. In most stations, the programming department is the most expensive unit in terms of salaries, program purchases, licensing fees, shipping costs, line charges, production expenses, and other program-related costs. A Bookkeeper may be assigned specifically to monitor all income and expenses of that department.

At some smaller stations, the Bookkeeper works in all areas of accounting, including payroll, accounts payable and receivable, and advertising billings.

Additionally, the Bookkeeper:

- keeps accurate backup records of all invoices or bills and related documents for reference purposes;
- maintains an ongoing file of past-due accounts receivable or payable for use by the Business Manager or Accountant;
- manages an inventory file of facilities, supplies, and equipment used or purchased by the station.

Salaries

Salaries for this position are somewhat low in commercial TV, ranging from $9,000 in a small market station to $15,000 in a major market operation, according to industry estimates. The average payment in 1980 was $10,500. Amounts at the top of the salary range are usually paid to beginners who have an undergraduate degree in accounting or to Bookkeepers who have been at the station for some time.

Salaries in public television averaged $11,800 in 1980, according to the Corporation for Public Broadcasting (CPB). This represented a 6.8 percent increase over the 1979 average. The range in 1980 was from $7,500 to $16,800. Salaries are higher at larger market stations in both public and commercial television.

Employment Prospects

This is frequently an entry-level position and job possibilities at a television station or in a related telecommunications or video operation are good. Most stations employ two or more Bookkeepers and with 1,000 television stations on the air, there are opportunities available for bright young business school graduates.

Positions open up frequently as the station's Bookkeepers are promoted or move to other television or telecommunications positions. The necessary skills are readily transferable to related jobs at multipoint distribution service (MDS) and subscription TV (STV) firms or cable TV companies, and a large number of jobs also exist in non-media business organizations.

Advancement Prospects

The possibilities for advancement are fair, and the competition is heavy. Once a beginning Bookkeeper learns the job or becomes a specialist in various accounting areas within the business department, promotion is possible. The promotion may be to a specific bookkeeping responsibility with a higher salary, or to the job of Bookkeeper in charge of a single departmental area within the station.

Some Bookkeepers, with further training and education, advance to the position of Accountant after apprenticing in all areas of the business department. Still others advance their careers by joining advertising agencies or other television-related organizations. Some use their business and accounting experience to assume more responsible positions outside the television and video industries.

Education

A minimum of one year in a business school, with the major emphasis of study in bookkeeping and accounting, is usually essential. An undergraduate degree in accounting, or in business administration with a major in accounting, is sometimes required at major or middle market stations. Courses in computer science are extremely valuable as well as some education in personnel management. A solid educational background in accounting and bookkeeping, however, is the prime requisite for the position.

Experience/Skills

While some stations recruit experienced individuals from other stations or telecommunications companies, extensive experience is not the primary requirement for bookkeeping. Many Bookkeepers are between the ages of 22 and 30 but some are much older and have been in the position for a number of years.

At least one year of experience, either part-time during school or college or full-time as a bookkeeper in another business, is very helpful. Successful Bookkeepers must be bright, accurate, reliable, and able to work rapidly. Above all, however, they must be good at and enjoy working with figures.

Minority/Women's Opportunities

The employment opportunities for minority-group members and, particularly, women in this position are good. Federal Communications Commission (FCC) statistics indicate that more than 89 percent of the office clerical employees (including Bookkeepers) in commercial television in 1980 were women. Minority-group employment amounted to 25.3 percent of all persons in this job category in 1980.

In public television, 93.5 percent of those employed as Bookkeepers in 1980 were women, while 25.2 percent were members of minorities, according to CPB statistics. Women and members of minorities showed gains in this position in both commercial and public television during the past four years.

Unions/Associations

There are no unions or professional organizations that represent Bookkeepers in commercial or public television.

PROGRAM MANAGER

CAREER PROFILE

Duties: Selecting and scheduling of all programming broadcast by a TV station

Alternate Title(s): Program Director; VP of Programming

Salary Range: $6,400 to $70,000

Employment Prospects: Poor

Advancement Prospects: Fair

Prerequisites:

Education—Undergraduate degree in communications, radio-TV, theater, journalism, business, or marketing

Experience—Minimum of five years in television production and sales

Special Skills—Creativity; extensive knowledge of TV programming; organizational ability; sound judgment; good taste

CAREER LADDER

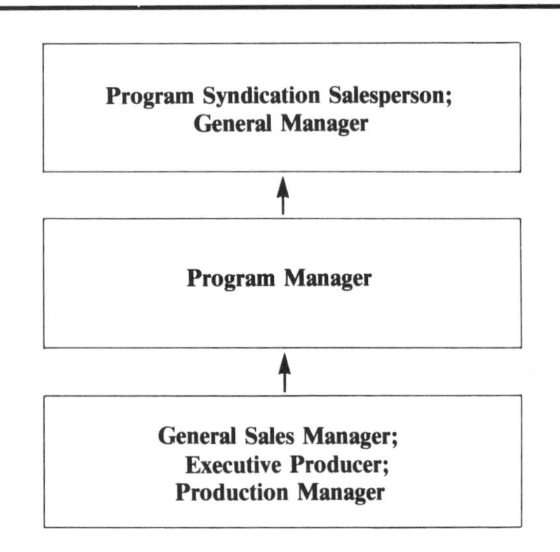

Position Description

The Program Manager, who reports to the General Manager, is responsible for all local, network, independent, and syndicated programming broadcast by a television station and must ensure that it is in compliance with all Federal Communications Commission (FCC) regulations. He or she selects and schedules all programs and tries to create a mix that will guarantee the largest possible viewing audience and highest ratings. At a commercial station, the Program Manager works closely with the sales department to determine the programming that is most marketable to advertisers during various times in the broadcast schedule.

He or she keeps informed about the needs of the audience, from sign-on to sign-off, and monitors the quality, type, and profitability of programs broadcast by the station. The Program Manager must be constantly aware of the competition for the audience from other local stations. He or she uses local and national research organizations, reports, and ratings services to define the demographics of the station's viewing audience at various times during the broadcast day, and to measure the success or failure of a particular program.

The Program Manager decides which film packages, syndicated shows, reruns of previously aired network programs, and locally produced programs will be selected, produced, or purchased to provide the overall programming balance. He or she also supervises the scheduling of spot announcements, commercials, public service announcements, and station breaks.

The Program Manager must also create and schedule local public affairs programs, to comply with the station's obligations to operate in the "public interest, convenience, and necessity" under FCC regulations. He or she must, therefore, understand the community's needs and problems, and select programs accordingly.

Additionally, the Program Manager:

- controls the programming budget and keeps within the station's affordable limits;
- negotiates with independent producers and program syndicators to purchase programming for the station;
- evaluates local program ideas for possible production.

Salaries

Despite the demands and responsibility of this position, Program Managers are relatively low-paid. Those at independent commercial stations are usually paid more than individuals at network-affiliated stations, because of their increased responsibility in selecting programs.

In its analysis of more the 278 Program Managers at commercial stations in 1980, the Broadcast Information Bureau (BIB) noted that salaries ranged from under $7,500 to more than $50,000. Most were in the $20,000 to $40,000 range, and the average was $29,900. Most commercial-station Program Managers are paid a combination of salary and bonus.

Salaries for Program Managers in Public TV are similar to those of their counterparts in commercial broadcasting even though they generally have more responsibility and are often the second in command at their stations. The 1980 average was $26,200, according to a Corporation for Public Broadcasting (CPB) survey. This represented an increase of 10.5 percent over the previous year. The range was between $6,400 and $70,000. Program Managers in public television in the top 50 markets whose stations were community-owned earned more than their peers in college- or school-licensed stations in smaller markets.

Employment Prospects

With only 1,000 television stations on the air, the opportunities for employment are poor. Cable TV and multipoint distribution service (MDS) and subscription television (STV) operations are increasing, however, so there will be additional opportunities for Program Managers. Still, the number of available positions is not expected to rise dramatically in the near future and the competition is severe.

Advancement Prospects

Opportunities for advancement are fair. Most Program Managers at both commercial and noncommercial stations move on to more responsible positions in broadcasting, or become Program Managers at larger stations at higher salaries. According to a survey conducted by Michael Fisher of Temple University in 1976, the typical Program Manager in commercial TV starts at a smaller station and moves to progressively larger stations in the same capacity. Almost 90 percent of those surveyed felt they had not reached the top of their profession.

Some commercial-station Program Managers join network program staffs, and many become salespersons of syndicated programs. Still others join or form their own TV production companies. In recent years, some Program Managers have moved to similar positions in cable TV to advance their careers.

In public television, Program Managers are usually next in line for the job of General Manager. Many Program Managers move to smaller stations to assume the top post, at higher salaries, while others wait until normal attrition makes the position available at their own stations.

Education

Today, more than 64 percent have an undergraduate degree, according to the Fisher survey, and many have graduate degrees—usually in mass communications.

An undergraduate degree in communications, radio-TV, theater, journalism, business administration, or marketing is recommended. Study in speech, drama, and writing, coupled with courses in sales and marketing, can offer very useful background for a Program Manager.

Experience/Skills

Prospective Program Managers should have at least five years of experience in television production and, in commercial broadcasting, sales. Most Program Managers now working in commercial stations have held the job of General Sales Manager and have had some on-air experience in radio or television. According to the Fisher survey, 58 percent have written programs for TV. Most have worked at their jobs for an average of seven years.

In public broadcasting, the majority of Program Managers have been Executive Producers and Production Managers at some time in their careers. Some have worked as Writers or Directors.

A Program Manager is expected to have creativity, taste, organizational ability, a firm knowledge of television programming, show business savvy, and sound business skills.

Minority/Women's Opportunities

Program Managers at both commercial and public television stations are usually white males between the ages of 35 and 45. In 1980, some 23.1 percent of the officials and managers (including Program Managers) in commercial TV were women, while 8.9 percent were members of minorities, according to FCC statistics.

In public television, 8 percent of the Program Managers in 1980 were minority-group members and 19.2 percent were women, according to the CPB.

Unions/Associations

There are no unions that serve as representatives for Program Managers. Most commercial and some public television Program Managers belong to the National Association of Television Program Executives (NATPE) to share mutual concerns and to learn about programs available for purchase.

FILM MANAGER

CAREER PROFILE

Duties: Acquiring and pre-screening commercials and film and tape programs for broadcast on a television station

Alternate Title(s): Film/Videotape Librarian; Film Specialist

Salary Range: $9,000 to $14,200

Employment Prospects: Fair

Advancement Prospects: Fair

Prerequisites:

Education—High school diploma; technical school training helpful

Experience—Operation of audio-visual equipment helpful

Special Skills—Attention to detail; timing ability; record keeping accuracy

CAREER LADDER

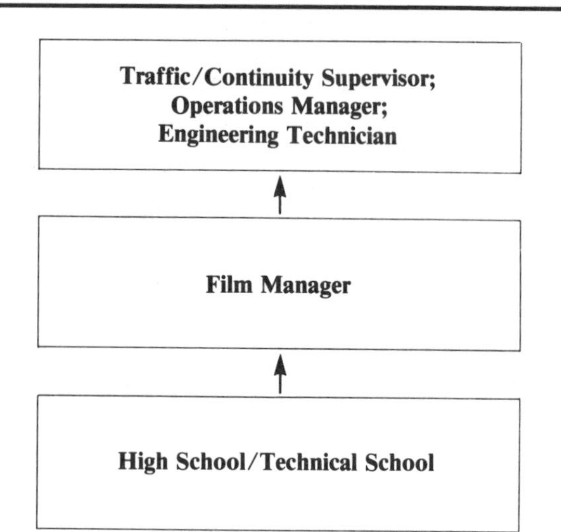

Position Description

The Film Manager keeps track of all commercials and film and videotape programs and prepares them for broadcast. He or she orders, receives, files, and ships films and tapes from and to a variety of sources. It is his or her job to get all on-air videotape or film to the proper location at the station in time for airing.

A major responsibility of the Film Manager is to pre-screen all incoming tape and film materials and evaluate their technical quality prior to their being broadcast. Television relies on split-second timing. A film or tape program that runs too long or too short can interfere with the entire station operation. A commercial on film that is aired scratched or dirty could cause the station to have to rerun it at no cost to the client. The Film Manager ensures that all films and tapes are correctly timed and ready for broadcast.

In some stations, the Film Manager also serves as a projectionist and checks, cleans, and loads the film chain (a TV projection system for film and slides) for programming and station breaks. In others, he or she receives and processes new, incoming blank videotapes (both in cassette and reel-to-reel formats), assigns control numbers, and readies them for technical evaluation by the engineering department. When they have been returned, the Film Manager inventories the tapes for use by the production department.

The Film Manager works closely with the Traffic/Continuity Supervisor, and reports to the Program Manager at many stations, or to the Operations Manager at others. In either case, he or she reviews the daily advance log of materials to be aired, prepared by the programming department, to determine that all scheduled tapes and films are in the house and ready for broadcast.

Additionally, the Film Manager:

- maintains accurate and up-to-date inventory of all pre-recorded tapes and films in the station and records their location and disposition;
- edits feature and syndicated films to conform with logged time and commercial breaks;
- edits and inserts commercials into film programs;
- assembles and labels all taped or film commercials scheduled for station breaks and edits them onto a single reel in advance of broadcast.

Salaries

Income for Film Managers in commercial and public television is quite low. Salaries in commercial TV ranged from $9,000 to $14,000 in 1980. The average for the related title of Film/Videotape Librarian was $9,500, according to industry sources.

In public TV, salaries for positions similar to Film Manager averaged $10,200 in 1980, according to the Corporation for Public Broadcasting (CPB). At major market community stations, people in the related position of Videotape Librarian earned from $10,400 to $14,200 in 1980.

Employment Prospects

Most TV stations employ only one or two Film Managers (or Film/Videotape Librarians). Since this is often considered an entry-level job openings do occur as people are promoted. A few cable TV systems also employ Film Managers, as do many large advertising agencies. The overall opportunities for employment, however, are only fair.

Advancement Prospects

Some Film Managers advance to become Traffic/Continuity Supervisors. Individuals with some engineering skill and talent sometimes obtain Second Class Radiotelephone Licenses from the Federal Communications Commission (FCC) and eventually move to more responsible positions as Operations Managers or Engineering Technicians. The opportunities for advancement, in any case, are only fair, and are dependent upon the individual's initiative, skills, and interests.

Education

Most employers require a high school diploma. Some additional training in electronics or film at a technical school is helpful.

Experience/Skills

Most employers provide on-the-job training for this entry-level position, under the supervision of the Program or Operations Manager. Some experience operating audio-visual equipment, including 8 and 16mm projectors, slide machines, and associated gear is helpful. Some knowledge of common transportation carriers and shipping procedures is useful.

A Film Manager must be careful and detail-oriented. It is essential that he or she be able to keep accurate records and time programs precisely.

Minority/Women's Opportunities

The majority of Film Managers at commercial and public television stations in 1980 were male, but women have found increasing opportunities in this area. During the past ten years, employment of minority-group members has also increased, according to industry sources, particularly in major urban centers.

Unions/Associations

There are no unions or professional organizations that represent Film Managers.

ANNOUNCER

CAREER PROFILE

Duties: Providing the spoken portion of station identifications, public service announcements, and programs at a television station

Alternate Title(s): None

Salary Range: $10,500 to $26,000+

Employment Prospects: Poor

Advancement Prospects: Poor

Prerequisites:

Education—Undergraduate degree in communications, drama, or speech preferable

Experience—Minimum of one year of varied radio or television announcing

Special Skills—Good control of voice and diction; acting ability; good personal appearance and style

CAREER LADDER

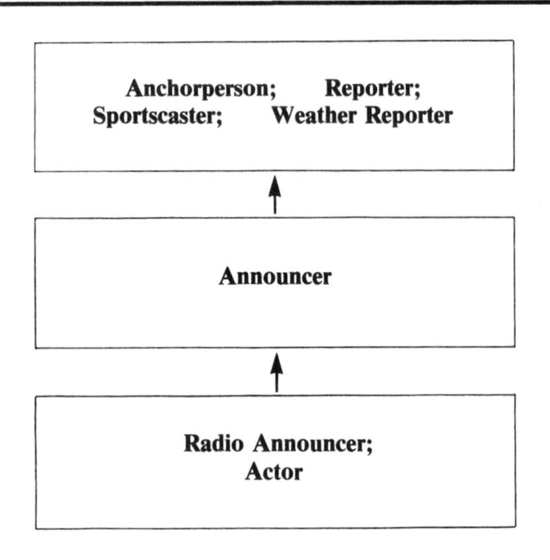

Position Description

An Announcer's voice is heard on station identifications and public service announcements at commercial or public television stations. It is his or her responsibility to provide artistically satisfying, technically sound, and professionally acceptable spoken portions of broadcasts. He or she handles a wide variety of assignments and usually reports to the Program Manager.

In addition to his or her other duties, an Announcer may also host or emcee programs, and sometimes appears on camera as a Performer. He or she might assume a character role for children's programs, or introduce horror movies, special films, or other station-produced programs. At some smaller stations, an Announcer does live or taped commercials for local station clients, often demonstrating products on camera. Sometimes, he or she fills in for an absent Sportscaster, Reporter, or Weather Reporter. Announcers also do freelance work on specific projects for advertising agencies, cable TV systems, and subscription television (STV) and multipoint distribution service (MDS) companies.

A good television Announcer learns to develop eye contact with the camera, and reads copy easily from a teleprompter or cue cards with a good sense of timing and at an appropriate pace. He or she marks up scripts, edits, and delivers written copy, rephrasing as necessary to avoid awkward word usage, and to provide a smooth and intelligent reading.

Most good Announcers try to read concepts, not just phrases, and groom their styles to provide a natural quality rather than a hard-sell delivery. They use a variety of speech techniques, including variation in the rate at which they speak, emphasis of key words and phrases, and grouping of words into meaningful sub-ideas. A good Announcer uses inflections appropriately, pronounces words clearly and accurately, and makes it all appear effortless.

Frequently, an Announcer is required to work without a prepared script and must, therefore, be quick-thinking, have a good vocabulary, charm, and a style appropriate to the situation. He or she must use body movement, voice qualities, gestures, and his or her personality to provide audience identification with the program.

Announcers are known by their voices. An individual's other qualities—wit, style, and knowledge—find expression through the delivery of spoken words.

Additionally, the Announcer:

• narrates opening or closing credits for locally-produced programs;

• conducts, on occasion, interviews for a newscast or documentary.

Salaries

Salaries in commercial television ranged from a low of $10,500 at small market stations to more than $26,000 for experienced personalities at major market stations in 1980, according to industry sources. The average salary was $17,000. Some personalities in the top 20 markets make considerably more. In public television, the average payment was $18,700 in 1980, according to the Corporation for Public Broadcasting (CPB).

Employment Prospects

There are more Announcers available than there are opportunities, and the prospects for employment are poor. The position of staff Announcer at a television station is not usually an entry-level job. Experience in radio announcing or as an actor is often required.

Most TV stations employ only one or two staff Announcers, and at many small outlets, Announcers from local radio stations handle the daily voice-over duties on a freelance basis.

Advancement Prospects

Opportunities for advancement are poor, and depend on the individual's talent and skill as well as on being in the right place at the right time.

Most staff Announcers seek higher-paid jobs as regular on-camera Anchorpersons, Reporters, Sportscasters, or Weather Reporters. The competition, however, is severe. Some Announcers advance their careers by moving to the same position at a larger market station, or by obtaining an on-camera position at a smaller station. Most freelance in commercials for advertising agencies and many also work at local radio stations owned by the TV company. A few move to larger stations that have special projects or programs, and some find freelance opportunities in cable TV systems and at STV and MDS operations.

Education

While a formal education is less important than a facility with the language, the majority of Announcers have an undergraduate degree in communications, drama, or speech. A broad liberal arts background is important, and courses in diction and phonetics are helpful.

Experience/Skills

At least one year of working in radio or television, in various announcing formats and under a number of conditions, is usually required. Many television stations prefer to employ experienced professionals with three or more years of announcing assignments in smaller markets.

Announcers must have a quick memory and an ability to work under distracting conditions. They must also have clear voices without regional characteristics for most assignments, but be capable of doing dialects and special character voices. Most Announcers are good actors, understand microphone techniques, and have a knowledge of television production. They are capable of precise timing, have excellent speech and language skills, a good personal appearance, style, and are able to project a youthful quality, authority, and enthusiasm.

Minority/Women's Opportunities

Opportunities for women and members of minorities have increased greatly during the past ten years. Although the majority of staff Announcers in commercial television are white males, more women and minority-group members are occupying these positions each year, according to industry observers. In public television, 12.5 percent of the individuals in the category of Announcers/on-air talent in 1980 were women, and 25 percent were members of minorities, according to the CPB.

Unions/Associations

In markets where Announcers are represented by unions for bargaining purposes, some individuals belong to the American Federation of Television and Radio Artists (AFTRA). Others are members of the Screen Actors Guild (SAG) and some of the American Guild of Variety Artists (AGVA).

DIRECTOR OF PUBLICITY/PROMOTION

CAREER PROFILE

Duties: Responsibility for all public relations and promotion at a television station

Alternate Title(s): Director of Information; Director of Public Relations

Salary Range: $7,800 to $33,000 +

Employment Prospects: Fair

Advancement Prospects: Poor

Prerequisites:

Education—Undergraduate degree in public relations, advertising, or communications

Experience—Minimum of two years in public relations, advertising, promotion, or publicity

Special Skills—Writing and editing ability; sales talent; interpersonal skills

CAREER LADDER

Director of Publicity/Promotion (Large Station, Network, Cable TV, or Non-Media Field)

↑

Director of Publicity/Promotion

↑

Publicity/Promotion Assistant

Position Description

A Director of Publicity/Promotion plans and executes comprehensive public relations campaigns to promote a television station's image, its programs, and its activities. He or she uses a variety of methods to help increase the audience for the station and to attract new advertisers. His or her duties normally evolve in response to current and projected broadcast schedules and to the promotion needs of the station.

On a continuing basis, a Director of Publicity/Promotion promotes the station, its programs, and its on-air personalities to current and potential audiences through imaginative campaigns, using many types of advertising and on-air and print promotional techniques. He or she conceives, plans, and edits printed materials, including press kits, news releases, posters, fliers, fact sheets, and feature articles. He or she also generates consumer advertising, using billboards, newspapers, radio, and consumer and trade magazines.

The Director of Publicity/Promotion ensures that program listings and schedule changes are promptly distributed to the media and that they are printed accurately. To further promote upcoming shows and events, he or she oversees the production and scheduling of promotional spot announcements to be aired on the station during the broadcast day.

At a commercial station, the Director of Publicity/Promotion works closely with the General Sales Manager to increase advertisers' awareness, and with the art department to produce print and promotional materials to attract new advertisers. With the Community Relations Director, he or she coordinates public functions and provides printed materials and other public relations support. If the station is affiliated with a network, the Director of Publicity/Promotion collaborates with the network's promotion and publicity department to advertise the station's programs.

At a public TV station, the Director of Publicity/Promotion works closely with the volunteer coordinator, the director of fund-raising and development, and other members of the staff, to promote the station and its image. He or she also interacts with the publicity and promotion staff at the Public Broadcasting Service (PBS).

The Director of Publicity/Promotion reports to the Program Manager or, at some stations, directly to the General Manager. Because of the importance and complexity of the job, he or she usually employs from one to three, or more, assistants who specialize in different media or techniques of promotion and publicity.

Additionally, the Director of Publicity/Promotion:

• supervises special promotion events, including parades and appearances of on-air talent at public functions;
• arranges interviews in other media for the station's personalities;
• provides advance screenings of unique and special programs;
• speaks at public or community functions on behalf of the station.

Salaries

Salaries in commercial television during 1980 ranged from $13,000 at small market stations to more than $33,000 at middle and major market stations, according to industry sources. The average was $18,400.

In public television, 1980 income ranged from $7,800 for beginners to $33,000 for experienced individuals at major market community stations, according to the Corporation for Public Broadcasting (CPB). The average was $16,700, an increase of 9.2 percent over the previous year.

Employment Prospects

Opportunities for moving into this position are only fair. Although all TV stations have promotion and publicity employees, there is not much movement out of the top position and openings do not occur that frequently. Talented and ambitious Publicity/Promotion Assistants can, however, move into the job at their own or other stations. Some opportunities also exist for qualified people in other public relations fields to transfer into television as Directors of Publicity/Promotion.

Advancement Prospects

Opportunities are poor in both public and commercial television. At a commercial station, this is often the top job in publicity and promotion, and there are few opportunities to advance within the station organization. Some individuals obtain higher-paid jobs in the same position at larger stations or at the network level. An increasing number find opportunities in cable TV. Some move into other public relations fields, or to more responsible public relations positions in other media-related organizations. In public television, some Directors of Publicity/Promotion move to high-level fund-raising positions at their own stations or at other stations in smaller markets.

Education

An undergraduate degree in public relations, advertising, or communications is usually mandatory. Courses in writing, photography, printing, and speech are valuable.

Experience/Skills

A minimum of two years of experience in all phases of public relations, advertising, promotion, or publicity is mandatory. Knowledge of print design is useful, and some experience with outside vendors of printing services is helpful.

A candidate for the job must have a vivid imagination and possess solid writing and editing skills. He or she should be a good salesperson who is poised, articulate, and capable of dealing with a variety of people in relatively sophisticated environments. He or she must also be creative, personable, persuasive, and enthusiastic.

Minority/Women's Opportunities

The prospects for women to obtain this position in commercial TV have improved dramatically over the past ten years, according to industry sources, and are now excellent. Opportunities for minority-group members are increasing at a slower pace. In public television, 72 percent of the Directors of Publicity/Promotion in 1980 were women, and 5 percent were members of minorities.

Unions/Associations

There are no unions that specifically represent Directors of Publicity/Promotion in commercial or public television. Many belong, however, to the Broadcasters' Promotion Association (BPA) to share mutual concerns and advance their careers.

PUBLICITY/PROMOTION ASSISTANT

CAREER PROFILE

Duties: Handling the details associated with public relations, promotion, and publicity at a television station

Alternate Title(s): Promotion Specialist; Assistant Promotion Director; Public Information Assistant

Salary Range: $7,500 to $28,000+

Employment Prospects: Good

Advancement Prospects: Good

Prerequisites:

Education—Undergraduate degree in public relations, journalism, English, advertising, or communications

Experience—Some promotion and public relations work helpful

Special Skills—Writing talent; language skills; typing ability

CAREER LADDER

Director of Publicity/Promotion

↑

Publicity/Promotion Assistant

↑

Media or Non-Media Promotion/ Public Relations Position; College; Secretary

Position Description

A Publicity/Promotion Assistant handles all the detail work involved in providing public relations, promotion, and publicity services for a television station in support of its programs and other activities. He or she reports to and assists the Director of Publicity/Promotion in assigned tasks. A similar position exists at many larger cable TV systems, and subscription television (STV) and multipoint distribution service (MDS) operations.

Small commercial stations often employ only one Promotion/Publicity Assistant who performs many and varied tasks. In middle market commercial and public TV stations, two or more Assistants are employed, and each usually has a specified major responsibility, such as promotion or public relations. Most publicity/promotion departments are relatively small, however, and the Assistant performs many duties to attract new viewers and advertisers to the station.

At commercial stations, the Publicity/Promotion Assistant works closely with the Community Relations Director, the traffic department, and the art department to implement promotion and public relations projects. At a public station, he or she works closely with the coordinator of volunteers and the director of fund raising and development to help raise money effectively.

Some of the Publicity/Promotion Assistant's tasks include writing press releases and feature articles, and compiling and distributing press kits. He or she also makes arrangements for special events and appearances, and conducts tours of the studio or station. It is up to this person to maintain up-to-date mailing lists of the station's advertisers, viewers, volunteers, and supporters.

A Publicity/Promotion Assistant establishes and is in daily touch with a wide variety of media contacts in order to publicize and promote station activities, and usually spends at least one-quarter of his or her time outside the station. When inside, he or she spends a considerable amount of time on the telephone.

Often, a Publicity/Promotion Assistant helps to develop and produce on-air promotion spots and slides that publicize the station's upcoming programming. He or she also writes and types boldface listings, program schedules, and schedule changes, and distributes them regularly to newspapers in the station's coverage area. The Publicity/Promotion Assistant is usually responsible for writing program announcements (overcrawls) to be superimposed on the video image during newscasts and other programs.

Additionally, the Publicity/Promotion Assistant:

- responds to viewer comments, complaints,

suggestions, and questions about programming by mail or telephone;

- maintains all promotion and publicity files, photographs, stories, national press releases, and taped promos;
- makes speeches and represents the station at community and business affairs.

Salaries

Earnings in commercial television in 1980 ranged from $8,500 for entry-level employees at small market stations to more than $28,000 for experienced individuals with seniority at major market operations. In public TV, salaries in 1980 ranged from $7,500 at small stations to $26,000 in community-owned stations in major markets, according to the Corporation for Public Broadcasting (CPB). The average salary in public television was $12,500, a slight decrease over the previous year.

Employment Prospects

Chances of getting a job as a Publicity/Promotion Assistant are generally good. There is some turnover in what is often considered an entry-level position. As such, there are openings for bright, alert college graduates with degrees in public relations or communications. Occasionally, a secretary in the publicity/promotion department of a commercial or public television station is promoted to the position.

Advancement Prospects

The opportunities for advancement for bright, capable Publicity/Promotion Assistants are good, particularly for those with imagination and style. Some become Directors of Publicity/Promotion at their own stations, after some years of successful experience, while others move to better-paid positions at major market stations at higher salaries. Some individuals move to cable TV or to subscription television (STV) or multipoint distribution service (MDS) operations as Directors of Publicity/Promotion. Many use their television experience to obtain more rewarding positions in public relations or promotion at companies or organizations outside the media field.

Education

An undergraduate degree in public relations, journalism, English, advertising, or communications is often required. Courses in all the liberal arts are helpful, but writing courses are mandatory.

Experience/Skills

Some experience in promotion and public relations at a media or non-media related organization is expected, and experience in promoting a non-profit organization, even on a part-time basis, is helpful. In many circumstances, however, an undergraduate degree and some background in writing (articles, college newspaper, or radio station) can replace full-time experience.

Candidates should be enthusiastic and creative individuals who are versatile, dependable, and adaptable to many situations and assignments. Being able to write short articles and press releases as well as having good language skills are required. Interpersonal skills and social talents are helpful. All Publicity/Promotion Assistants must be competent typists.

Minority/Women's Opportunities

Opportunities for women are excellent. The majority of Publicity/Promotion Assistants at commercial stations in 1980 were women, according to industry estimates. In public television in 1980, 76 percent of individuals with the related title of Public Information Assistant were women, and 20 percent were members of minorities, according to the CPB. These represent increases of 14 and 6.2 percent, respectively, over the previous year.

Unions/Associations

There are no unions that act as bargaining agents for Publicity/Promotion Assistants. Many, however, belong to the Broadcast Promotion Association (BPA) to improve their skills and advance their careers.

COMMUNITY RELATIONS DIRECTOR

CAREER PROFILE

Duties: Determining local public needs and creating programs and announcements to meet them

Alternate Title(s): Public Service Director; Director of Community Affairs

Salary Range: $12,000 to $28,000+

Employment Prospects: Poor

Advancement Prospects: Fair

Prerequisites:

Education—Undergraduate degree in government, urban affairs, sociology, or communications

Experience—Minimum of two years in community, civic, cultural, or service organizations

Special Skills—Sensitivity; excellent writing and speaking abilities; service orientation; interpersonal skills

CAREER LADDER

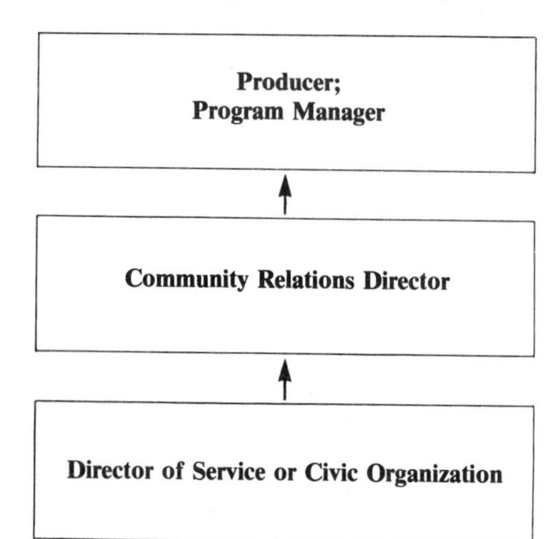

Position Description

A Community Relations Director is directly responsible for developing a continuing program of station services in response to the needs and interests of the local community. He or she is usually the station's primary liaison with public and private agencies, businesses, and professional organizations, in the station's viewing area.

All commercial and public television stations are licensed by the Federal Communications Commission (FCC) to operate "in the public interest, convenience, and necessity." Stations meet this responsibility by broadcasting local programs, public service announcements (PSAs), and special-events coverage that address local needs and interests. Much of the Community Relations Director's time is spent in representing the station at public meetings in order to determine these needs and interests.

As the direct communications link with business associations, civic organizations, social service agencies, and youth organizations, the Community Relations Director is responsible for the effective use of specialized station programming to meet the interests of these groups.

Most Community Relations Directors are responsible for planning and producing the public affairs panel programs, interviews, or talk shows that deal with community-oriented issues. In many instances, he or she serves as the host for these programs. He or she also develops the necessary informal and contractual relationships to involve individual organizations in station programming.

The Community Relations Director oversees the production of all local public service announcements. In addition, he or she screens those that are produced elsewhere to determine their quality and appropriateness for broadcasting by the station.

Community Service Directors deal with a variety of groups daily. They must sympathetically evaluate requests for promotion and help from many worthy organizations. They are service-oriented people concerned with using television to deal effectively with community needs.

A Community Relations Director is usually employed only at middle or major market commercial stations where he or she generally reports to the Program Manager. At smaller market stations, the Program Manager and the General Manager often share the duties. In public television, a station's entire programming operation is designed to be responsive to community needs. The duties of a Community Relations Director are, therefore, performed by a number of people, including the General Manager, the Program Manager, the Director of Publicity/Promotion, and the person in charge of fund raising and development.

Additionally, the Community Relations Director:

- administers all ascertainment files (comments by community members about the station's programming), as required by the FCC;
- arranges and conducts station tours for organizations and groups;
- establishes and trains a speaker's bureau from the station's employees and schedules their appearances at public or organization meetings;
- plans and organizes public service events to discuss the station's programming with community members.

Salaries

Salaries ranged from $12,000 at middle market commercial stations to more than $28,000 at major market stations in 1980, according to industry sources. The average was $18,500. The smaller salaries are usually paid to individuals who have fewer responsibilities.

Employment Prospects

Opportunities are poor. There are relatively few openings in commercial or public television specifically for a Community Relations Director. There were only 16 such positions in public television in 1980, according to the Corporation for Public Broadcasting (CPB).

At many stations the position is filled by a director of a service or civic group with an understanding of local problems and of the organizations that deal with them, or by another individual experienced in community relations.

Advancement Prospects

Opportunities for advancement to other broadcast-related positions are fair. Some Community Relations Directors become full-time Producers of public affairs or news programs. A few use their television skills and knowledge of programming to become Program Managers. Still others use their knowledge to obtain better-paid public relations positions at non-profit organizations or government agencies.

Education

An undergraduate degree in government, urban affairs, or sociology is usually necessary, although a degree in communications or public relations is some times required. A broad liberal arts background with courses in the social sciences is helpful.

Experience/Skills

Most Community Relations Directors have at least two years of previous experience as a director or a public information officer for a community, civic, cultural, or service organization. They should have experience in dealing with a wide variety of social problems and a familiarity with the structure of local business, community, and governmental groups.

Candidates must be sensitive individuals with a strong desire to assist in alleviating problems. They should possess excellent speaking and writing skills and be able to appear on camera with poise. Good social skills are also required. Above all, a Community Relations Director should be service-oriented and capable of dealing with a large number of organizations and people in an understanding manner.

Minority/Women's Opportunities

Opportunities for women and members of minority groups are excellent. The job is relatively new in television, created in response to the influence of the minority and women's movements in the past 15 years. As a result, the majority of Community Relations Directors in commercial television during 1980 were women or minority-group members, according to industry sources. Surveys by the Corporation for Public Broadcasting (CPB) indicate that of the 16 Community Relations Directors in public television in 1980, eight were women and five were members of minorities.

Unions/Associations

There are no unions that serve as bargaining agents for Community Relations Directors. Many individuals join a variety of civic and service organizations in the course of performing their duties.

PRODUCTION MANAGER

CAREER PROFILE

Duties: Responsibility for all local studio and remote productions at a television station

Alternate Title(s): Production Director; Production Supervisor; Production Coordinator

Salary Range: $10,000 to $47,000

Employment Prospects: Poor

Advancement Prospects: Good

Prerequisites:

Education—Undergraduate degree in radio-TV, communications, or theater; graduate degree preferable at some stations

Experience—Minimum of five years in TV production

Special Skills—Creativity; organizational ability; leadership qualities; taste

CAREER LADDER

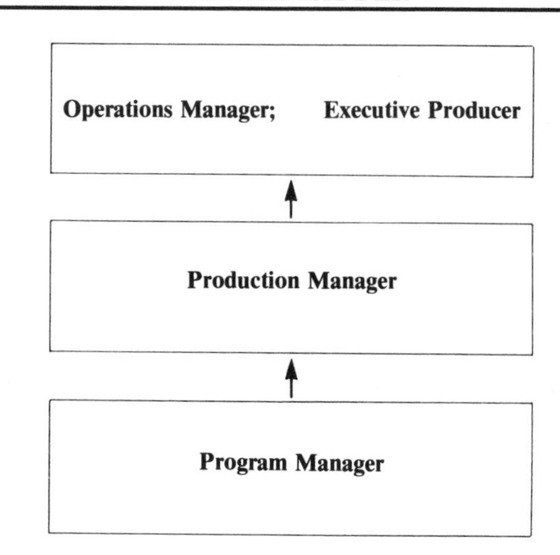

Position Description

The Production Manager's primary responsibility is to conceive, design, and develop ideas for local television productions that result in entertaining and informative programs. He or she is in charge of developing, and complying with, detailed budgets for all local productions, including costs for materials, capital equipment, supplies, and talent.

As overall manager of the station's production schedule, he or she coordinates the creation of locally produced commercial spots, public affairs programs, entertainment shows, specials, and remote coverage of sporting events. He or she oversees the design of sets, art, and prop elements to ensure quality design and workability in a studio or at a remote location. This involves directing and coordinating the creative efforts of many people during the planning and execution stages of television productions.

The Production Manager often supervises Executive Producers, Directors, Lighting Directors, Production Secretaries, the Art Director, Graphic Artists, and other production support personnel, including Camera Operators, Production Assistants, and Floor Managers. At many stations, he or she also supervises the Operations Manager. The Production Manager also negotiates with many individuals from outside the station, including Performers, Writers, and Musicians.

At a smaller commercial station that has few local productions, the Production Manager typically reports to the Program Manager. In many public television stations, where local productions are more frequent, he or she is in charge of a separate department.

Additionally, the Production Manager:

- monitors standards of quality on all local productions;
- coordinates equipment and facilities requirements for local productions.

Salaries

Salaries for Production Managers at both commercial and public television stations are relatively low.

At some small market commercial stations, the position is viewed merely as a glorified Operations Manager, with scheduling duties added, and the salary is correspondingly low. Production Managers' salaries in commercial TV ranged from a low of $10,000 at small market stations to $35,000 in major market outlets with heavy local-origination schedules. The average salary was $18,900 in 1980, according to industry sources.

At most public television stations, however, this is a more important position because of the greater involvement with local productions, and the Production Manager is better paid.

According to a Corporation for Public Broadcasting (CPB) survey, Production Managers' salaries ranged from $15,000 to $47,000 at non-commercial TV stations in 1980. The average was $24,100 a 9.2 percent increase over the previous year. In general, the higher-paid positions are in the top markets, which place an emphasis on locally originated programs. Production Managers' salaries at community stations (those not affiliated with schools or colleges) exceed those of their peers.

Employment Prospects

The job of Production Manager is not an entry-level position. Opportunities for employment are generally poor and competition is heavy. There are fewer than 1,000 Production Managers in television and most have been promoted from within the ranks of the production department of their stations.

As a result of some turnover, there are a few positions open for the more talented Operations Managers or Executive Producers. A typical route to the Production Manager's position at a local station is from Production Assistant to Camera Operator to Floor Manager to Assistant Director to Director to Executive Producer to Production Manager.

The skills of a Production Manager are transferable to cable TV systems, subscription TV (STV), and multipoint distribution service (MDS) operations, and independent production companies, and some Production Managers move into a job with these organizations, thus creating some additional openings.

Advancement Prospects

The opportunities for advancement are good. Most Production Managers at commercial or public TV stations move on to more responsible positions, either within the station or at a smaller market. Some assume similar positions at larger stations. Still others form their own production companies or join established commercial or independent production companies.

At public television stations, Production Managers are often promoted to Program Manager or move to that position at another public station. Occasionally, a Production Manager moves to a higher-paying position as Director of Instructional Television at an educational telecommunications center, as Director of Media Services in private industry, or as a Media Resource Manager or Media Specialist in health organizations.

Education

While it is not essential for the job, most Production Managers have a college degree in radio-TV, communications, theater, or a related liberal arts field. A few have graduate degrees. This is particularly true in public television, where increased emphasis is placed on the role and responsibilities of a Production Manager. A degree is less important at commercial television stations, but some education other than on-the-job experience is usually required.

Experience/Skills

In general, Production Managers are between the ages of 25 and 35. Most have had extensive prior experience in television production. Having spent time at all levels in a production crew, a Production Manager is better able to coordinate all the elements and individuals that are involved in a station's locally originated program schedule.

Experience in cinematography or theater staging is helpful. In most cases, however, the Production Manager is hired from within the industry, and has had at least five years of experience in television production.

Creativity, organizational ability, leadership, and taste are important talents in a Production Manager. He or she must combine a sense of esthetics with pragmatism and logic, and be excellent at working with a variety of people in highly complex tasks. He or she must be able to transform elusive ideas and concepts into visually exciting programs.

Minority/Women's Opportunities

The opportunities for employment as a Production Manager at a local television station for women and minority-group members are limited. In public TV, 5.4 percent of the Production Managers were women, and 5.4 percent were members of minorities, according to a CPB survey in 1980. At both local commerical and non-commercial stations, the position of Production Manager is usually held today by a white male.

Unions/Associations

Some Production Managers at commercial stations have belonged to the National Association of Broadcast Employees and Technicians AFL-CIO (NABET) and a few have been members of the American Federation of Radio and Television Artists (AFTRA), but most become inactive when they reach this level. There is no union or professional organization that specifically represents Production Managers.

EXECUTIVE PRODUCER

CAREER PROFILE

Duties: Conceiving, developing, and producing a television series, complex program, or special production

Alternate Title(s): None

Salary Range: $8,200 to $100,000 +

Employment Prospects: Poor

Advancement Prospects: Poor

Prerequisites:

Education—Minimum of undergraduate degree in mass communications, radio-TV, or specialized subject area

Experience—Three years as a television Producer

Special Skills—Creativity; leadership qualities; interpersonal skills; organizational ability; business acumen; financial orientation

CAREER LADDER

Production Manager; Program Manager

↑

Executive Producer

↑

Producer

Position Description

An Executive Producer is responsible for conceiving and developing ideas for television series, individual complex programs, or special productions that meet defined audience objectives. He or she formulates the rationale for the project, determines the format, frames the budget, arranges the financing, and oversees the promotion.

Many Executive Producers work with independent production companies that create programs for network broadcast or syndication. Some are staff members at major market stations in charge of all news, documentary, or public affairs programs. Some are employed at the networks in charge of developing shows for specific audiences (children) or in specific subjects (religion or sports).

The responsibilities of the Executive Producer last through the life of the project. He or she seeks and identifies ideas that can be creatively developed into entertaining or informative shows, researches them, determines whether they can be produced within the available technical resources and facilities, and writes, edits, or approves the program scripts.

The overall budget is established by the Executive Producer who carefully monitors all expenses concerned with the project. He or she often finds and convinces sponsors to fund the broadcast. The Executive Producer exploits all of the resources of television, arranges for facilities, equipment, and personnel, and supervises every aspect of the project to ensure smooth, interesting, and entertaining broadcasts.

One or more Producers are selected by the Executive Producer to interpret and execute the concepts of individual shows or segments. The Executive Producer, however, acts as final arbiter in the selection of Writers, Directors, Performers, and special technical/craft and production staff, including Music Directors, Costume Designers, and Scenic Designers. An Executive Producer has the ultimate responsibility for the success or failure of the project as it is seen on the air.

Additionally, the Executive Producer:

- approves the completed program(s) prior to broadcast;
- monitors all promotion and publicity for the program(s);
- evaluates audience response to the show(s).

Salaries

In keeping with the talent and versatility needed for the position, Executive Producers are well paid. In commercial television, salaries in 1980 ranged from $25,000 to $45,000 at major market stations, according to industry sources. Some Executive Producers of network programs earn well over $100,000 yearly. In public television, 1980 salaries ranged from a low of $8,200 at the smaller university- or school-owned stations to $75,000 for Executive Producers of programs designed for broadcast on the Public Broadcasting Service (PBS) system, according to the Corporation for Public Broadcasting (CPB).

Employment Prospects

The opportunities for employment as an Executive Producer are generally poor. Only the brightest and most capable achieve the position and competition for the job is heavy. While some major market public and commercial TV stations employ two or more Executive Producers, there is little opportunity in small market stations or in government, education, or health telecommunication units, where Producers often perform the roles of Executive Producers.

An Executive Producer is usually promoted from within the station's ranks. Having produced a variety of successful productions, a bright and talented individual may be given overall responsibility for high-budgeted productions. Some entrepreneurial individuals with great initiative, exciting ideas (and often a sponsor) are hired from outside the station to be an Executive Producer for a series or special.

Advancement Prospects

The possibilities of advancement from Executive Producer to Production Manager or Program Manager are poor. Some individuals, however, obtain those positions in smaller market stations, although the assumption of these largely administrative posts reduces their creative activities as Executive Producers. Still others are able to get better paying positions at independent production companies. Some experienced people form their own companies and develop programs for syndicated sale to stations or to networks. A very few are skilled and talented enough to obtain a staff Executive Producer position at one of the commercial networks.

Education

Most employers require that candidates have a minimum of an undergraduate degree in mass communications or radio-TV prior to promoting or hiring an Executive Producer. Many Executive Producers have training and education in the subjects for which they develop programs. A degree in a particular area is often necessary for an Executive Producer of specific kinds of broadcasts, such as political science for news programs, or theater arts for dramatic productions.

Experience/Skills

Most Production Managers or Program Managers expect candidates to have at least three years of experience as a television Producer. The individual must have a thorough practical knowledge of all stages involved in producing a TV program and an understanding of television or film technology.

The Executive Producer is an idea person. He or she must be creative and able to develop concepts, ideas, and formats for television programming. The Executive Producer must have leadership capabilities and great interpersonal skills to manage creative and technical people effectively. He or she must be well-organized and a sound business manager. An Executive Producer should be able to estimate costs and develop budgets for television productions.

Minority/Women's Opportunities

Few Executive Producer positions in commercial television were occupied by women or members of minorities in 1980. Although their representation has increased in recent years, the job is most often held by white males. Public TV, however, has experienced an increase in programming for both women and minorities in the past ten years, with an accompanying growth in their opportunities to become Executive Producers. In 1980, 35 percent of Executive Producers in public television were women, and 10.7 percent were minority-group members, according to the CPB.

Unions/Associations

There are no unions or professional organizations that represent Executive Producers.

PRODUCER

CAREER PROFILE

Duties: Creating and developing individual television productions

Alternate Title(s): Producer/Director

Salary Range: Under $18,900 to $50,000+

Employment Prospects: Poor

Advancement Prospects: Poor

Prerequisites:

Education—Undergraduate degree in radio-TV, communications, theater, or journalism

Experience—Minimum of three years as a Director or Associate Producer

Special Skills—Leadership qualities; creativity; interpersonal skills; organizational abilities; orientation to detail; financial expertise; business acumen; television judgment

CAREER LADDER

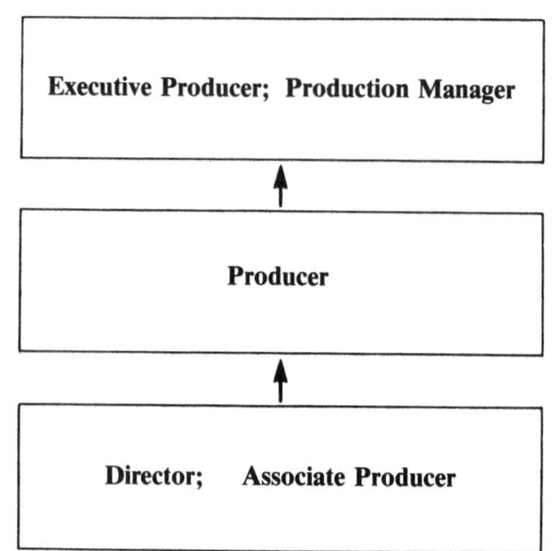

Position Description

A Producer is the overall manager and arranger of an individual television production or of one that is part of a series. He or she conceives and develops ideas for a program, consistent with the theme established by an Executive Producer or Production Manager. He or she ensures that the program meets all of the station's objectives and is compatible with audience interests.

A Producer is a skilled professional who is often selected by an Executive Producer to generate a single program or segment of a larger project. Producers are usually staff members of a TV station, a network, or an independent production company that creates shows for networks or syndication. Most commercial and public stations employ several full-time Producers. At network and independent production companies, two or three are often responsible for individual programs within a series and may have alternating responsibilities on a daily or weekly basis, depending on the complexity of the productions.

Virtually all types of programs require Producers, including variety shows, soap operas, talk shows, dramas, situation comedies, news, public affairs, sports, and all other categories of TV programming. Within the parameters of a series' or program's goals, the Producer has considerable latitude in developing ideas, selecting Performers, creating scripts, choosing Directors, and determining the specific approach and format of each production.

It is the job of the Producer to keep a program within budget and on schedule. He or she takes care of all business arrangements, and is responsible for contracts with Performers, handles clearances, orders technical facilities and equipment, and schedules rehearsals and performances. The Producer supervises negotiations with personalities and other Performers as well as with professional and technical people needed for a production. He or she selects Performers who will appear on camera, including guests, hosts, and musicians. He or she also coordinates production assignments and supervises the Director. On occasion, a Producer acts as on-air talent, that is, as a host, interviewer, or Reporter.

For newscasts and documentaries, a Producer is frequently involved in selecting the stories and special events to be covered. He or she often decides on the issues to be discussed, supervises background research, and helps to choose individuals to be interviewed.

Additionally, the Producer:

- supervises the development of promotional material for the program;
- ensures that the program meets quality production standards.

Salaries

In commercial broadcasting, Producers of news programs earned an average of $18,900 in 1980, according to industry sources. Some Producers of sports, special events, or entertainment programs at major market stations make more than $30,000 per year. Producers at networks and at independent production companies frequently earn $50,000 or more.

In public television, the average 1980 salary for a Producer was $23,700 according to the Corporation for Public Broadcasting (CPB). Producers of news and public affairs programs at major market community-owned stations earned from $25,200 to $37,800, and at state-owned public stations from $21,000 to $27,600 in 1980.

Producers of training, public information, instructional, and health programs in private industry, government, health and educational organizations usually receive slightly lower salaries.

Employment Prospects

The Producer's position is highly coveted at most stations and competition for the job is heavy. Most television stations employ from two to eight Producers who work on specific assignments, or on special programs during a given period. In addition, some Producers move to similar positions at independent production companies, advertising agencies, film production operations, or to larger market stations, and therefore, jobs become available at some stations. Because of the expertise and training required and the extensive competition, employment prospects as a Producer in TV broadcasting are generally poor.

Government, health, private industry, and educational video production operations also employ Producers who are often specialists in a particular subject area. The opportunities for employment in these non-broadcast settings, although usually better than at TV stations, are only fair.

Advancement Prospects

Many Producers view the position as the final step in their careers. Some, however, become Executive Producers or Production Managers at smaller market stations. Still others seek administrative, production, or program responsibilities at independent production companies or at educational, health, and governmental telecommunications organizations. The competition for such positions, however, is heavy.

A number of Producers specialize in news and public affairs, commercials, sports, or entertainment programs and advance their careers by working on higher budgeted or more prestigious shows where they have expertise in the subject covered. A very few Producers with great talent and ability obtain positions at the network level, or with independent production companies that produce programs for network or cable TV.

Education

Most employers require that a Producer have a degree in radio-TV, communications, theater, or journalism. He or she should have a broad exposure to the liberal arts and some specific education in the subject area of the assigned program.

Experience/Skills

A candidate for a Producer's job should have a minimum of three years of experience as a Director or Associate Producer. He or she needs a thorough practical knowledge of TV production and should be familiar with budget estimation. A Producer should have experience in all stages of program development from conception to videotape editing.

Producers must demonstrate leadership qualities and be creative and original in developing new formats and techniques for TV production. They must be able to coordinate a number of individuals, including Performers and technical and production people. Producers must be well-organized, detail-oriented, financially astute, and have good business and television judgment.

Minority/Women's Opportunities

According to industry sources, the majority of Producers at commercial stations in 1980 were white males, although women and minority-group members are increasingly employed in the position, particularly at smaller market stations. In public television, 23.2 percent of the Producers were female and 13.1 percent were members of minorities in 1980, according to the CPB. This represented an increase of 1.5 and 1.2 percent, respectively, over the previous year. As programming for minorities and women increases, industry observers predict a growth in opportunities for them as Producers.

Unions/Associations

Some Producers who also work as on-air talent are members of the American Federation of Television and Radio Artists (AFTRA). A few are members of the Directors Guild of America (DGA). In the majority of instances, however, Producers are not represented by a union or professional organization.

ASSOCIATE PRODUCER

CAREER PROFILE

Duties: Providing administrative and professional support to a television Producer

Alternate Title(s): Assistant Producer

Salary Range: $6,500 to $33,700

Employment Prospects: Fair

Advancement Prospects: Fair

Prerequisites:

Education—Undergraduate degree in mass communications or radio-TV

Experience—Minimum of two years in a variety of production positions

Special Skills—Creativity; organizational aptitude; leadership qualities; writing ability; cooperativeness

CAREER LADDER

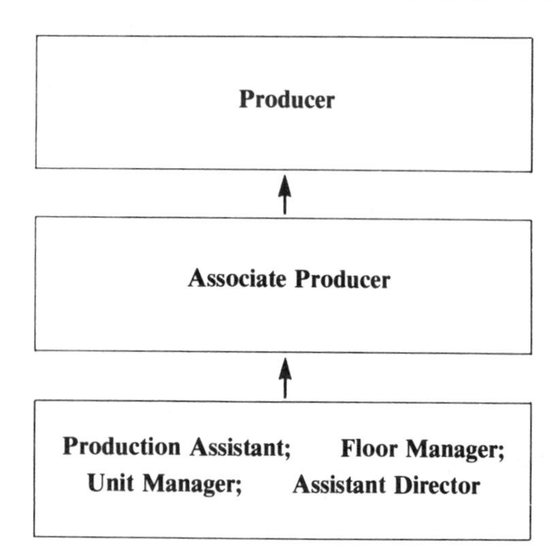

Position Description

An Associate Producer provides general administrative and professional support to the Producer of a TV program. He or she acts as the Producer's chief aide in conceiving, developing, and producing a TV show recorded on tape, or broadcast live from a studio or remote location. In the Producer's absence, he or she assumes full responsibility.

Some complex TV productions employ several Associate Producers, each of whom has a specific function during various production stages. In addition to helping others work out their concepts, an Associate Producer also develops his or her own original story lines and ideas, and, often, is given full responsibility to create taped elements or segments to be inserted into the finished show.

An Associate Producer helps to prepare the budget, organizes the production schedule, and occasionally supervises Camera Operators, Cinematographers, Videotape Engineers, and other members of the production team. He or she also evaluates scripts and program proposals, helps design set and lighting plots, and, sometimes, selects music for bridges and inserts. He or she often negotiates rights for film, slide, or tape inserts, Musician or Performer "step-up" fees (residual payments for wider distribution of a program), and, on occasion, the rights for cable TV or audio-visual distribution.

As a rule, an Associate Producer is given wide latitude in helping to develop the production. He or she usually reports to a Production Manager and is assigned to one Producer for specific projects or programs. He or she can work on news, sports, documentaries, talk shows, and entertainment specials.

The Associate Producer can be an important member of a station's news department. He or she may be called upon to assist a Producer or Reporter in developing a story or documentary by providing some of the research and doing part of the writing. The Associate Producer will also sometimes conduct background interviews and supervise the filming or videotaping of segments to be used in the finished program.

Additionally, the Associate Producer:

- works with the Operations Manager in scheduling facilities and equipment;
- assists in film or videotape editing of a final production.

Salaries

Salaries for Associate Producers range from $12,000 for beginners in smaller market commercial stations or production companies to more than $30,000 for experienced individuals who work on major TV programs or films for network broadcast.

In public television, Associate Producers' earnings went from $6,500 for entry-level employees at smaller school or university stations, to $33,700 at major market community-owned stations for those working on major program projects for PBS system distribution in 1980, according to the Corporation for Public Broadcasting (CPB). The average 1980 salary was $16,900.

Employment Prospects

Although some smaller market stations view the Associate Producer to be a general all-around person involved in detail work, most TV production operations require more experienced individuals. To become an Associate Producer, therefore, one needs a thorough knowledge of TV production which can usually be gained as a Production Assistant, Floor Manager, Unit Manager, or Assistant Director.

Although many stations or independent production companies employ more than one Associate Producer, competition for the position is usually quite strong. There is, however, considerable mobility among Associate Producers, and through attrition a number of positions become available. Advertising agencies also employ Associate Producers. The overall opportunities for employment are only fair.

Advancement Prospects

Most Associate Producers are ambitious, have considerable experience in TV production, and actively seek more responsible positions, usually as Producers. The more creative and talented move to a Producer position at a smaller market station or become a Producer of news, public affairs, or sports programs at their own station. The competition for these jobs is great, however, and only the best and the brightest make the advancement. Some Associate Producers move to advertising agencies or to education, health, or government TV operations as Producers of specific or specialized programs. The opportunities for advancement are generally only fair.

Education

Television stations most often hire Associate Producers who have an undergraduate degree in mass communications or radio-TV. Some theater training is desirable, and courses in writing are also helpful. An Associate Producer who wishes to work in a specialized area such as news, sports, or music, has a better chance if he or she has some educational background in that, or a related, discipline.

Experience/Skills

Most employers require a minimum of two years of varied experience in television production. An Associate Producer usually has worked as a Production Assistant, Floor Manager, or Unit Manager and has operated film and tape editing equipment. The majority have had some experience as Assistant Directors.

Employers seek bright and aggressive people with original ideas for programs. An Associate Producer must be creative, well organized, and should demonstrate leadership abilities. He or she should also be versatile, persistent, a good writer, and capable of working well with a variety of people.

Minority/Women's Opportunities

The opportunities for women and members of minority groups in gaining employment as Associate Producers are increasing yearly. This is particularly true at middle and major market stations, although advertising agencies, government, health, and educational production units are also employing women and minority members as Associate Producers in greater numbers.

Unions/Associations

There are no unions or professional groups that represent Associate Producers.

DIRECTOR

CAREER PROFILE

Duties: Orchestrating all technical and creative elements of a television production

Alternate Title(s): Producer/Director

Salary Range: $7,500 to $65,000+

Employment Prospects: Poor

Advancement Prospects: Fair

Prerequisites:

Education—Undergraduate degree in radio-TV or communications

Experience—Minimum of two years in television production

Special Skills—Creativity; technical knowledge; organizational and leadership qualities; motivational skills

CAREER LADDER

Producer

↑

Director

↑

Assistant Director; Unit Manager; Floor Manager

Position Description

A Director plays a critical role in any television production. He or she turns a concept or script into a cohesive and interesting program, creatively, effectively, and within budget. Working in a complex, pressure-filled environment, he or she is the unifying force during the planning, shooting, and completion of a TV program. The show's final success is largely due to the Director's talent and ingenuity.

The Director coordinates all elements, facilities, and people during rehearsal and actual production. He or she gives significant and detailed instruction to the studio or location crew and talent, including the production team, technical staff, and Performers. A Director is usually assigned to a Producer for a particular show or program, but sometimes he or she also assumes the Producer's role.

The work of a Director begins in pre-production planning stages. He or she meets frequently with the Producer, and sometimes the Writer, to develop a workable program. The Director collaborates with important members of the team to determine production elements, including equipment and engineering requirements, lighting needs, the number of cameras necessary, music, costumes, and sets. He or she also hires or approves of the cast. Working with a floor plan, the Director plots camera shots, placement of equipment, and blocking of Performers.

During rehearsals and production, the Director oversees all production and engineering personnel from the control room, and communicates with the studio via headphones. He or she selects all camera shots and movements that will be seen eventually by the audience. He or she combines all pre-planned creative and technical elements into a smooth and polished show. In the post-production phase, the Director supervises the videotape editing of the program.

A Director's role is somewhat different in news and documentaries. The Producer chooses what stories or events will be covered and selects the film or videotape segments to be shown. The Director then organizes the pre-edited pieces and studio elements into a unified broadcast.

Directors are employed on staff by all networks and virtually every station. At some smaller outlets, one or two staff Directors may be responsible for all locally-produced programs. At others, one Director may be in charge of a continuing program or series, or may specialize in a particular format such as news or sports. Independent production companies usually employ Directors on staff and, occasionally, may hire freelance Directors for specific projects.

Additionally, the Director:

- assists in the design of the set and lighting plot;
- reviews the script with Performers and conducts rehearsals;

• selects slides, art work, graphics, and tape and film inserts for the program.

Salaries

1980 salaries in commercial television for the related position of Producer/Director (the one often used in surveys) averaged $14,500, according to industry sources. Some experienced people at major market stations earned over $30,000. The few Directors of major TV programs for network release make considerably more. Directors received a minimum of $2,566 per week for such projects in 1980, according to the Directors Guild of America (DGA). Most such individuals work on one or two productions annually, but can earn over $65,000 if they work six months or more a year.

In public TV, 1980 salaries for Directors ranged from a low of $7,500 in small university- or school-owned stations to $34,900 for experienced individuals with seniority at major market community-owned stations, according to the Corporation for Public Broadcasting (CPB). The average was $17,100, an increase of 18 percent over the previous year.

Employment Prospects

While most commercial and public stations employ from two to six Directors, the opportunities for employment are very limited. The job requires considerable experience and skill, and competition for the position is strong. Most Directors are promoted to their positions after working as Assistant Directors, Floor Managers, or Unit Managers.

Some additional opportunities exist at networks and independent production companies, as well as at advertising agency, educational, government, health, and corporate TV operations. The overall chances of becoming a Director are, however, poor.

Advancement Prospects

For some, the position of Director is considered the peak of a successful career. A considerable number, however, become Producers and others assume more responsibility as producer/directors. Usually, a Director obtains one of these positions at a smaller market station at more salary, although some move into comparable positions at independent production houses or ad agencies, at higher salaries. Some Directors specialize in a particular kind of program, such as news, sports, variety, or cultural shows and command higher salaries for more complex productions. In general, opportunities for advancement are fair.

Education

An undergraduate degree in radio-TV or communications is usually required to be hired as a Director. Many individuals have some education in theater and others come from a journalism background. While a degree is not an absolute requirement, a broad education, usually in the liberal arts, is extremely useful for Directors, who must work with programs that involve various subjects, styles, and formats. Few Directors are employed straight out of college or broadcast training school.

Experience/Skills

Many television Directors are between the ages of 25 and 30, and have had at least two years of production experience. They often apprentice in all positions on a production crew, including Camera Operator, Floor Manager, Unit Manager, and Assistant Director. A well-known actor may also be given the opportunity to direct a show. Occasionally a successful stage or film Director may use his or her skills to direct a TV show.

Directors must have fast reflexes to give instructions to a number of people working in a variety of tasks under severe time constraints. They should have a thorough knowledge of the medium, should be well-organized, methodical planners, with excellent leadership qualities, and an ability to motivate others. They must act decisively under pressure, be creative and flexible, and have a good eye for the composition of images.

Minority/Women's Opportunities

The majority of Directors in commercial and public TV are white males, although women and minority-group men have reached the position more often in the past few years. During 1980, 9.2 percent of the Directors in public television were women, and 18.3 percent were members of minorities, according to the CPB.

Unions/Associations

Directors working for networks, independent production companies, and some major market stations are usually represented by the Directors Guild of American (DGA). The majority of Directors, however, are not represented by a union or professional organization.

ASSISTANT DIRECTOR

CAREER PROFILE

Duties: Assisting the Director in television productions

Alternate Title(s): Associate Director

Salary Range: $12,000 to $28,000+

Employment Prospects: Fair

Advancement Prospects: Fair

Prerequisites:

Education—Undergraduate degree in mass communications, radio-TV, or theater

Experience—Minimum of two years as a Unit Manager, Floor Manager, or Production Assistant

Special Skills—Firm knowledge of TV production; organizational ability; facility with details; creativity

CAREER LADDER

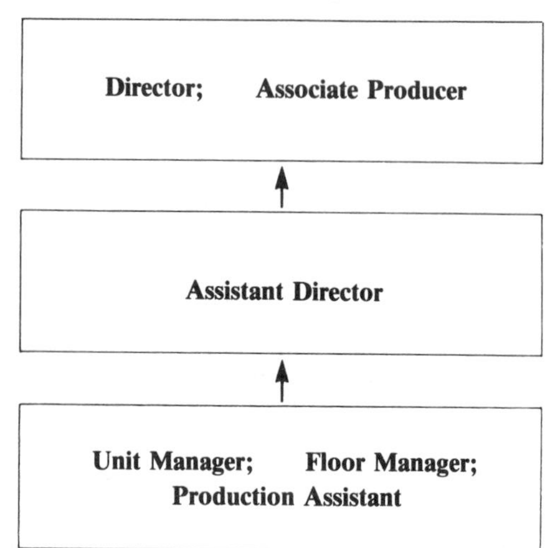

Position Description

The Assistant Director helps the Director to coordinate the production elements of a TV program and makes certain that Performers, equipment, sets, and staff are ready for rehearsals and shooting. He or she serves as the Director's right hand prior to, during, and immediately following the production. Usually, the Assistant Director reports to a Production Manager and is assigned to a Director on a project or program basis.

A major responsibility of the Assistant Director is to make sure that all film segments, slides, tapes, inserts, and electronic titles have been assembled and timed in the pre-production stage. He or she assists in the development and preparation of scripts, times rehearsals, and prompts Performers. The Assistant Director helps the Director block camera shots and Performer movement on the floor plan.

For a remote production or a complex dramatic film, the Assistant Director scouts locations and supervises the Unit Manager in organizing arrangements for transportation, lodging, facilities, and personnel. For studio rehearsals, he or she takes notes and transmits the Director's production requirements to Production Assistants, Technical Directors, Floor Managers, Makeup Artists, Lighting Directors, and other production team members.

During rehearsals and actual production, the Assistant Director is in the control room and is in charge of timing the various elements of the program and relaying time cues to the Floor Manager and other production personnel. In the absence of a Technical Director, he or she is responsible for presetting and readying camera shots and cueing film, tape, and slide inserts. He or she also helps the Director in post-production editing of a videotaped or filmed TV show.

Additionally, the Assistant Director:

- coordinates all deadlines related to the production;
- handles immediate emergencies pertaining to sets or Performers;
- ensures that all elements of the production blend into a cohesive, smooth process.

Salaries

Many small market stations, both commercial and public, with relatively simple news or interview programs, offer low annual salaries of $12,000 to $15,000, according to industry estimates. The few Assistant Directors working on major TV films, complex entertainment specials, or major market productions, command much higher salaries. Minimum payments for such Assistant Directors were $1,100 a week in 1980, according to the Directors Guild of America (DGA). These individuals work on assignment for one or two specials a year as contract personnel, and are not part of the regular station staff. A non-staff person who works six months a year at union minimum, earns more than $28,000 annually.

Employment Prospects

The Assistant Director job is a coveted one in most commercial or public TV stations and telecommunications operations because it often leads to higher production positions. The competition among Floor Managers, Unit Managers, and other production workers is, therefore, very heavy.

Most complex TV or film productions require one Assistant Director on a project-by-project basis. Only major market stations, networks, and independent production companies hire them as full-time staff members. A few advertising agencies, cable TV, health, government, and educational TV operations also use them. Since many good Assistant Directors move on to more responsible positions at smaller market stations, positions are usually available for qualified people. Because of the competition, however, employment prospects are considered to be only fair.

Advancement Prospects

Most Assistant Directors view the job as a stepping-stone to becoming Directors, and some obtain that position at smaller market stations. Still others become Associate Producers at their stations, or take similar positions at independent production companies.

There are not as many openings for either Director or Associate Producer positions as there are qualified candidates and only the most creative Assistant Directors are likely to be promoted. Because of the general competition, overall prospects for advancement for Assistant Directors are only fair.

Education

Production Managers usually require an undergraduate degree in radio-TV, mass communications or theater in promoting or hiring a person as an Assistant Director. Although a degree is not required at smaller market stations, most employers select applicants who have had some post-high school training in television production or in the theater.

Experience/Skills

Most employers in major market stations require a minimum of two years of experience as a Unit Manager or Floor Manager for promotion to Assistant Director. In some stations, however, an experienced Production Assistant may be promoted to the job.

In addition to having a sound background in all aspects of TV production, an Assistant Director must be well organized and detail-oriented in order to follow through on the complex elements involved in working on a program. Above all, he or she must be creative and be capable of dealing with a variety of other creative as well as technical people in pressure situations.

Minority/Women's Opportunities

The opportunities for women to become Assistant Directors are excellent, according to industry sources. At smaller market commercial and public TV stations, over one-half of the people in this position were female in 1980. Additionally, there are an increasing number of opportunities for minority-group members to obtain this job.

Unions/Associations

In the majority of instances, there are no unions or professional organizations that represent Assistant Directors. For many major made-for-TV dramatic films or taped specials, however, the DGA serves as the bargaining agent.

UNIT MANAGER

CAREER PROFILE

Duties: Overseeing logistics and budget control of a television production or series, usually on location

Alternate Title(s): Studio Supervisor; Remote Supervisor; Unit Production Manager; Unit Supervisor

Salary Range: $8,200 to $27,500

Employment Prospects: Fair

Advancement Prospects: Fair

Prerequisites:

Education—High school diploma; TV or film production training; some college preferable

Experience—Minimum of two years TV/film production

Special Skills—Technical aptitude; organizational and administrative skill; financial management ability

CAREER LADDER

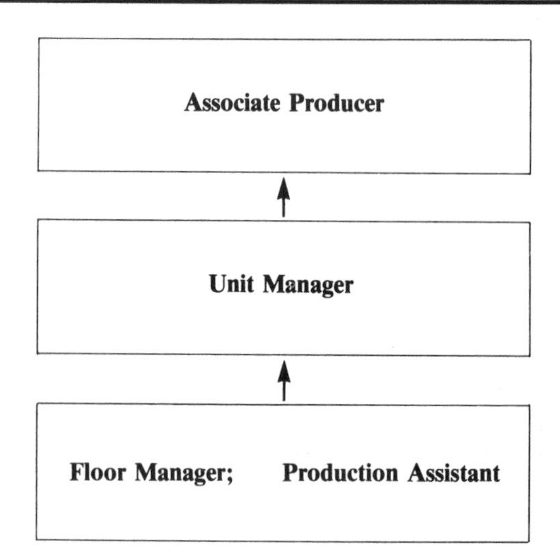

Position Description

A Unit Manager is the primary logistics organizer of all elements of a television production. He or she also serves as the person immediately in charge of budget expenditures and ensures that all production costs are within budgetary limitations.

Usually, a Unit Manager is immediately responsible for all pre-production scheduling and for the setup, maintenance, and operation of all facilities and equipment. He or she organizes staging and lighting personnel, Camera Operators, Production Assistants and other production employees in the preparation of rehearsals and in pre-production activities. A Unit Manager usually reports to an Assistant Director or Associate Producer, and is normally employed only at middle or major market stations.

A Unit Manager must be able to manage the diverse technical and practical elements of a television production efficiently and effectively.

Although Unit Managers are most often responsible for on-location film or remote television productions, the title is sometimes used in major market stations for the people in charge of a daily in-studio television production. They oversee the budget and physical organization of a continuing series of television productions.

Additionally, the Unit Manager:

- scouts and selects locations for remote productions;
- schedules transportation, lodging, food, and other requirements for a TV remote;
- organizes the delivery and setup of all equipment, including cameras, lights, audio equipment, monitors, sets and scenery, easels, props, and other items necessary for a production;
- checks and monitors all facilities and crew during production to ensure a smooth operation;
- organizes the strike (dismantling) of a remote production and the return of all equipment and personnel.

Salaries

Unit Managers in commerical television were paid from $13,000 in smaller markets to more than $25,000 for experienced personnel in the major markets and production companies in 1980, according to industry sources.

The salaries at public TV stations for employees functioning as Unit Managers were approximately the same. Payments for these people, often titled Unit Supervisors or Studio Supervisors, ranged from $8,200 to $27,500 in major market public TV stations in 1980, according to the Corporation for

Public Broadcasting (CPB). The average salary was $16,600 in 1980, an increase of 6.2 percent over the previous year.

Employment Prospects

The position of Unit Manager is not normally an entry-level job and opportunities are only fair for diligent and ambitious production employees at a station. Unit Managers are often assigned from the production staff to specific remotes, or to a continuing series of in-studio programs. Because of this, there are sometimes opportunities for a production employee to be assigned periodically as a Unit Manager to various projects during a year. In addition, many major market stations maintain standing production units for continuing series, and efficient Floor Managers or Production Assistants are sometimes promoted to Unit Managers on a permanent basis. Other opportunities for employment are in film production or with commercial TV production companies, where on-location shooting is routine.

Advancement Prospects

Most Unit Managers are aggressive and have a penchant for organization. In large stations, the opportunities for advancement to Associate Producer or to other responsible production posts are fair. A series of successful assignments often impresses a superior enough to promote the Unit Manager to a more responsible position. Some Unit Managers seek advancement in smaller stations as Production Managers. Still others join independent film or television commercial production organizations to further their careers.

Education

A high school diploma and some evidence of technical training in television or film production is helpful for employment in or promotion to the position. Some major market stations prefer an undergraduate degree in radio-TV, mass communications, or theater. Courses in theater staging and lighting can also be useful. Experience in television production, however, is often more important than formal education.

Experience/Skills

New Unit Managers are usually expected to have at least two years of experience in television or film production and a thorough knowledge of TV equipment and facilities.

The position also requires an ability to work well with people on both an individual and a collective basis, and to organize personnel for complex and interrelated tasks in pressure situations. The Unit Manager must understand budgets, be highly organized, and have a good mind for detail.

Minority/Women's Opportunities

The majority of Unit Managers in commercial television in 1980 were white males, according to industry estimates. The number of women holding the job has increased somewhat and should continue to do so.

At public television stations in 1980, 21.6 percent of the Unit Managers and Studio Supervisors were white females and 15.3 percent were minority-group members, according to the CPB.

Unions/Associations

Some Unit Managers working on television films (60 minutes or more) for network release are classified as Unit Production Managers, and as such are represented by the Directors Guild of America (DGA). Most Unit Managers in local commercial or public television stations, however, do not belong to a union.

FLOOR MANAGER

CAREER PROFILE

Duties: Coordination of Director's instructions with studio and remote production activities

Alternate Title(s): Stage Manager; Crew Chief

Salary Range: $6,700 to $24,300

Employment Prospects: Fair

Advancement Prospects: Fair

Prerequisites:

Education—High school diploma; some college preferable

Experience—Minimum of one year in television production

Special Skills—Versatility; cooperativeness; organizational ability; cool-headedness

CAREER LADDER

Assistant Director

↑

Floor Manager

↑

Camera Operator; Production Assistant

Position Description

The Floor Manager is responsible for coordinating the Director's instructions with all crew and talent activities on the studio floor or on remote location during rehearsals and actual production. A major function of the Floor Manager is to follow along with the script and cue Performers, both in rehearsal and, silently, during production. It is one of the most important positions on the production team, especially during actual production.

Prior to the production, the Floor Manager supervises all staging activities for each program, including the setup of scenery and all production-related equipment and devices. He or she works closely with the Art Director or Scenic Designer for on-the-spot construction, painting, and modifications of the set. He or she verifies that all props and costumes are accounted for and are on hand.

During actual production, the Floor Manager positions easels and graphics, operates the teleprompter, places props, and gives timing signals and other cues to Performers. He or she wears headphones and acts as the on-site extension of the Director, relaying instructions by hand signals and cue boards to the crew and Performers.

The Floor Manager makes some independent, immediate decisions regarding the solution of production problems that occur in the studio during rehearsal or actual production. As the Director's on-site representative, he or she must behave calmly and efficiently so that any tensions related to the production are not transmitted to the Performers or crew.

Additionally, the Floor Manager:

- assists the Lighting Director in the transport, setup, and placement of lights and accessories;
- positions video monitors during rehearsal and production;
- directly supervises the dismantling and storage of set pieces and production equipment.

Salaries

In keeping with the necessary experience and skills required for the position of Floor Manager, it is a relatively well-paid job in commercial and public TV. At commercial stations, salaries ranged in 1980 from $9,000 at smaller market stations to more than $20,000 for individuals with years of experience and seniority. At unionized, major market stations, salaries ranged from $14,900 for people with less than six months of experience to $21,300 for those with more than two and a half years of experience.

In public television in 1980, Floor Managers' salaries went from $6,700 for beginners to $24,300 for experienced, senior people, according to the Corporation for Public Broadcasting (CPB). The average was $14,500 in 1980, an increase of 11.1 percent over the previous year.

Employment Prospects

The chances of employment are only fair. The position of Floor Manager is not an entry-level job in either commercial or public television.

Most employers require candidates to have had a minimum of one year of experience in a production crew, usually as a Camera Operator or Production Assistant. Many stations promote only their most experienced crew members to the job. While some major market stations employ more than one Floor Manager for various programs or shifts, most small and middle market stations have only one individual for whom this job is his or her primary responsibility.

Some Floor Manager positions are available in larger production firms, at cable television systems, and in health, government, and education television units where a significant amount of production is scheduled.

Advancement Prospects

Opportunities for advancement are fair for competent, talented individuals. The position of Floor Manager is often a stepping-stone for ambitious, professional younger people employed in television production. The majority of individuals in such positions were between the ages of 25 and 30 in 1980, according to industry estimates. A bright Floor Manager often moves up to Assistant Director after a year or two in the job. Some Floor Managers switch to smaller market stations or to independent production companies for more responsible positions, such as Unit Manager.

Education

A high school diploma and some college training in theater or radio-TV are often required. An undergraduate degree in mass communications or radio-TV is particularly useful in obtaining the job at major market commercial or public stations.

Experience/Skills

A minimum of one year of experience as a member of a production crew is usually required. A Floor Manager needs a good knowledge of all aspects of television production, including lighting, staging, make-up, and camerawork. An ability to organize many elements into a smoothly running operation is a necessity. In addition, the Floor Manager must be resourceful, have initiative, and be capable of independent action in managing the complex aspects of a studio production. Above all, the Floor Manager must appear calm and confident during a hectic television production.

Minority/Women's Opportunities

While the majority of Floor Managers in commercial television are white males, women are occupying the position in growing numbers, according to industry estimates. Members of minorities are also increasingly being hired for the position in commercial television.

In public television, 14.7 percent of the Floor Managers in 1980 were women, and 20.6 percent were minority-group members, according to the Corporation for Public Broadcasting (CPB). These figures represent decreases of 3 and 5 percent, respectively, over the previous year.

Unions/Associations

The majority of Floor Managers in commercial and public TV are considered members of the production department and thus are not represented by a union. At some major market stations, they are members of and are represented by the National Association of Broadcast Employees and Technicians AFL-CIO (NABET) or the International Brotherhood of Electrical Workers (IBEW). Floor Managers working on long-form made-for-television movies are sometimes represented by the Directors Guild of America (DGA).

PRODUCTION ASSISTANT

CAREER PROFILE

Duties: Assisting in the production of studio and remote television programs

Alternate Title(s): Staging Assistant; Floor Assistant

Salary Range: $4,800 to $17,100

Employment Prospects: Good

Advancement Prospects: Good

Prerequisites:

Education—High school diploma; undergraduate degree preferable

Experience—Some theater, photography, or film experience

Special Skills—Resourcefulness; organizational ability; cooperativeness; initiative; typing skills

CAREER LADDER

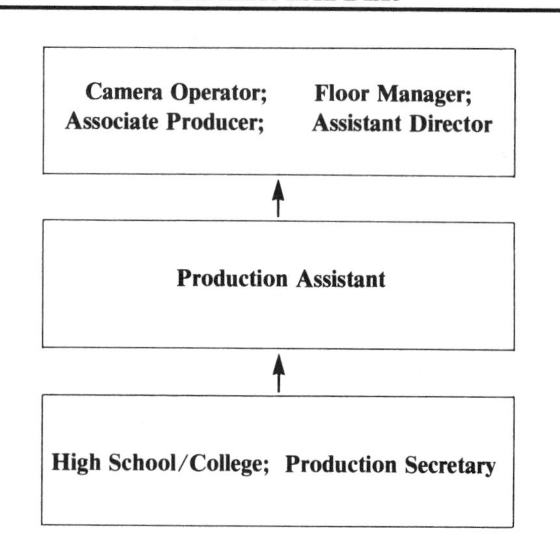

Position Description

The Production Assistant aids in the creation of a local, network, or independent television production in a variety of ways to help achieve a smooth, polished, and professional program. He or she helps the Floor Manager, Unit Manager, Producer, or Director to schedule, coordinate, and set up a TV production. As an apprentice, he or she has the opportunity to gain experience in a wide assortment of TV programming, including news, documentaries, talk shows, and entertainment specials.

The major responsibility of this position is to assist the production staff in non-engineering matters by providing effective and timely support in all phases of a production. The Production Assistant often acts as a principal message and communications liaison between various staff members, and, as such, works with the Camera Operators, Assistant Director, Associate Producer, Director or Producer, and the Floor Manager. During the preparation of a program, he or she often reports to the Producer or Director. During actual production, however, he or she reports to the Assistant Director, Floor Manager, or Unit Manager.

Although Production Assistants usually work on assignment in the studio, or on remote on a specific program, some also assist in the planning, scheduling, coordination, and daily operation of the production department or unit.

The role of Production Assistant is one of continuing importance to the orderly flow of a production. It is a jack-of-all-trades job in which the individual may do research or copywriting, assist in casting or scheduling guests, and work on sets, costumes, and makeup. He or she helps in the control room or studio by placing monitors, easels, and prompting devices, holding cue cards and slate cards, and performing other production duties. The Production Assistant is a vital part of the production team.

Additionally, the Production Assistant:

- prepares and distributes daily shooting schedules and notifies the crew of all script changes and production arrangements;
- records all production shot sheets, detailing the timing of various program segments;
- assists in the setting up, lighting, and eventual strike (dismantling) of the set;
- assists in the research, development, coordination, and finalization of program scripts;
- collects files and records upon completion of a production, including visual materials, photos, personnel work sheets, on-air talent releases, and other related material.

Salaries

Wages for Production Assistants are rather low. The average payment in commercial TV in 1979 was $10,600, according to industry sources. At some major market stations, where the Production Assistant is represented by a union, the salaries ranged from $11,100 for a person with less than a year of experience, to $15,800 for an individual with more than three years.

In public TV, 1980 earnings went from $4,800 for part-time work to $16,000 for an experienced Production Assistant, according to the Corporation for Public Broadcasting (CPB). The average was $10,100 in 1980, an increase of 9.5 percent over the previous year. At two major market public TV stations in the same general geographic area, the 1980 salaries ranged from $9,600 to $12,500 at a state-owned facility, and from $11,900 to $17,100 at a community station.

Employment Prospects

Chances of getting this job are good. It is generally considered an entry-level position in most commercial and public TV stations. Production Secretaries, however, are often promoted from within the station to Production Assistant. Opportunities for employment often depend on being in the right place at the right time.

Most stations employ from three to eight Production Assistants, depending on the size of the market and on the amount of production work. In addition to the opportunities at TV stations, all independent TV production companies employ Production Assistants. Some cable and closed circuit TV systems, as well as some government, health, and media centers that produce programs offer possibilities for employment.

Advancement Prospects

The opportunity for advancement from the position of Production Assistant to other production and programming positions is good. A bright and enthusiastic person may be promoted to Camera Operator (in non-union shops), or to Floor Manager, Associate Producer, or Assistant Director. The experience gained in all aspects of television production, coupled with resourcefulness, often leads to promotion.

Competition for more responsible production positions is heavy, and while there is considerable turnover among Production Assistants, only the more experienced and reliable are promoted. Some move to smaller stations or production units in order to gain more responsible production positions.

Education

A high school diploma is usually a minimum requirement. Many large market public and commercial stations prefer to hire the recent college graduate who has majored in mass communications, radio-TV, or theater.

Experience/Skills

Since Production Assistant is an entry-level position, most employers do not require extensive experience in television production. Some background in theater staging and lighting, photography, or film work is helpful.

A Production Assistant must be bright, resourceful, organized, be able to plan, and to think intuitively and effectively. The individual must also be able to work under pressure with a variety of people and get a job done cooperatively and creatively. Some typing ability is also usually required.

Minority/Women's Opportunities

Opportunities for women and members of minorities as Production Assistants are excellent. In commercial television, more than 50 percent of the Production Assistants in 1980 were women, according to industry estimates. While those in union commercial and public stations are often white males, minority-group members are occupying the position in increasing numbers.

In public television, 45.6 percent of the Production Assistants in 1980 were women according to the CPB. Members of minorities held 23.7 percent of the jobs in 1980, an increase of 2.8 percent over the previous year.

Unions/Associations

The majority of Production Assistants are considered part of the production department at many stations and are not represented by unions. In some major market commercial and public TV stations, however, they are represented by the National Association of Broadcast Employees and Technicians AFL-CIO (NABET).

PRODUCTION SECRETARY

CAREER PROFILE

Duties: Providing clerical support for a television production department or unit

Alternate Title(s): None

Salary Range: $4,700 to $21,200

Employment Prospects: Good

Advancement Prospects: Good

Prerequisites:

Education—High school diploma; minimum of one year of business school preferable

Experience—Minimum of one year as a secretary in TV or an ad agency

Special Skills—Typing ability; shorthand or speed writing; familiarity with office equipment

CAREER LADDER

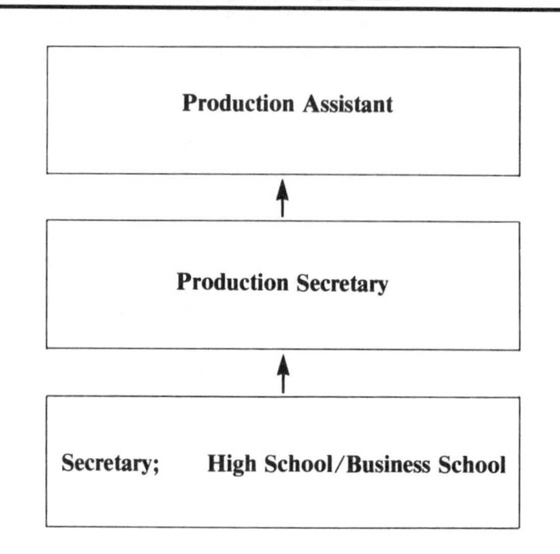

Position Description

A Production Secretary is an integral part of the production team and is responsible for the office support of the production department at a television station or independent production company. He or she coordinates much of the work within the department and provides general secretarial assistance, including typing, filing, transcribing, taking dictation, and general office work.

As a rule, the Production Secretary reports directly to the Production Manager and, in small and middle market stations, he or she also provides general clerical support for Producers, Directors, and other production people. While the position requires standard secretarial skills, the Production Secretary must also be knowledgeable about current practices, terms, and nomenclature used in TV. He or she must be resourceful and organized in dealing with staff, talent, guests, and others involved in a production.

The Production Secretary often works independently in researching and developing the basic material and information needed for the timely development of a program script. He or she prepares and organizes the script from draft through final form, then duplicates, collates, and distributes it to the production staff.

Sometimes, the Production Secretary is assigned on a temporary basis to perform the duties of a Production Assistant.

Additionally, the Production Secretary:

- assists in obtaining clearances for remote locations;
- schedules talent and guests;
- monitors the production schedule;
- maintains the production files containing scripts, program information data, contracts, business records, talent sheets, and other records.

Salaries

Production Secretaries are considered support personnel and, as such, are relatively low paid in comparison with other television positions. In commercial TV, salaries ranged from $7,000 for beginners at small market stations to $14,000 in major markets for individuals with experience and seniority, according to industry estimates.

In public TV, salaries for all secretaries ranged from $4,700 (part-time) to $21,200 for experienced full-time employees in 1980, according to the Corporation for Public Broadcasting (CPB). The average salary was $10,500 in 1980, an increase of 8.2 percent over the previous year.

Employment Prospects

The opportunities for employment are generally

good. Although the position is not usually considered an entry-level job, smaller market stations sometimes employ beginners and provide on-the-job training. At some large market commercial or public stations, various program units within the production department employ Production Secretaries. In addition, all independent production companies and many government, health, and education television operations that are engaged in production have them on staff.

Advancement Prospects

The prospects for advancement for bright, alert people are good. The responsibilities of the position require that the individual learn all aspects of television production, and many Production Secretaries become Production Assistants or move to other, more responsible positions within the production department. Some Production Secretaries are promoted to Assistant Director or Associate Producer, usually within the station or unit, while others use their experience to obtain more responsible positions at other production organizations and at advertising agencies.

Education

The minimum educational requirement for the position of Production Secretary is a high school diploma. Most employers prefer at least one year of training in a business school. Some major market commercial and public TV stations usually employ individuals who have had some education in mass communications or who have an undergraduate degree in theater or radio-TV.

Experience/Skills

Employers in some middle and major market commercial stations require at least one year of experience as a secretary in a television station or advertising agency. In many such stations, an undergraduate degree in radio-TV is acceptable in lieu of experience.

In smaller market commercial and public television stations or independent production companies, the position is sometimes considered an entry-level job and, as such, little experience is required. Although the individual must possess good secretarial skills.

A knowledge of standard office equipment, including photocopy, mimeograph, transcribing, postal meter, telex, and teletype machines is necessary. A Production Secretary must also be able to type at at least 50 words per minute and have some experience in shorthand or speed writing.

Minority/Women's Opportunities

The vast majority of Production Secretaries in commercial TV in 1980 were female, according to industry estimates. Minority women represent a significant percentage. In public TV, 98 percent of all secretaries were females in 1980, and 20 percent were members of minorities, according to the CPB.

Unions/Associations

There are no unions or professional organizations that represent Production Secretaries.

LIGHTING DIRECTOR

CAREER PROFILE

Duties: Lighting studio and remote television productions

Alternate Title(s): Staging/Lighting Supervisor; Lighting Direction Engineer

Salary Range: $11,000 to 27,800

Employment Prospects: Fair

Advancement Prospects: Good

Prerequisites:

Education—High school diploma; Second Class FCC License helpful

Experience—Two years as a Camera Operator or Engineering Technician

Special Skills—Technical aptitude; creativity; physical strength; cooperativeness

CAREER LADDER

Floor Manager; Technical Director; Lighting Director (Large Station or Independent Production Company)

↑

Lighting Director

↑

Camera Operator; Engineering Technician

Position Description

The Lighting Director designs and executes the lighting for television productions, either in the studio or on remote locations. Since the transmission of a TV picture is dependent upon the illumination of a subject, the job is a vital one. He or she must balance the technical limitations of TV with creativity to achieve sparkle and brilliance in lighting effects, and complement the overall purpose and design of a production.

Incorporating theatrical lighting techniques to sharpen, balance, and add dimension, the Lighting Director uses spotlights, floodlights, filters, a light meter, and other equipment and accessories to provide the lighting. He or she must ensure that the tonal qualities and color needs of faces, furniture, backdrops, props, and other set elements work together to achieve the desired lighting effects.

To obtain the correct results, the Lighting Director works with a floor plan, created by the Scenic Designer and Director, and a copy of the script which he or she annotates to indicate lighting needs. The Lighting Director then designs and lays out a lighting plot for the program. This is a complex and detailed version of the floor plan, showing how and where each lighting instrument will be placed.

The Lighting Director usually supervises a part-time or on-assignment crew of two to five people to install and position lighting instruments. He or she makes certain that the lighting is properly focused and balanced and sets up a cue sheet of instructions. Adjustments are made in rehearsal and, as needed, during the actual production.

Virtually all TV stations employ individuals whose primary responsibility is lighting, but at some smaller stations, they often have additional responsibilities in staging, engineering, camera work, or other production activities. In some cases, the duties of the Lighting Director are handled by members of the production crew.

Some stations consider the Lighting Director to be part of the engineering staff, while others consider it a production job. Depending on the station, a Lighting Director reports either to an Engineering Supervisor, or to a Production Manager. During the actual production, he or she reports to the Director of the program.

Additionally, the Lighting Director:

- cues the lighting control board operator during production to alter lighting as necessary;
- oversees the striking (dismantling) of lighting equipment at the completion of a program;
- maintains the inventory of lights and accessories, and purchases replacement parts and supplies;
- performs routine maintenance on lights, electrical equipment, and accessories.

Salaries

At smaller commercial TV stations, the salaries for Lighting Directors range from $11,000 to $15,000, depending on whether additional responsibilities are required and if the position is considered to be part of the engineering or the production staff. In major market stations where the position of Lighting Director is considered part of the engineering department and is represented by a union, the 1980 salaries ranged from $15,900 for people with one year in the position to $27,800 for those with more than four years of seniority.

In public TV, the duties of the Lighting Director are often combined with others into a position called Staging/Lighting Supervisor. Salaries for such employees are generally lower than those of their commercial counterparts.

Employment Prospects

This position is seldom an entry-level job at a station or other production-related organization. Openings at small public and commercial TV stations, where the Lighting Director might have additional production duties, occur infrequently. Even at small independent production firms, subscription TV (STV) and multipoint distribution service (MDS) companies, or cable TV systems where programs are produced, the opportunities are limited since these duties are often handled by members of the production crew.

The possibilities for getting a job as a Lighting Director are only fair and largely confined to middle and major market stations or large production organizations where a full-time person is employed in this position. Many larger stations employ more than one Lighting Director for various units or production teams.

Advancement Prospects

In spite of the somewhat ambiguous nature of the position at smaller stations, the opportunities for advancement are generally good. An alert Lighting Director has the opportunity for promotion to Floor Manager in the production department, or to Technical Director, or other middle-management positions in the engineering department. Good Lighting Directors often move to larger market stations or to independent film or TV production organizations in order to advance their careers in this specific job.

Education

A high school diploma is a minimum prerequisite. At some stations, a Second Class Radiotelephone License from the Federal Communications Commission (FCC) is preferred. Obtaining this license indicates that the individual has a basic knowledge of the electronic and technical aspects of television transmission and operation.

Experience/Skills

Two years of previous experience as a Camera Operator or Engineering Technician are usually required. The experience gained in actual TV production of in-studio and remote programs is invaluable in obtaining the position. A background in theater staging and lighting is very helpful, and many supervisors seek such experience when hiring a Lighting Director.

The Lighting Director must be able to combine a thorough knowledge of the technical requirements of television with a creative and esthetic sense. He or she must be detail-oriented and well organized. He or she must also possess some physical strength to handle lighting equipment, and have the ability to work cooperatively with production and engineering personnel in pressured situations.

Minority/Women's Opportunities

The vast majority of Lighting Directors in commercial or public broadcasting are men. The physical demands of the job have been a factor in keeping the representation by women low. The percentage of minority males in the position is increasing, according to industry estimates.

Unions/Associations

In some major market commercial stations, Lighting Directors are part of the engineering department and are represented by the National Association of Broadcast Employees and Technicians AFL-CIO (NABET) or by the International Brotherhood of Electrical Workers (IBEW). In many smaller commercial and public television stations where the individual reports to the production department, such employees are not represented by a union.

ART DIRECTOR

CAREER PROFILE

Duties: Conceiving and shaping the visual elements of a TV production and the visual style of a TV station

Alternate Title(s): Broadcast Designer; Graphic Arts Supervisor; Senior Artist

Salary Range: $7,600 to $32,500

Employment Prospects: Fair

Advancement Prospects: Poor

Prerequisites:

Education—Minimum of one year of art school; undergraduate degree in commercial art, fine art, or design preferable;

Experience—Minimum of five years as a Graphic Artist in TV production

Special Skills—Creativity; sense of esthetics; administrative abilities; technical aptitude

CAREER LADDER

Art Director (Ad Agency); Owner (Art/Design Company)

↑

Art Director

↑

Graphic Artist

Position Description

The Art Director is responsible for the design of all television programs produced and broadcast by a station. While much of the work is directly related to TV production, he or she is often in charge of the visual representation of a station in printed material. This would include advertising rate cards and displays, promotion/publicity brochures, booklets, pamphlets, and stationery that project the station's corporate image.

The Art Director conceives ways of visually portraying television production ideas and supervises the execution of all design, layout, and art materials for the station. As an administrator, the Art Director supervises a staff of from two to eight full- and part-time employees who produce all art and visual materials in various formats. He or she must, in association with other Graphic Artists and the station's production, promotion, and technical staffs, combine talent and technical knowledge to complete creative projects on time and within budget.

The Art Director is the final authority on the appropriateness of all materials produced by the art department. He or she must create, develop, and maintain the station's overall visual representation in all types of internal and external presentations.

Additionally, the Art Director:

- conceives and supervises the execution of all visual elements of photography, scenic design, and animation;
- develops story boards, layouts, and rough sketches for television, print, and film productions;
- consults with Producers, Directors, and other production and technical personnel on graphic design for local productions;
- supervises the design and construction of new sets and the modification of existing scenery;
- develops and administers the art department budget and allocates funds for specific projects;
- orders, installs, operates, and maintains all art equipment and supplies.

Salaries

Art Directors are not very well compensated in the field of television. Salaries in commercial TV range from $13,000 in some small market stations to more than $29,000 in most of the larger ones. A survey by the Broadcast Designers Association (BDA) found the range to be $14,000 at stations in the above-100 (smaller) markets to $30,000 in the top ten. The average was $15,000 in 1980. In union shops where a senior artist sometimes performs the functions of Art Director, the minimum 1980 payment required by one union was $25,800.

In public television, Art Directors were paid between $7,600 and $32,500 in 1980, according to the

Corporation for Public Broadcasting (CPB). The average was $14,100, an increase of 7.9 percent over the previous year.

Employment Prospects

Overall, the chances for employment are fair. The position of Art Director at a television station requires considerable knowledge and experience in all phases of commercial art, particularly in TV production. According to the BDA survey, the typical Art Director had an average of nine years of television art experience. With 1,000 operating television stations, the opportunities are somewhat limited. Some very small market stations use only free-lance Graphic Artists as the occasion demands. Cable television systems and multipoint distribution service (MDS) and subscription television (STV) companies occasionally employ Art Directors for specific projects, but few are large enough to require a full-time person. Positions are often available, however, in health, government, and educational telecommunications and video operations.

Advancement Prospects

The possibilities for advancement are quite limited in television. An Art Director is generally confined to the creation and execution of various formats of the visual arts. While the position requires a knowledge of production, the opportunity for promotion within a station to Production Manager or other supervisory job (Producer or Director) is seldom available. Some Art Directors in smaller television stations move to similar positions in larger markets and thus gain greater responsibility and a larger salary.

Art Directors often advance their careers by joining advertising agencies in supervisory art positions. Some use their experience to form their own commercial art/design firms, or join independent television production agencies that create art work for specific clients.

Education

Most Art Directors have an undergraduate degree in commercial art, fine art, or design. While a degree is less important than taste, ability, and visual style, the Art Director's background usually includes some college level courses in the liberal arts.

In lieu of such education, a year or two of training at a commercial art school is usually a requisite for obtaining the position of Graphic Artist and for eventual promotion to Art Director. Courses and education in all the visual art media are helpful.

Experience/Skills

When hiring an Art Director, most employees require at least five years of experience in television production art work. In addition, wide experience in various art forms (layout, sculpture, scenic design, cartooning, drawing, etc.) is usually required.

Art Directors should demonstrate a clean style that is visually appealing and that conforms to the image desired by the station. Sophisticated taste in the visual arts, strong technical skills, and an ability to organize and manage a diverse group of creative subordinates are required. In addition, an Art Director must be able to work in cooperation with other production and technical people, on-air talent, and promotion and public relations personnel.

Minority/Women's Opportunities

In commercial television, about 25 percent of the Art Directors in 1980 were women, according to most industry estimates. The figures for members of minorities are not as high, although their representation in the position is increasing yearly.

In public television, 20.4 percent of Graphic Arts Supervisors (an equivalent title) were women in 1980 and 8.2 percent were minority-group members, according to the Corporation for Public Broadcasting (CPB).

Unions/Associations

Some Art Directors are members of the National Association of Broadcast Employees and Technicians AFL-CIO (NABET). A few, at major market stations, are represented by United Scenic Artists for bargaining purposes. In both commercial and public TV, however, most are not represented by a union. Some commercial television Art Directors are members of the Broadcast Designers Association in order to share mutual professional concerns.

GRAPHIC ARTIST

CAREER PROFILE

Duties: Design and creation of television art and related visual materials

Alternate Title(s): Television Artist

Salary Range: $8,000 to $21,300

Employment Prospects: Good

Advancement Prospects: Fair

Prerequisites:

Education—Minimum of one year of training in commercial art; undergraduate art degree preferable

Experience—Minimum of one year as a commercial artist

Special Skills—Creativity; versatility; talent for illustration

CAREER LADDER

Art Director

↑

Graphic Artist

↑

School; Commercial Artist

Position Description

The Graphic Artist in a commercial or public television station creates, designs, and executes a variety of visual art forms to enhance, and often serve as the focal point of, a television production. This includes the creation of charts, graphs, title cards, maps, and three-dimensional objects, as well as set design and construction.

In addition, he or she may be responsible for the design, layout, and execution of art for newspaper advertising, publicity brochures, program guides, billboards, and other advertising and promotional displays. In these areas, the Graphic Artist may be involved with photography, photo laboratory developing, color separations, typography, and other print-oriented art requirements.

The Graphic Artist is primarily an illustrator of ideas and concepts. He or she must translate the basic thoughts and perceptions of a variety of people into finished art that improves the visual image of a program or print campaign and clearly communicates ideas to the viewer or reader. The Graphic Artist works in a variety of styles, including cartooning, realistic renderings, decorative background painting, and sculpting. He or she is a versatile executor of a variety of art techniques within the framework of the television medium.

Additionally, the Graphic Artist:

- performs the layout and pasteup of printed materials;
- operates various types of art equipment, including typesetting units, air brushes, and dry mount machines;
- designs and creates backdrops, set pieces, and props for television productions.

Salaries

Salaries for Graphic Artists in commercial television stations range from $9,000 at smaller market stations to $21,000 in major markets, according to industry sources. The average was $12,200 in 1980.

Where Graphic Artists were represented by a union, the 1980 salary scale ranged from $14,900 for people with six months or less tenure to $21,300 for those with more than thirty months of service.

In public television, 1980 salaries ranged from $8,000 at small market school or university stations to more than $20,500 at the larger metropolitan stations. In general, salaries for Graphic Artists are lower in public TV, regardless of location.

Employment Prospects

The position of Graphic Artist is often an entry-level job in the art department of a television station. The chances for employment are quite good. Most middle market stations employ at least two Graphic Artists and major market stations may have more than eight on staff. According to a survey by the Broadcast Designers Association (BDA), the average commercial station employed two or three in 1981.

Graphic Artist positions are also available at cable TV systems and subscription television (STV) and multipoint distribution service (MDS) companies. Other opportunities exist in instructional and closed circuit television (ITV and CCTV). In addition, TV operations in business, health, and government settings usually have at least one Graphic Artist.

Advancement Prospects

The opportunities for advancement for creative and diligent Graphic Artists at a commercial or public TV station are only fair. The relative scarcity of Art Director positions and the competition tend to limit advancement. In addition, the skills and talents required by an artist are not readily transferable to other production or technical positions within the station.

Positions in other TV-related industries, however, are often available for an experienced Graphic Artist. Many utilize the experience gained in television art, as well as in print media, to excellent advantage in obtaining positions in advertising agencies or public relations firms. Still other Graphic Artists advance their careers by moving to larger stations at higher salaries or by becoming Art Directors at smaller market operations.

Education

Most stations require a minimum of a high school diploma and at least one year of training in commercial art when employing a Graphic Artist. Some employers require an undergraduate degree in commercial or fine art, as well as some evidence of private study in specific art techniques and styles.

Experience/Skills

An extensive background in television art is often not essential although some middle or major market stations do require at least one year of experience.

Facility in a variety of commercial art techniques and styles, as well as considerable versatility in print art, is often a prerequisite to obtaining the position. The Graphic Artist should display a variety of different projects in his or her portfolio. A clean, clear style is important. Some experience with basic art equipment and an ability to work rapidly and meet deadlines are added advantages.

Minority/Women's Opportunities

This is one of the areas in television in which opportunities for women are excellent. In 1980, more than 50 percent of the Graphic Artists in commercial TV were female, according to industry estimates. While the employment rate for women was not as high in public television, they held 40.8 percent of the Graphic Artist jobs in 1980, according to Corporation for Public Broadcasting (CPB) statistics. Minority-group members were not as well represented in public TV; they filled 12.4 percent of the positions that year.

Unions/Associations

Some Graphic Artists belong to and are represented by the National Association of Broadcast Employees and Technicians AFL-CIO (NABET). In most instances, however, Graphic Artists in commercial and public television are not members of a union. Some people in commercial TV belong to the Broadcast Designers Association (BDA) to share mutual concerns and to improve their skills.

CINEMATOGRAPHER

CAREER PROFILE

Duties: Shooting and editing of film for broadcast at a television station

Alternate Title(s): Film Cameraman; News Photographer; Film Camera Operator; Photographer

Salary Range: $8,800 to $24,500

Employment Prospects: Fair

Advancement Prospects: Fair

Prerequisites:

Education—Undergraduate degree in film

Experience—Minimum of one year of shooting and editing film

Special Skills—Creativity; technical skill; originality; resourcefulness

CAREER LADDER

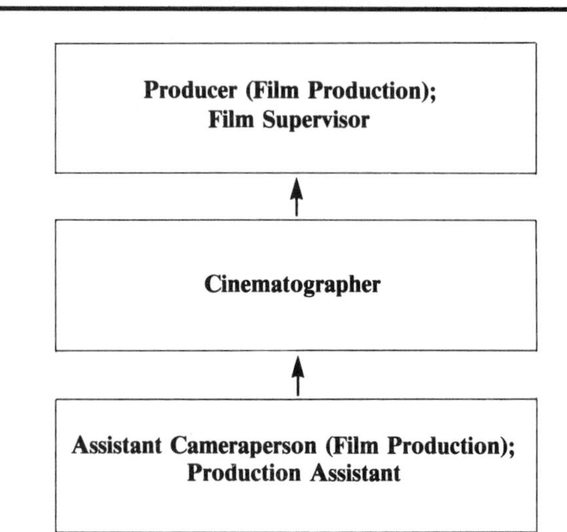

Position Description

A Cinematographer is responsible for shooting and editing film inserts to be used as elements of the news or special-events programs broadcast on television. He or she may also shoot commercials for a station's clients. On occasion, a Cinematographer works on major productions, including documentaries, sports, or other special programming created for TV.

Depending upon their involvement with film as a production tool, most stations employ two or more Cinematographers. During the past ten years, however, many stations have changed over to portable Electronic News Gathering (ENG) equipment for most of their short, on-location productions. This equipment is based on video technology, rather than film, and is used for the majority of news coverage and interviews shot outside of a studio. In many stations Cinematographers have switched from shooting film and routinely use ENG systems. (In some cases Camera Operators operate ENG equipment on news assignments.) Cinematographers are also often responsible for a station's black and white and color still photography.

Most commercial and public television stations still employ Cinematographers for some news coverage, commercial production, and special programs. At many commercial stations, they are part of special units that provide production services for advertising clients. Often they are members of the production or programming departments. Still other Cinematographers are assigned permanently to a station's news department.

The Cinematographer's overall responsibility is to produce quality film images for use in various types of television productions. He or she interprets the needs of the Director, Producer, or Reporter and assists in developing cinematic approaches to a script or news story. The Cinematographer chooses the cameras, lenses, accessories, and film stock appropriate for each shooting assignment.

Additionally, the Cinematographer:

- selects angles and shot composition to achieve desired film effects;
- directs the selection and placement of lights;
- supervises the processing of film within the station or at outside laboratories;
- performs minor repairs on film equipment;
- maintains an inventory of raw film stock and other photographic supplies.

Salaries

Income for Cinematographers is relatively low. At commercial stations, salaries ranged from $10,000 to $18,000 in 1980 while the average payment was $13,000, according to industry sources.

In public television, 1980 salaries ranged from $8,800 for beginners at small stations to $24,500 for experienced individuals at major market community stations, according to the Corporation for Public Broadcasting (CPB). The average was $14,200, which represented a slight reduction from the previous year.

Employment Prospects

Overall opportunities for employment are only fair. An assistant cameraperson in commercial film production or a talented and knowledgeable Production Assistant can sometimes become a Cinematographer. The use of film, however, is diminishing at commercial stations. They are using ENG equipment more and more frequently for location shooting. Fortunately, the techniques and skills required of a Cinematographer are easily transferable to the newer technology. Some stations even use the term cinevideographer for an individual who uses the techniques of film production in operating portable video camera equipment.

Opportunities for Cinematographers are usually better at public television stations, where many documentaries are shot on film. In addition, most commercial production firms and advertising agencies prefer film and continue to use it to shoot commercials. There are some additional opportunities for Cinematographers in government, education, and health video settings where film is still often used.

Advancement Prospects

The opportunities for Cinematographers to advance in film production are fair. Some seek to create longer-form feature, artistic, or documentary films for theatrical presentation and use their TV experience at traditional film production firms as Producers. In television stations that have large film departments, Cinematographers can be promoted to film supervisor in charge of other Cinematographers, or to the position of Producer specializing in film.

Education

An undergraduate degree in film with some training in film production techniques is often required. Courses in film production, editing, film history, and film as an art form are helpful.

Experience/Skills

Most employers require at least one year of experience in shooting and editing motion picture film. Familiarity with many types of cameras and related audio tape recorders is helpful. A background in still camera equipment and techniques, processing, and film chemistry is also useful.

Cinematographers are often given considerable latitude in shooting film for television. Individuals who are imaginative and have a feel for composition and the esthetics of shooting and editing film are good employment prospects. A Cinematographer must be original and resourceful in dealing with a variety of different circumstances during film production.

Minority/Women's Opportunities

While minority representation has grown, the majority of Cinematographers in commercial television were white males in 1980. In 1980, 4.4 percent of the Cinematographers in public television were women, and 23.5 percent were members of minorities, according to the Corporation for Public Broadcasting (CPB).

Unions/Associations

Some Cinematographers in commercial television are represented by the National Association of Broadcast Employees and Technicians AFL-CIO (NABET) or the International Alliance of Theatrical Stage Employees (IATSE), which act as bargaining agents. Many Cinematographers belong to the American Film Institute (AFI), the Association of Independent Video and Film Makers (AIVF), and other local, state, and national film councils, to share ideas and advance their careers.

OPERATIONS MANAGER

CAREER PROFILE

Duties: Scheduling and coordinating all technical and production facilities at a television station

Alternate Title(s): Operations Supervisor; Operations Director

Salary Range: $9,600 to $40,000 +

Employment Prospects: Poor

Advancement Prospects: Fair

Prerequisites:

Education—Minimum of high school diploma; technical school or college training preferable

Experience—Minimum of two years in various TV station positions

Special Skills—Organizational ability; logical mind; detail orientation; scheduling talent; cooperativeness; leadership qualities

CAREER LADDER

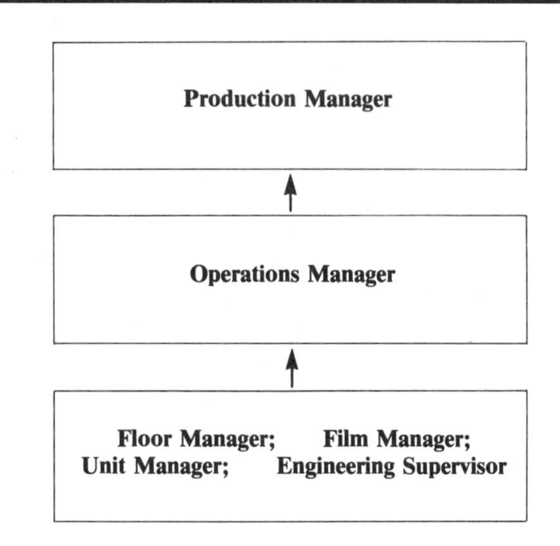

Position Description

The Operations Manager allocates the facilities and resources at a television station to achieve and maintain a smooth and professional operation and maximize production opportunities. He or she schedules all production, engineering, and technical facilities from sign-on to sign-off and coordinates the station's on-air activities. He or she also arranges for the leasing of telephone lines for the distribution of programs to and from networks and other locations.

The Operations Manager's responsibility is to schedule production studios, control rooms, videotape machines, film projectors, and other equipment, and, in some cases, operating staff. He or she periodically coordinates the activities of the traffic/continuity, program, film, production, and engineering departments. At some stations, the Operations Manager oversees the activities of the Film Manager.

In addition, the Operations Manager acquires, processes, and distributes all program information including the station logs, and slides, films, or tapes for station breaks. He or she establishes the procedures for the prompt procurement, scheduling, and delivery of commercials for broadcast and the recording of programs or program elements from network and other sources.

The duties of an Operations Manager vary from station to station. In some small settings, the job is considered a glorified traffic coordinator whose main functions are the delivery of tapes and films, and the setup of slides, films, and commercials for station breaks. In most middle and major market stations with heavy production schedules, however, he or she has considerable responsibility. At some stations, the Operations Manager reports to the Program Manager or Production Manager. In others, his or her immediate supervisor is the Assistant Chief Engineer.

Additionally, the Operations Manager:

- supervises the station's library of videotapes, films, and records and oversees cataloging procedures;
- schedules staff announcers.

Salaries

Operations Managers in commercial television earned from $15,000 to more than $35,000 in 1980, according to industry sources. The average salary was $24,900.

In public television, 1980 salaries ranged from $9,600 at small market stations to more than $40,000 for a few experienced personnel at major market stations with heavy national and local production schedules, according to the Corporation for Public

Broadcasting (CPB). The average was $21,400 in 1980, an increase of 8.9 percent from the previous year.

Employment Prospects

The chances of getting a position as an Operations Manager are poor. This is not an entry-level job. In most stations, the position is held by an experienced individual who has a broad knowledge of station operations and a thorough understanding of each department's functions. A talented Floor Manager, Film Manager, or Unit Manager already on the station's staff is likely to be promoted to the position should an opening occur.

Most cable television, subscription television (STV), multipoint distribution service (MDS), and other production organizations do not maintain operations that are as complex as those at broadcasting stations. As a result, opportunities for employment as an Operations Manager in those settings are extremely limited.

Advancement Prospects

Possibilities for advancement are only fair. Diligent Operations Managers who are efficient and experienced sometimes move to the position of Production Manager. Some obtain similar positions at larger stations with more responsibility and better salaries. Still others advance their careers by moving to smaller stations in more responsible positions. Since each station operates in a slightly different manner, the skills acquired at one station are not automatically transferable. In the majority of instances, Operations Managers are not highly mobile and they seek promotion within their own stations.

Education

All employers require a high school diploma. Some call for technical school training or college courses in radio-TV or mass communications. A few prefer an undergraduate degree. Some education and study leading to a First Class Radiotelephone License from the Federal Communications Commission (FCC) is desirable.

Employers seek alert and responsible individuals with a thorough understanding of the operation of every department in the station as well as a knowledge of production and engineering equipment. The candidate should be well organized, logical, and able to create and manage a detailed schedule of people and equipment. Most important, the individual must be cooperative and have leadership qualities to deal with a variety of people in day-to-day pressure situations.

Minority/Women's Opportunities

While the majority of Operations Managers are white males, the number of women in this position in commercial TV has increased over the past few years. Minority-group member representation has also increased, according to industry sources.

In public TV, 23.3 percent of the Operations Managers were women, and 10 percent were members of minority groups in 1980, according to the Corporation for Public Broadcasting (CPB).

Unions/Associations

A few Operations Managers are considered part of the engineering department at major market stations and are members of the National Association of Broadcast Employees and Technicians AFL-CIO (NABET), or of the International Brotherhood of Electrical Workers (IBEW). In the majority of stations, however, they are considered members of the programming or production department and are not represented by a union or professional organization.

Experience/Skills

Most employers require a minimum of two years of experience in the engineering, production, or programming departments within the particular station when promoting an individual to Operations Manager. Many Operations Managers gain experience as Floor Managers, Film Managers, or Unit Managers, while others earn valuable experience in the traffic/continuity department. Some serve as Engineering Supervisors and acquire experience in scheduling and supervising operating and technical people.

NEWS DIRECTOR

CAREER PROFILE

Duties: Directing of all activities of a television station's news department

Alternate Title(s): News and Public Affairs Director

Salary Range: $7,800 to $110,000

Employment Prospects: Poor

Advancement Prospects: Poor

Prerequisites:

Education—Undergraduate or graduate degree in political science, journalism, or mass communications

Experience—Several years in other television news positions or in print or radio news

Special Skills—Sound news judgment; inquisitive mind; imagination; administrative abilities; objectivity; integrity

CAREER LADDER

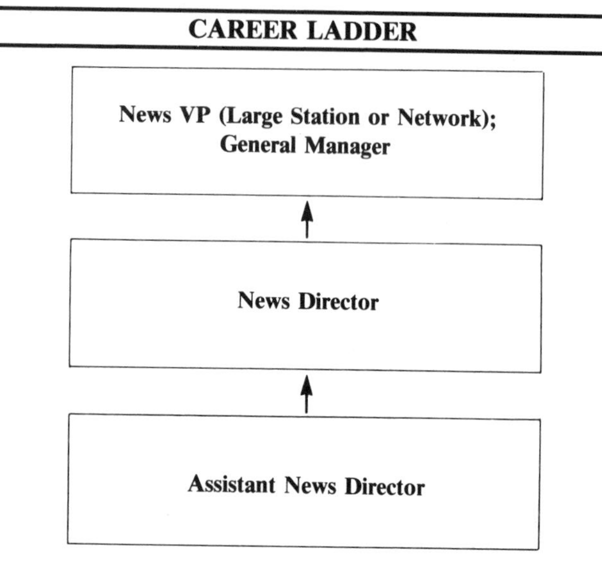

Position Description

The News Director is the senior executive in charge of a television station's news department. He or she serves as the final authority for the choice of all news, interviews, documentaries, and special event programs broadcast by a station. The decisions of the News Director determine what events will be covered, which stories will be broadcast, and how and when they will be presented.

Supervision, coordination, and evaluation of the performances of the news staff are the prime responsibilities of a News Director. Most television station news departments have staffs of 10 or 12 people. In some major market stations, however, the news staff includes 40 or more Reporters, Anchorpersons, Sportscasters, Weather Reporters, News Writers, film and video Camera Operators, and a sizeable number of other professionals and support personnel. The News Director must, therefore, be able to supervise and communicate with a news team that is very often composed of individuals with strong personalities.

When selecting particular stories and news events to be covered, the News Director must rely on strong journalistic instincts. He or she must be able to assess the importance of a story and decide how best to utilize the available talent and technical resources of the station. The individual must also be capable of making quick decisions about fast-breaking events. At smaller market television stations, the News Director may also be involved in actual on-the-air reporting.

Additionally, the News Director:

- develops and administers the news department budget;
- monitors the progress of in-depth investigative research and reporting and of special events coverage;
- reviews and approves news film, videotape stories, and edited news copy from all sources;
- resolves production and technical problems with the production and engineering departments;
- coordinates news department activities with the traffic/continuity and programming departments.

Salaries

News Directors' salaries vary considerably, depending on the size of the station, its geographical location, and the extent of local news coverage.

A 1980 salary survey by the Radio-Television News Directors Association (RTNDA) reported that TV News Directors in small markets averaged $18,700 per year, while in major markets, the figure was $34,400. The average annual salary for all markets in 1980 was $25,700, an increase of 15 percent from the previous year. This number is made up of salaries as low as $7,800 and as high as $110,000.

In public television, the average salary for the related position of News and Public Affairs Director was $21,900 in 1980, according to the Corporation for Public Broadcasting (CPB).

While these salaries are respectable, News Directors are generally paid less than the super star of local and network news—the Anchorperson.

Employment Prospects

Television News Director positions are limited to commercial stations and the few public stations that broadcast news. The news profession is filled with individuals who are capable and aggressive, and the competition is fierce. Only exceptionally talented people become News Directors, usually with the help of some breaks. Occasionally, a very successful Assistant News Director is promoted to the top position. There is, however, a high turnover rate among News Directors. *The New York Times* has reported that not one of the News Directors at the three network-owned television stations in New York City has been in that position for more than two and a half years.

The demand for higher ratings in a highly charged competitive marketplace and the economic value of news to a station's income create a continuing search for new formats, different air personalities, and new News Directors.

Advancement Prospects

Although there is a fairly high turnover rate among News Directors, there are few positions to which they advance. Some move to print journalism to advance their careers. A few become news vice presidents at the very large TV stations or the networks, and in some recent cases, News Directors have been promoted to the top general management position at stations and networks.

To many individuals, the position is the climax of a successful career. Because of the relatively few opportunities currently available, advancement from this position is presently considered poor. However, with the growing popularity of and increase in television news, and with the expansion of the new cable news networks, opportunities for News Directors will increase in the future.

Education

An undergraduate degree is essential in attaining the position of News Director, and a graduate degree is extremely helpful. The major areas of study should be in political science, journalism, or mass communications. A solid liberal arts background and courses in law, economics, and sociology are helpful.

Experience/Skills

Several years of hard work, imagination, and initiative in lower-level news department positions are essential for promotion to News Director. Previous employment and experience in radio news or in print journalism or as a Reporter for a wire service is also good training.

Experience, energy, an inquisitive mind, and sound journalistic judgment are expected in a News Director, as are good communication skills, objectivity, and a willingness to assume responsibilities. Also necessary are a thorough knowledge of Federal Communications Commission (FCC) regulations involving fairness and equal time; an understanding of libel, slander, and copyright law; and familiarity with the intricacies of the Freedom of Information Act.

Minority/Women's Opportunities

A 1980 survey by the RTNDA found that only 5 percent of the News Director positions in commercial television were held by women, and a 1980 report in the *Washington Post* indicated that opportunities for women were increasing at a faster rate than for minority groups.

In public television, minority-group members and women were equally represented as News Directors in 1980, each holding 20 percent of the positions, according to the Corporation for Public Broadcasting (CPB). However, the number of public television stations employing a News Director is so small that the comparative representation is not a reliable indication of trends.

Unions/Associations

As members of management, News Directors are not represented by a union for bargaining purposes.

RTNDA is the professional membership association for News Directors and other professionals in the field. The Association sets standards for its members, encourages college students who are preparing for broadcast news careers, operates a placement service, and acts to secure and protect the right to report the news.

ASSISTANT NEWS DIRECTOR

CAREER PROFILE

Duties: Making news coverage assignments; supervising newsroom operations

Alternate Title(s): TV Managing Editor; Assignment Editor; Desk Editor

Salary Range: $16,000 to $31,500

Employment Prospects: Poor

Advancement Prospects: Good

Prerequisites:

Education—Undergraduate degree in journalism, political science, or mass communications

Experience—Several years as a Reporter and in other news functions

Special Skills—Leadership qualities; sound news judgment; communication skills; objectivity; organizational ability

CAREER LADDER

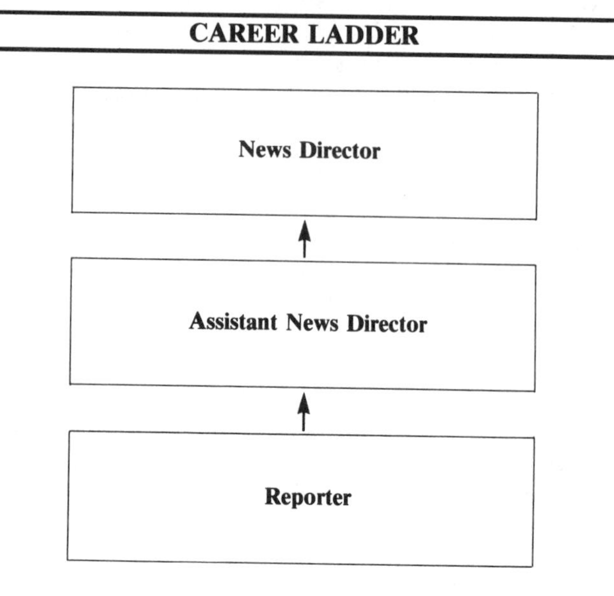

Position Description

Most commercial and some public television stations employ an Assistant News Director. The responsibilities vary from station to station, but nearly every Assistant News Director selects and assigns Reporters, News Writers, and commentators to cover specific news and special events.

The Assistant News Director oversees the daily operation of the newsroom, coordinating wire service reports, network feeds, taped or filmed inserts, and stories from individual News Writers and Reporters. He or she monitors all of the assignments and activities leading to each newscast and arranges work shifts that keep the newsroom functioning at all times. At some stations, the Assistant News Director also has the responsibility for designating the technical crews for each assignment as well as for selecting the Producer of each newscast segment.

The Assistant News Director often has direct supervisory control over the video and photographic news units, and develops procedures to improve and strengthen the technical aspects of news gathering.

Most of the duties that are outside the normal daily routine of the Assistant News Director are carried out in consultation with the News Director. The Assistant usually assumes complete supervision of the news staff when the News Director is away from the station and when fast-breaking stories are developing.

Additionally, the Assistant News Director:

- analyzes news reports, determines their significance and the extent of coverage necessary for an individual story;
- evaluates finished stories, film inserts, and taped reports and selects significant stories for use on a newscast;
- searches for any reportorial mistakes or misinterpretations in order to provide creditable and balanced news reporting for the station.

Salaries

Industry sources place the average salary for Assistant News Directors at major market stations between $16,000 and $31,500. Earnings vary depending on geographic location, the ranking of markets, and the relative size of individual stations. Salaries at smaller market stations and in public television are generally lower.

Employment Prospects

Some major market stations employ more than one Assistant News Director with each responsible for a specific news area. There are, however, many ambitious and able Reporters competing for the job at stations of all sizes and the opportunities for employment are generally poor. Some experienced Reporters do find middle management positions at cable television news operations.

Advancement Prospects

With television news enjoying a greater popularity and higher ratings than ever before, with both local stations and networks seeking to expand their news broadcasts, and with technological advances that include all-news cable television channels, the opportunities for advancement are good.

The relatively high turnover rate of News Directors increases the possibility of direct promotion. The wire services, government, and information and public relations fields also represent employment possibilities for people who have had some management experience in broadcast news. Some Assistant News Directors move to magazine or newspaper positions to advance their careers.

Education

An undergraduate degree in journalism, political science, or mass communications is essential in obtaining the job. Equally important is a strong background in liberal arts. While history courses contribute a sense of perspective and studies in English improve language skills, many employers also look for some educational background in business administration or economics.

Experience/Skills

The Assistant News Director job is not an entry-level one. It is most frequently filled by promotion from within the news department at a television station. Several years of experience as a Reporter and in various other newsroom positions is usually required.

Some Assistant News Directors are recruited from a TV station in a smaller market or from print journalism. In the latter case, some prior experience in broadcast news is usually necessary.

The position of Assistant News Director is demanding and challenging, and requires an inquiring mind and qualities of leadership and communication, as well as the awareness and initiative of a good Reporter.

Employers seek well-organized and experienced people who can bring a balanced perspective to television journalism and who are capable of good judgment under pressure. An ability to coordinate many activities on a tight schedule is also required.

Minority/Women's Opportunities

While employment for women and minority-group members in reportorial, writing, and on-air news positions has increased during the past decade, the middle management position of Assistant News Director is still most commonly held by white males.

Unions/Associations

If their duties involve any news writing or news reporting on the air, Assistant News Directors may be members of the Writers Guild of America (WGA) or the American Federation of Television and Radio Artists (AFTRA). This is particularly true at some large market stations or at the network level.

Membership in the Radio-Television News Directors Association (RTNDA) is open to Assistant News Directors. Some also belong to the Society of Professional Journalists, Sigma Delta Chi.

NEWS WRITER

CAREER PROFILE

Duties: Writing and editing of news stories and continuity for newscasts

Alternate Title(s): None

Salary Range: $15,000 to $30,000

Employment Prospects: Good

Advancement Prospects: Good

Prerequisites:

Education—Undergraduate degree in journalism or mass communications with strong liberal arts background

Experience—Occasionally entry level; minimum of one year as a Desk Assistant or related position preferable

Special Skills—Good, concise writing style; talent for research; ability to meet deadlines

CAREER LADDER

Reporter

↑

News Writer

↑

College; Desk Assistant

Position Description

A News Writer writes and edits the news stories, commentaries, continuity (transitional phrases and sentences), introductions, and descriptions that constitute the oral portions of a regularly scheduled newscast.

At most television stations, News Writers do not write advertising or commercial copy, the continuity for station breaks, or other locally produced television programming; they write primarily for newscasts and, occasionally, for public affairs and documentary programs produced by the news department. They do not write all of the news stories; very often, the Reporter who gathers the information and delivers the report on the air writes or assists in writing the story.

The News Writer must be able to use language effectively, conveying a thought in concise phrases and sentences. He or she must know where to acquire information and how to translate that information into clear, understandable prose.

Additionally, a News Writer:

- utilizes such resources as teletype and picture transmitting machines, printed material, electronic storage and retrieval systems, and personal and telephone interviews, in gathering information;
- verifies the accuracy of questionable facts for news broadcasts;
- obtains additional relevant details of news events by research and questioning.

Salaries

News Writers are generally paid less than the more visible on-air members of the news team, but all news salaries have increased annually at a rate of 10 to 15 percent in recent years. In a non-union station, News Writers may expect to start at around $15,000, and as they gain seniority, salaries will increase to the mid-$20,000 range. Union scale in 1980 was about $5,000 higher.

Employment Prospects

Large television stations employ several News Writers. The best of these people often advance to more prestigious positions, opening up entry-level jobs for the fledgling writer. The networks also employ News Writers in large numbers, and the trend towards increasing the length and frequency of newscasts, and the emergence of all-news cable television networks, indicate additional opportunities in news. Prospects for employment are good for potential News Writers.

Advancement Prospects

Industry predictions about the growth potential of the entire news profession are so optimistic that News Writers can look forward to a number of opportunities in the future, either at the station or network level, in cable TV news, or with one of the wire services. Today, some ambitious News Writers with a desire to move in front of the cameras, are becoming Reporters at their own stations or in larger markets.

Education

An undergraduate degree in journalism or mass communications is almost a necessity, particularly if the News Writer wishes to grow and advance in the profession. A strong background in English, economics, political science, and the humanities is also recommended.

Experience/Skills

Talented people are occasionally hired as News Writers right out of college. A candidate with at least one year as a Desk Assistant or similar news function, however, stands a better chance for employment. This experience could be acquired outside of television broadcasting, but a knowledge of TV news is helpful.

A crisp journalistic writing style and the ability to translate complex ideas into meaningful copy are the primary skills needed by a News Writer. Research skills and the ability to work well under the pressure of regular deadlines are also most important.

Minority/Women's Opportunities

Both qualified women and minority-group members have an excellent chance of success in television news. In 1980, 29 percent of all broadcast journalists (including News Writers) were women and 15 percent were members of minorities, according to the Radio-Television News Directors Association (RTNDA). Regardless of sex or race, recent graduates of university broadcast journalism programs have been much in demand, according to the RTNDA.

Unions/Associations

The National Association of Broadcast Employees and Technicians AFL-CIO (NABET) and the Writers Guild of America (WGA) represent News Writers for bargaining purposes at a few major market stations and at the network level. The RTNDA is also open to News Writers, and promotes professional growth and offers support.

DESK ASSISTANT

CAREER PROFILE

Duties: Providing general assistance to the news department of a television station

Alternate Title(s): News Assistant

Salary Range: $8,800 to $12,300

Employment Prospects: Fair

Advancement Prospects: Good

Prerequisites:

Education—High school diploma; some college or business school preferable

Experience—Any news-related activities helpful

Special Skills—Typing ability; writing aptitude; organizational skills

CAREER LADDER

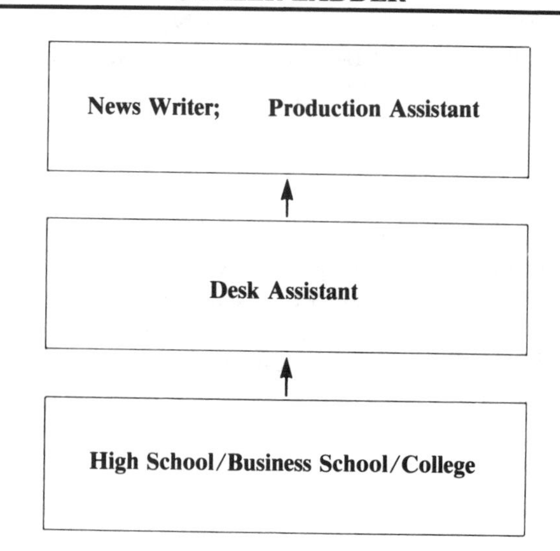

Position Description

The Desk Assistant is an all around helper in a television news department. He or she is usually an entry-level apprentice who has a specific interest in a broadcast news career.

The Desk Assistant performs general office duties in the newsroom. These include answering telephones and taking messages for staff members, opening and distributing mail, delivering newspapers and magazines, ordering office supplies, and filing scripts and correspondence. He or she is also responsible for refilling wire service machines with paper and collecting wire service copy and distributing it to News Writers and Reporters.

The Desk Assistant is usually considered to be a general messenger. He or she may have to pick up food for newsroom employees and deliver packages both inside and outside the station. He or she can also be called on to deliver film or videotape to and from editors, the library, the videotape room, laboratories, and studios. Often the Desk Assistant is involved in logging incoming and outgoing news film and tape and seeing that it is delivered to the proper location for shipping or for use at the station. On occasion, he or she may go to an airport, train depot, or bus terminal to ship or pick up news film and tape.

Helping with script preparation can also be part of the job. The Desk Assistant may type updated sections of the script for a newscast, special report, or documentary and deliver them to appropriate staff members. He or she may also be called on to transcribe the audio portion of a film or videotape interview into typewritten form so that a Reporter or Producer can select the segments to be included in the final production.

The Desk Assistant sometimes collects routine data such as sports scores and weather information for inclusion in a newscast. He or she also does occasional background research for a particular project and locates library film or videotape footage for use in a report or documentary.

In larger stations and at the networks, the Desk Assistant can be assigned on a regular basis to a specific news show or special unit. In most outlets, however, he or she is involved in the entire newsroom operation.

The Desk Assistant position can offer an excellent introduction to the news profession. The capable, aggressive individual will find it a challenge and an invaluable opportunity to become familiar with many aspects of television news.

Additionally, the Desk Assistant:

- assists studio and control room personnel during production;
- accompanies a Reporter to an on-location as-

signment to rush film or videotape back to the station or to a laboratory.

Salaries

Salaries are comparable to those of many other entry-level positions. In 1980, they ranged from $8,800 to $12,300, according to industry estimates. Seniority, the size of the station and market, and geographical differences all affect the salary scale. The union requirement for part-time (up to twenty hours per week) Desk Assistants in 1980 was $5.65 an hour.

Employment Prospects

Opportunities are currently only fair. Positions are limited to the middle and large market commercial television stations, but there is relatively rapid turnover in the job as individuals are promoted or decide to pursue careers in other fields. Openings will be increasingly available in the cable television news industry as the popularity and expansion of news programming continues. The job-seeker has to search for the openings, but a qualified, ambitious candidate will find the opportunities.

Advancement Prospects

Industry sources are enthusiastic about the future of news personnel, and indicate that the need for qualified professionals at all levels will increase markedly over the coming years. A bright, ambitious Desk Assistant should find the opportunities good for promotion to a higher-ranking position. Some, with the necessary skills, become News Writers. Others, move into news production as Production Assistants. Still others find advancement opportunities at cable TV news networks.

Education

A high school diploma is required, and a year or two of college or business school education is helpful, for obtaining a Desk Assistant position. In order to prepare for a better-paying, higher-level television news position, however, candidates should consider the educational requirements for News Writers, Reporters, Anchorpersons, and News Directors. An undergraduate degree in journalism or political science, with a heavy liberal arts emphasis, is recommended.

Experience/Skills

Part-time or college extracurricular experience in a news-related field is extremely useful in obtaining a Desk Assistant position. Typing, writing, and organizational skills are important. Initiative, an inquisitive mind, and an enthusiastic manner are also helpful.

Minority/Women's Opportunities

Neither women nor members of minority groups are finding discrimination in the hiring practices of television news departments today. This is particularly true for entry-level positions, such as Desk Assistant. Both groups were well represented in 1980 surveys within the television news industry.

Unions/Associations

At network owned-and-operated stations and certain other major market stations, Desk Assistants are represented, for bargaining purposes, by the National Association of Broadcast Employees and Technicians AFL-CIO (NABET). At most outlets, however, they are not represented by a union or professional organization.

ANCHORPERSON

CAREER PROFILE

Duties: Reporting the news and introducing reports on the air; serving as focal point for a newscast

Alternate Title(s): TV Newscaster

Salary Range: Under $17,000 to $1,000,000+

Employment Prospects: Poor

Advancement Prospects: Poor

Prerequisites:

Education—Undergraduate degree in mass communications, journalism, or political science, with a strong liberal arts background

Experience—Many years in various TV newsroom positions

Special Skills—Attractive appearance; good speaking voice; ability to communicate authoritatively; integrity; pleasant personality

CAREER LADDER

Anchorperson (Large Station or Network)

↑

Anchorperson

↑

Reporter

Position Description

The main functions of an Anchorperson are to host regularly scheduled newscasts, report some of the major news stories, provide lead-ins for other stories, and serve as the personality around whom the entire newscast revolves. The Anchorperson has extremely high visibility in a community and can even achieve star status.

As one of the most coveted jobs in television newscasting, the Anchorperson's position appears to be high-paying, glamorous, and easy. In fact, it is sometimes high-paying, occasionally glamorous, and almost never easy. In smaller and middle market stations, the Anchorperson often does the legwork necessary to research news stories, or constructs reports from wire service copy or network feeds. Even at stations with large news staffs and at the network level, he or she usually writes or rewrites some material.

The person holding this position must have a particular appeal to viewers and be able to speak to them in a satisfying and authoritative manner. The Anchorperson must have a thorough understanding of all news developments locally, nationally, and internationally. He or she must be able to identify trends and significant events, analyze and interpret them, and convey such information clearly to the viewing audience.

With news escalating in importance and stations involved in ratings wars, the Anchorperson has become the most important element of a newscast. An attractive and popular individual brings more viewers to the newscast, and thus a larger consumer audience to the station or network.

Additionally, the Anchorperson:

- takes part in determining the order of stories and the content of a newscast;
- provides voice-over commentary for filmed or taped clips of a particular story or event;
- conducts in-studio or on-location interviews;
- hosts documentary productions and special news reports.

Salaries

The salary range is very broad, with small market stations paying near the bottom of the scale and the major networks near the top. Industry sources report that Barbara Walters is earning $1 million a year and Dan Rather between $1.6 and $2 million. *Broadcasting* magazine reports that in New York, Chicago, Los Angeles, and Washington, DC, a few salaries are in the $150,000 to $350,000 range.

Apart from the super stars, Anchorpersons in 1980 earned an industry-wide average of $28,500 (up 12 percent from 1979), $36,000 in the top 50 markets, and $17,000 in the smallest markets, according to a survey by the Radio-Television News Directors Association (RTNDA). Other industry polls show the national average to have been $25,000 in 1980.

Employment Prospects

The anchor position at any television station or network is a very attractive post, and is considered the height of television Reporter's career. It is well paid, highly visible, and frequently accompanied by substantial benefits. Competition is extremely fierce, and only a small percentage of aspiring Reporters become Anchorpersons.

Advancement Prospects

Until recently, the networks were considered the ultimate goal for Anchorpersons, but with current trends toward higher salaries and better quality in local news, the rush to join a network has diminished. With the proliferation of cable TV news networks, opportunities for Anchorpersons will increase in the future. Current advancement possibilities, however, are quite poor.

Education

An undergraduate degree is necessary, and a graduate degree is preferable, for the position of Anchorperson. The major field of study should be in mass communications, journalism, or political science, but with a very strong liberal arts education. Many News Directors recommend that up to three-quarters of an aspiring Anchorperson's college course work should be in such areas as English, speech, government, sociology, and the humanities.

Experience/Skills

A considerable number of years of newsroom experience at smaller market stations is usually required before promotion to Anchorperson, and many more years of news work are necessary as an individual moves up the ladder to larger market stations. This experience includes news writing and reporting in all content areas and in various market sizes. Anchorpersons often have some experience in radio news or, less often, in print journalism.

A large part of an Anchorperson's success is based on appearance and delivery. He or she must look attractive on camera and speak in a clear and authoritative manner. In addition, he or she should project integrity and a pleasant personality.

Minority/Women's Opportunities

As with most on-camera news positions, opportunities for women and minority-group members have improved in recent years. Television News Research of New York reported in 1980 that the number of female Anchorpersons was increasing but there was a significant turnover. RTNDA found that in 1980, 29 percent of all broadcast journalists were women and 15 percent were members of minority groups.

Unions/Associations

When represented by a union, Anchorpersons may be members of the Writers Guild of America (WGA) or the American Federation of Television and Radio Artists (AFTRA).

The RTNDA is very active in promoting the professionalism of broadcast news, and many Anchorpersons belong to that organization. Some are also members of the Society of Professional Journalists, Sigma Delta Chi.

REPORTER

CAREER PROFILE

Duties: Gathering of news from various sources; analyzing and preparing news and features for television broadcast; on-air reporting

Alternate Title(s): Correspondent

Salary Range: $12,100 to $100,000

Employment Prospects: Good

Advancement Prospects: Fair

Prerequisites:

Education—Undergraduate degree in political science or journalism with strong liberal arts background

Experience—Minimum of two years as a News Writer in television or radio or in print journalism

Special Skills—Good writing and speaking ability; inquiring mind; dependability; persistence

CAREER LADDER

Assistant News Director; Anchorperson

↑

Reporter

↑

News Writer

Position Description

A television Reporter is a working journalist who gathers news from many different sources, organizes each report, sometimes writes it, and reports it on the air at a local station or network. At the major market station and network level, a Reporter assigned to an outlying area or overseas is usually called a correspondent or, occasionally, a foreign correspondent. In many stations, Reporters serve as their own Producers for specific television news coverage.

The day-to-day responsibilities of a Reporter vary considerably from station to station and from market to market. In a small station that does not employ News Writers, the Reporter will write his or her stories and deliver them on the air. In a larger station with several Reporters and News Writers on staff, each Reporter may specialize in particular areas such as politics, economics, health, or consumer information. In those instances, News Writers develop the Reporter's stories for delivery on the air.

Reporters' assignments may vary in importance, from covering a civic luncheon or a local fire to in-depth investigative reporting on corruption in local government or the significance of national economic policies to a local community. Reporters acquire information through library research, telephone inquiries, interviews with key people, observation, and questioning. Covering news conferences and press briefings, researching feature stories, and reporting on special one-time-only events are part of the Reporter's responsibilities.

A Reporter is skilled in investigating and gathering the news; in developing the information into a factual, understandable report; and in broadcasting it in a clear, concise and credible manner.

Additionally, the Reporter:

- determines the slant or emphasis of a particular news story;
- examines significant news items to determine ideas for features and news reports and evaluates leads and tips in developing story ideas;
- conducts live, taped, or filmed interviews in the studio and presents stand-up live, taped, or voice-over reports from the site of a news event;
- follows through on the developments of a previously reported story for news updates and additional reports.

Salaries

All news position salaries have increased in recent years, and industry sources believe they will continue to do so. Reporters, however, are paid substantially less than the top Anchorpersons. There is also a wide variance in Reporters' salaries between small local stations and the top market stations and networks, where salaries may reach $50,000 to $100,000.

According to the Radio-Television News Directors Association (RTNDA), top Reporters at commercial stations averaged $17,900 per year in 1980. A market-by-market analysis indicates that earnings at small market stations averaged $13,500 and in the top 50 markets, about $24,500. Other industry sources indicate that Reporters' earnings ranged from $12,100 to $18,400 in commercial television in 1980.

Employment Prospects

Although the position is sometimes considered an entry-level job at small market stations, most Reporters have had experience as television News Writers. Middle-size and large market stations employ several Reporters, as do the networks, and many more opportunities are opening up in cable television news operations. Industry sources believe that more and more people will be needed in key news positions during this period of expansion in the news industry. There is also a high turnover rate in this position at TV stations, and opportunities are good for employment as a replacement Reporter.

Advancement Prospects

The career goal of nearly every television Reporter is to become an Anchorperson. Prospects for that particular promotion, however, are poor because of the limited number of anchor positions available, the reluctance of Anchorpersons to vacate those positions, and the extremely heavy competition for the few openings that do occur.

Yet, it is possible to advance by moving to a larger market station, and thus to a higher salary and more prestige. Promotion to Assistant News Director and ultimately to News Director is also a possible career track.

For many Reporters, the position represents the peak of their careers. Some large stations in the major markets have Reporters on staff who have retained their jobs for up to 14 years, according to *Television/Radio Age.*

Education

A television Reporter must have broad knowledge in many areas. An undergraduate degree in political science or journalism, with considerable course work in all of the liberal arts is a necessity. Graduate study is useful for those who specialize in reporting on political or economic affairs.

Experience/Skills

Two or three years of experience in a television station's news department as a Desk Assistant or News Writer is usually preferred by employers, although some entry-level reporting positions do exist at smaller market stations. Actual news experience in college at a student newspaper or radio station is helpful. Some Reporters move to television reporting from a successful career in newspaper reporting.

Employers look for a bright, alert individual with an inquiring mind, a considerable degree of curiosity, and a great deal of persistence. A Reporter must be dependable, have an excellent knowledge of the English language, a reasonably good speaking voice, and good diction.

Minority/Women's Opportunities

As with other high visibility news positions, opportunities for minority-group members and for women have improved in recent years. The RTNDA reports that in 1980, 29 percent of all broadcast journalists were women and 15 percent were members of minority groups. The RTNDA report adds that audience surveys have shown that the sex of an on-air Reporter makes little or no difference to most viewers.

Unions/Associations

News personnel who write or perform are sometimes members of the Writers Guild of America (WGA) or the American Federation of Television and Radio Artists (AFTRA).

Television Reporters are also eligible for membership in RTNDA and other professional associations. Those who cover Congress can belong to the Radio and Television Correspondents Association (RTCA). Many Reporters have belonged to Phi Delta Epsilon, an honorary college journalism fraternity; Kappa Tau Alpha, a journalism scholastic society; or Pi Gamma Kappa or Alpha Epsilon Rho, student broadcasting fraternities.

WEATHER REPORTER

CAREER PROFILE

Duties: Reporting of weather conditions and forecasts as a part of a regularly scheduled TV newscast

Alternate Title(s): None

Salary Range: $13,000 to $100,000+

Employment Prospects: Poor

Advancement Prospects: Poor

Prerequisites:

Education—High school diploma; undergraduate degree in meteorology required at major market stations and networks

Experience—Some public speaking

Special Skills—Pleasant appearance and manner; good verbal ability; distinctive style

CAREER LADDER

Weather Reporter (Large Station or Network)

↑

Weather Reporter

↑

Announcer

Position Description

Weather Reporters range from the personable, attractive high school graduate at some local stations to the highly qualified, knowledgeable meteorologist at most major market stations and networks. Many television stations fill this position with an experienced staff Announcer who is permanently assigned to the post and who is responsible for the subject regularly.

The Weather Reporter reports to the Assistant News Director on a daily basis with information about the day's weather conditions. In some parts of the country, the weather is an important factor in the economy, and in those locations, it is reported on regular newscasts as well as in short summaries several times during the day. Extensive weather coverage of the entire country is usually reported twice a day, during major newscasts. Industry studies have indicated that most viewers are interested in weather conditions in other locations only if something unusual has occurred.

A Weather Reporter often uses a variety of visual devices to illustrate weather conditions, including radarscopes to show storm conditions; dials and charts that indicate temperature, humidity, wind velocity, barometric pressure, and pollen count; magnetic boards and chalkboards; transparent plastic overlays; and satellite photographs of the continental United States. A Weather Reporter works with the station's production team and Art Director to develop new techniques of making weather information more interesting and informative for the viewer.

Additionally, the Weather Reporter:

• works with the production crew and Director in setting up and changing or modifying weather visuals for a newscast;

• maintains communications with the U.S. Weather Bureau and other weather reporting stations;

• checks with contacts in different parts of the station's coverage area for information on weather conditions in various locations.

Salaries

Weather Reporters' salaries vary with the size of a station, its geographic location, the importance with which weather is viewed by management as an audience-builder, and the qualifications of the person holding the position. Industry sources place the salary range between $13,000 at small stations for relatively inexperienced individuals to $26,600 at major market stations for those with experience and seniority. The average salary was $22,300 in 1980. Some Weather Reporters who are well known and have a distinctive style that attracts viewers are paid considerably more. Network Weather Reporters often earn more than $100,000 per year.

Employment Prospects

The potential for employment is very limited. Most stations employ only one full-time Weather Reporter, with an Announcer filling in on early-morning and some weekend newscasts. Some major market stations that also operate AM and FM radio stations employ two or more Weather Reporters. Most industry sources expect, however, that as the news industry continues to expand and new communications media become available, there will be increasing opportunities for Weather Reporters.

Advancement Prospects

For the professional Weather Reporter, the most likely career path for advancement is to a larger market station or possibly to a network. Some Weather Reporters move to cable TV systems in larger markets to advance their careers. For the staff Announcer who is assigned the job, a good performance can help pave the way to promotion within the News Department. In general, however, most successful Weather Reporters remain in that position and the chances for advancement to another market or to a different television career are quite limited.

Education

Educational requirements vary from station to station and from market to market. At some smaller stations, a high school diploma is sufficient; at others and at the network level, an undergraduate or graduate degree in meteorology is required. Courses in public speaking and mass communications are helpful.

Experience/Skills

The weather segment of a newscast is usually delivered without a script, and with almost constant eye contact with the camera. Some public speaking experience and an ability to ad lib with style, wit, and enthusiasm is generally required of all Weather Reporters. Most News Directors seek articulate, attractive people who have pleasant voices and proper diction. An ability to relate to and attract a viewing audience is the primary requirement for the position.

Minority/Women's Opportunities

As with many on-camera positions in television news, both women and minority-group members are finding more opportunities open to them as Weather Reporters.

Unions/Associations

Some Weather Reporters are members of the American Federation of Television and Radio Artists (AFTRA). Most, however, are not represented by a union. For meteorologists, membership in the American Meteorological Society is a prestigious affiliation that can enhance their career opportunities.

SPORTSCASTER

CAREER PROFILE

Duties: Reporting of athletic and sports events as part of a regularly scheduled TV newscast

Alternate Title(s): Sports Director; Sports Reporter

Salary Range: $18,900 to $100,000+

Employment Prospects: Poor

Advancement Prospects: Poor

Prerequisites:

Education—Undergraduate degree in journalism or mass communications with heavy exposure to the sports scene

Experience—Some newsroom and sports desk experience in television, radio, or newspapers

Special Skills—Extensive knowledge of all sports; writing and verbal ability; pleasant personality

CAREER LADDER

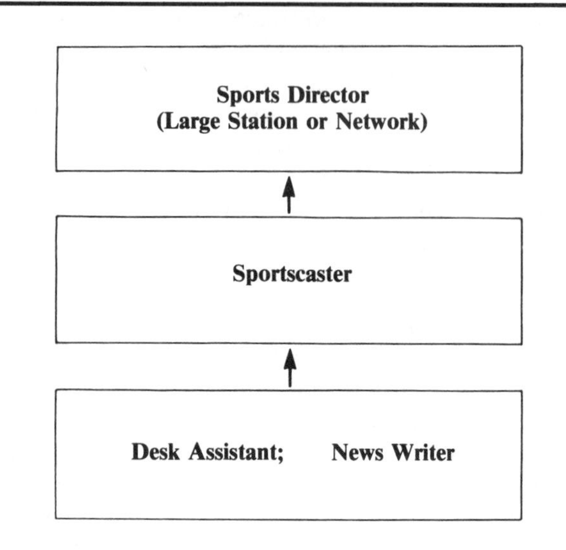

Position Description

A Sportscaster's duties and responsibilities are similar to those of a Reporter, but in a highly specialized area—the sports world.

At small stations, the Sportscaster may be a department of one, reporting to the Assistant News Director. In this situation he or she is allowed considerable autonomy in selecting, writing, and preparing, as well as delivering, the sports news for each newscast. At major market stations and at the network level, there may be several Sportscasters on staff, all of whom report to a sports director. Each person may specialize in a particular sport or event.

The duties of a Sportscaster go beyond merely reporting the outcomes of local games and contests, to include coverage of such diverse subjects as a baseball strike, the major professional sports' annual drafts, local personalities who appear headed for national prominence, legislation that affects the athletic scene, professional players' contracts and salary arrangements, record-breaking events, and more. Generally, the Sportscaster covers any event that involves or affects the sports scene.

Additionally, the Sportscaster:

- reviews network feeds, syndicated sports clips, and wire stories on national and international sports news;
- selects video material for use on a newscast;
- provides play-by-play (description of action in an event as it occurs) or color (informative commentary) coverage for games broadcast by the station;
- interviews local and visiting sports personalities;
- maintains continuing contact and encourages good station relationships with local and regional sports figures—coaches, managers, athletic directors, and players;
- originates and develops special sports commentaries, features, and documentaries.

Salaries

Salaries for all news personnel have risen over recent years, particularly at smaller market stations. Sportscasters, however, are still paid substantially less than Anchorpersons. *Broadcasting* magazine reports that in 1980, in a market approximately 50th in size, an Anchorperson may have been paid $50,000, while a Sportscaster earned only about $25,000 to $30,000. Industry studies indicate that the 1980 salary range for Sportscasters was between $18,900 and $29,400, with the average at $20,000.

Exceptions to the above range can occur at major market stations and networks where popular Sportscasters can be paid well over $100,000 per year.

Employment Prospects

Competition for sportscasting jobs in commercial television is fierce and employment opportunities are poor. The position requires extensive background in athletics and solid reporting and writing skills. It is not considered an entry-level position, and most stations employ only one or two Sportscasters. There is, however, a high degree of interest in sports in this country, and the current expansion of news programs includes sports along with general news. According to industry reports, technological advances in satellite transmission are creating more opportunities in sports reporting, both at the station level and in the cable television industry.

Advancement Prospects

Sportscasting is generally not a stepping-stone to bigger and better positions within the news department. A Sportscaster may look forward to promotion to sports director at his or her station, if such a position exists, or at the network level, where salaries and prestige can be higher. Expertise and specialization in a particular sport is usually necessary.

Cable television news operations also offer advancement possibilities for the best and the brightest Sportscasters, and industry estimates indicate that these opportunities will continue to increase. The overall outlook for advancement, however, is considered poor.

Education

Sportscasters can no longer get by on a "dumb-but-charming-jock" image. They must be articulate and knowledgeable. Their writing skills must be highly developed since Sportscasters often write their own copy. Most employers require an undergraduate degree in journalism or mass communications with course work in speech, English, sociology, and psychology. Extensive exposure to, or training in, virtually all aspects of sports is essential.

Experience/Skills

Many Sportscasters gain general news experience as a Desk Assistant or News Writer before moving to sports reporting. Writing and speaking experience in television or radio is usually required. Newspaper sports reporting experience also provides background for the job. Part-time work in television or radio or at a newspaper while in college is useful training for those wishing to enter the field.

A Sportscaster should have an enthusiastic interest in sports and broad experience in a wide variety of athletic events. He or she should also have an appealing voice, good intonation and diction, a personal style, and a distinctive on-air personality.

Minority/Women's Opportunities

Long a male-dominated profession, sportscasting is beginning to open up to women. At the network level, Jayne Kennedy and Phyllis George have both contributed to the acceptance of women in this role with their knowledgeable contributions, poise, and professionalism. Some of this influence is noted at the local station level, although sports reporting in the early 1980's is largely dominated by male reporters. Male members of minority groups achieved acceptance as Sportscasters earlier than women, and there is little racial discrimination in hiring practices at all levels of sportscasting.

Unions/Associations

Some Sportscasters may belong to the Writers Guild of America (WGA) or to the American Federation of Television and Radio Artists (AFTRA) at certain networks and at some network-owned stations in major markets. At small market stations, however, Sportscasters are not usually represented by a union.

Sportscasters may hold membership in the American Sportscasters Association (ASA), the Radio Television News Directors Association (RTNDA), and other journalism organizations.

CHIEF ENGINEER

CAREER PROFILE

Duties: Overseeing of all engineering and technical operations at a television station

Alternate Title(s): Director of Engineering; Vice President of Engineering

Salary Range: $15,000 to $70,000+

Employment Prospects: Fair

Advancement Prospects: Fair

Prerequisites:

Education—Undergraduate degree in electrical engineering preferable; training in broadcast equipment; FCC First Class License

Experience—Minimum of five years as an Assistant Chief Engineer

Special Skills—Technical aptitude; administrative and organizational ability; leadership qualities

CAREER LADDER

General Manager; Engineering Management (TV Network)

↑

Chief Engineer

↑

Assistant Chief Engineer

Position Description

The Chief Engineer is ultimately responsible for all of a television station's technical facilities, equipment, and services necessary to conduct its broadcast and programming activities. He or she reports to the General Manager and is responsible for the administration and supervision of all engineering and operations personnel, and for keeping the station on the air during all broadcast periods.

The Chief Engineer is in charge of all long-range facilities planning, of systems design, and of the budgeting and purchasing of technical equipment. He or she must also make sure that station operations are in compliance with Federal Communications Commission (FCC) regulations and applicable local, state, and federal laws.

He or she is instrumental in the design, construction, and installation of all engineering equipment, develops preventive maintenance programs for station facilities, and ensures the efficient and effective use of a station's equipment.

Additionally, the Chief Engineer:

- coordinates and negotiates the use of outside telecommunications facilities with appropriate utility companies and private agencies;
- prepares all technical applications, including FCC construction authorization permits, licenses, and the renewal or modification of licenses.

Salaries

In spite of their many responsibilities, Chief Engineers are not as well paid as other management personnel in broadcast television. Salaries in commercial TV ranged from $20,000 at small market stations with limited facilities to more than $70,000 at major market stations in 1980, according to industry sources.

In public television, salaries ranged from $15,000 in small markets to $58,000 at some major market community stations in 1980, according to the Corporation for Public Broadcasting (CPB). The average was $27,000 in 1980, an increase of 9.4 percent over the previous year.

Employment Prospects

The opportunity to become a Chief Engineer at any of the 1,000 commercial and public television stations is somewhat limited. Although qualified engineers are increasingly difficult to find, a minimum of five years of experience in all aspects of broadcast engineering is essential. Because of the high degree of technical expertise required, the position is often viewed as a climax to a successful career in engineering. There is relatively little turnover in this area.

Positions as a Chief Technician at a cable television system or at a subscription television (STV) or multipoint distribution service (MDS) company, however, are available in increasing numbers. Other opportunities exist in government, health, or educational production organizations. Because of the lack of qualified broadcast engineers, ambitious and diligent individuals who have the necessary experience and education should find additional opportunities in these newer areas.

Advancement Prospects

As mentioned above, many Chief Engineers view the position as a successful completion of a career. Some, however, move onward to the same position at higher salaries at larger stations. Still others transfer to networks or large production companies to advance their careers. A few obtain the position of General Manager at smaller market stations. Recently, a number of individuals have advanced by obtaining more financially rewarding positions of a similar nature in large cable TV, STV, or MDS operations.

Education

Although an undergraduate degree is not required in the position, many Chief Engineers have degrees in electrical engineering, physics, or a related science. A few have some graduate work or a graduate degree. All have some technical training in broadcast engineering, at a school or technical center, and have passed the examination for a First Class Radiotelephone License from the FCC.

Experience/Skills

Television engineering is not an exact science. The possibilities of breakdowns in transmission or technical equipment are high. Considerable experience with all types and models of electronic gear, a thorough understanding of the principles of electronics, and an ability to rapidly scan and decipher schematic drawings are necessities in this position. Most TV stations require a Chief Engineer to have at least five years of experience as an Assistant Chief Engineer.

The Chief Engineer must thoroughly understand all FCC regulations and be current on all state-of-the-art television equipment. In addition, he or she must be capable of designing various technical systems using interrelated components to improve engineering facilities. He or she must also demonstrate leadership qualities, be budget conscious, and be a capable administrator.

Minority/Women's Opportunities

As of this writing, there are no women Chief Engineers reported at any commercial TV station, and only one white female held the related title of Director of Engineering at a public television station in 1980. The vast majority of positions are held by white males. Less than 2 percent of such positions in commercial TV are held by minority males, according to industry sources, and 3 percent of the Chief Engineers and Directors of Engineering at public stations in 1980 were members of minority groups, according to the CPB.

Unions/Associations

Chief Engineers are considered part of the management of a station and are not represented by a union or bargaining agent. Some have been members of the International Brotherhood of Electrical Workers (IBEW) or the National Association of Broadcast Employees and Technicians AFL-CIO (NABET). Most become inactive, however, upon becoming Chief Engineers.

Many belong to the Society of Broadcast Engineers (SBE) or the Society of Motion Picture and Television Engineers (SMPTE), sharing mutual concerns and increasing their technical knowledge.

ASSISTANT CHIEF ENGINEER

CAREER PROFILE

Duties: Day-to-day scheduling, maintenance, and operation of a television station's equipment and facilities

Alternate Title(s): Assistant Director of Engineering

Salary Range: $12,000 to $43,000

Employment Prospects: Fair

Advancement Prospects: Good

Prerequisites:

Education—Undergraduate degree in electrical engineering or related field preferable; training in broadcast equipment; First Class FCC License

Experience—Five years as an Engineering Supervisor

Special Skills—Technical aptitude; administrative and organizational ability; leadership qualities

CAREER LADDER

Chief Engineer

↑

Assistant Chief Engineer

↑

Engineering Supervisor

Position Description

The Assistant Chief Engineer is usually the second in command in the engineering department at a commercial or public television station. As the senior technical staff member, he or she is directly responsible to the Chief Engineer for the day-to-day operation of the station and all scheduling of engineering employees and facilities. The person in this job also implements the department's budget and policies.

The Assistant Chief Engineer advises program and production personnel on the technical aspects of a TV program during the planning stages. He or she also implements the support systems necessary to mount a production. Other important functions include assisting the Chief Engineer in developing technical designs and specifications and creating (or supervising the creation) of the documentation necessary for installing or modifying equipment.

The Assistant Chief Engineer also helps to determine the connecting services required for program distribution and directly orders them from the telephone company and other carriers. He or she keeps abreast of state-of-the-art broadcast equipment specifications and of the vendors, prices, and availability of such equipment.

As the second in command, he or she directly supervises master control, editing, and transmitter operations, and conducts tests to ensure that the station is operating in accordance with Federal Communications Commission (FCC) regulations.

Additionally, the Assistant Chief Engineer:

- anticipates all possible contingencies in the studio or at the transmitter and develops procedures to minimize technical down-time;
- supervises equipment replacement and modification;
- assists in the installation of equipment;
- purchases engineering supplies and equipment;
- conducts transmission tests;
- oversees inventory of parts and components;
- serves as backup for other engineering supervisory employees and for the Chief Engineer.

Salaries

Salaries for Assistant Chief Engineers are relatively good in broadcasting. They ranged from $18,000 to more than $40,000 at major market commercial stations in 1980, according to industry sources.

In public TV, 1980 salaries went from $12,000 at small market stations to $43,000 in the larger markets, according to Corporation for Public Broadcasting (CPB) surveys. The average for that year was $22,700.

Employment Prospects

The opportunities for employment as or advancement to Assistant Chief Engineer in commercial or public television are fair. There is relatively little turnover in the position, and in many small market stations, the Engineering Supervisor in charge of transmission performs the duties of an Assistant Chief Engineer.

Positions in cable TV, and in subscription television (STV) or multipoint distribution service (MDS) operations, however, are available in increasing numbers, and most government, health, or educational production units require management-level technical workers who have some broadcast experience. Because of the increasing need for engineering personnel in all areas of broadcast and non-broadcast telecommunications, the opportunities for employment in this or in a comparable position are improving for ambitious and qualified people.

Advancement Prospects

Most Assistant Chief Engineers seek advancement to the position of Chief Engineer. Some advance their careers by moving to a smaller market in a more responsible role and at a higher salary. Still others join independent production organizations or become Chief Technicians in cable TV or at subscription television (STV) or multipoint distribution service (MDS) companies, which actively recruit broadcast-quality engineering managers. Because of the relative lack of experienced personnel, Assistant Chief Engineers have a good chance of advancing their careers in a variety of technical settings.

Education

While an undergraduate degree is not a prerequisite for the position, most Assistant Chief Engineers, have had some college education in electrical engineering, physics, or a related science. Some training by equipment manufacturers or at technical schools is required, and an individual must have passed the examination for a First Class Radiotelephone License from the FCC.

Experience/Skills

A minimum of five years of experience involving all aspects of broadcast technology (usually as an Engineering Supervisor) is frequently required for employment as or promotion to Assistant Chief Engineer. In order to reach that position, a person must have a complete knowledge of currently used television equipment, systems, and components and a thorough understanding of FCC technical regulations.

In addition, he or she must possess supervisory and leadership capabilities, must be organized and budget-conscious, and must be capable of dealing with a wide variety of individuals from various departments within a station. An Assistant Chief Engineer should also be experienced in dealing with manufacturers' representatives and electronic supply companies.

Minority/Women's Opportunities

There were few women Assistant Chief Engineers at commercial television stations in 1980. Employment of minority-group members in the position has increased during the past few years, but the job has usually been held by a white male. No white or minority women, and only four minority men, held the position in public TV in 1980 according to the Corporation for Public Broadcasting (CPB).

Unions/Associations

Assistant Chief Engineers are usually considered part of management, and as such are not represented by a union. Most people in the position, however, belong to one or more of the professional organizations mentioned in the Chief Engineer's entry.

ENGINEERING SUPERVISOR

CAREER PROFILE

Duties: Supervision of Audio/Video, Maintenance, Master Control, Transmitter, or Videotape Engineers at a television station

Alternate Title(s): Supervising Engineer

Salary Range: $14,500 to $35,000+

Employment Prospects: Fair

Advancement Prospects: Good

Prerequisites:

Education—High school diploma; one to two years of technical school training; FCC First Class License

Experience—Minimum of two years as an Audio/Video, Maintenance, Master Control, Transmitter, or Videotape Engineer

Special Skills—Technical aptitude; supervisory skills; leadership qualities

CAREER LADDER

Assistant Chief Engineer

↑

Engineering Supervisor

↑

Audio/Video, Maintenance, Master Control, Transmitter, or Videotape Engineer

Position Description

An Engineering Supervisor directs the day-to-day operation of electronic and technical equipment and of operating engineers at a television station or production company. In smaller market stations, one or two engineers supervise the entire engineering staff. In middle and major market stations, there are often four or more Engineering Supervisors, each of whom directs the activities of Audio/Video, Maintenance, Master Control, Transmitter, or Videotape Engineers. In some small and middle market stations, the Engineering Supervisor of transmitter operations often serves as the Assistant Chief Engineer.

One of the major responsibilities of the position is to ensure the proper maintenance and operation of electronic equipment. An Engineering Supervisor is held responsible for the safe, efficient, and effective operation of all electronic and technical equipment under his or her jurisdiction.

The Engineering Supervisor often oversees the performance of five or more engineers. He or she serves as the day-to-day manager of various shifts, schedules personnel, and determines assignments within each shift. Depending on the size of the station and of the department, the Supervisor may also be responsible for the actual running of equipment. In most cases, however, he or she acts as a backup and troubleshooter for a variety of operations.

Additionally, the Engineering Supervisor:

- makes tests and calibrations to ensure the station's conformance to Federal Communications Commission (FCC) rules and regulations;
- supervises preventive and emergency maintenance and repair of all equipment under his or her supervision;
- maintains and reviews all discrepancy and technical reports to spot and correct technical problems;
- plans for and installs new component parts and equipment.

Salaries

In keeping with their responsibility, Engineering Supervisors are relatively well paid. Salaries in commercial TV ranged from $16,000 at smaller market stations to more than $35,000 in major markets for people with experience and seniority.

In public television, salaries went from $14,500 to $32,400 at major market stations in 1980, according to the Corporation for Public Broadcasting (CPB). The average was $20,400 in 1980, an increase of 5.7 percent over the previous year.

Employment Prospects

The opportunities for employment as an Engineering Supervisor are fair. The position is not an entry-level job, and requires considerable experience and knowledge of broadcast equipment. Although most stations and the larger televison production organizations often employ more than four Engineering Supervisors, competition for the job is heavy. The position is usually acquired by the more diligent, competent, and experienced engineers from within the station. More opportunities for a similar position, however, exist at larger cable TV systems and subscription television (STV) and multipoint distribution service (MDS) companies. Some jobs are also available at government, health, and educational television production organizations.

Advancement Prospects

The possibilities for advancement to the position of Assistant Chief Engineer or in some cases directly to Chief Engineer are generally good. There is a lack of people with sufficient qualifications to fill these positions at most public and commercial television stations.

The bright, alert, and diligent Engineering Supervisor who has earned a reputation in a specific area (such as videotape recording and editing, maintenance, or transmission) and who has shown interest in other engineering disciplines, often finds promotional opportunities at his or her own station. Some use their skills and experience to obtain better-paying positions at larger market stations, while others advance by assuming more responsible engineering posts at smaller stations. Many experienced Engineering Supervisors obtain better positions at higher salaries in cable TV, STV, and MDS operations, which actively recruit broadcast-quality engineers.

Education

A minimum of a high school education and a year or two of training at a technical school is required for the position of Engineering Supervisor. Some college training in electrical engineering, physics, or a related field is also helpful. Engineering Supervisors have usually obtained a First Class Radiotelephone License from the FCC.

Experience/Skills

Most candidates for the position are expected to have a minimum of two years of experience as an Audio/Video, Maintenance, Master Control, Transmitter, or Videotape Engineer. A complete knowledge of the particular engineering specialty that the individual will supervise is required, as well as some understanding of other engineering functions. Most Supervisors have earned a reputation among their peers for skilled troubleshooting.

Engineering Supervisors must have the ability to supervise other engineers in the installation, operation, and maintenance of complicated electronic equipment, and must be organized and efficient. They should also be capable of dealing with a wide variety of technical problems and of determining effective solutions.

Minority/Women's Opportunities

There were but a handful of women Engineering Supervisors at commercial stations in 1980. The vast majority of the positions were held by white males. The number of minority-group members, however, has increased in the past few years, particularly in the major urban areas, acccording to industry observers. In public television, there were no white or minority females in the position in 1980, and only 5.6 percent of those holding the job were minority males, according to the Corporation for Public Broadcasting (CPB).

Unions/Associations

Engineering Supervisors are usually considered to be management personnel and, as such, are not represented by a union. Some advance their knowledge and their careers by joining the associations listed in the Chief Engineer entry.

MAINTENANCE ENGINEER

CAREER PROFILE

Duties: Maintenance and repair of equipment at a TV station

Alternate Title(s): Maintenance Technician

Salary Range: $9,000 to $29,300

Employment Prospects: Good

Advancement Prospects: Good

Prerequisites:

Education—High school diploma; minimum of FCC Second Class License

Experience—Minimum of one year as an Engineering Technician

Special Skills—Technical aptitude; troubleshooting and problem-solving ability; intuition; curiosity

CAREER LADDER

Engineering Supervisor

↑

Maintenance Engineer

↑

Engineering Technician

Position Description

A Maintenance Engineer is responsible for the repair and maintenance of all broadcast-related equipment at a television station. Under the direction of an Engineering Supervisor, he or she performs preventive maintenance on cameras, switchers, audio consoles, video monitors, microphones, videotape recorders, film chains, etc. He or she also designs, modifies, and repairs component parts, systems, and elements used in all production and transmission equipment.

A Maintenance Engineer is a problem-solver and troubleshooter. He or she handles major repair problems and often reconstructs existing equipment to upgrade it to conform to state-of-the-art technology.

It is his or her job to see that equipment meets all design specifications and that there is as little down time as possible. He or she usually works in the engineering shop, which is often adjacent to the master control room at a TV station.

The Maintenance Engineer is usually employed for a regular shift, but is often on call to do major and complex repairs that require immediate attention.

Additionally, the Maintenance Engineer:

- uses various testing monitors and meters to gauge the performance of equipment;
- maintains an inventory of parts and supplies needed to repair electronic equipment;
- keeps complete records of all equipment maintenance;
- recommends the replacement of equipment that is outdated or that cannot be repaired.

Salaries

The salaries for Maintenance Engineers are usually higher than for those of their peers who are involved in audio, video, or transmitter engineering. In commercial television, they ranged from $14,000 to $25,000 for experienced individuals at major market stations in 1980, according to industry sources. In union shops at major market network-owned stations, salaries ranged from $18,400 to $29,300 for people with four or more years of experience in 1980.

In public television, engineering maintenance workers were paid between $9,000 and $28,100 in 1980, according to the Corporation for Public Broadcasting (CPB). The average salary was $16,200 in 1980, an increase of 11 percent over the previous year.

Employment Prospects

Opportunities in all areas of television, video, and telecommunications are good. Qualified Maintenance Engineers are extremely difficult to find. Most commercial and public television stations employ four or more such people who are experienced electronic troubleshooters. Today's TV equipment consists of hundreds of complex, interrelated components. The ability to determine which element in the system has malfunctioned, and how best to repair it, is a rare talent. As a result, Maintenance Engineers are sought by commercial television stations as well as by educational, government, health, and corporate production organizations. Subscription TV (STV) and multipoint distribution service (MDS) companies and cable TV systems also offer employment opportunities. Some Engineering Technicians with the necessary skills, and with further study and training, can become Maintenance Engineers.

Advancement Prospects

Creative and innovative Maintenance Engineers have numerous opportunities to advance their careers. Many assume more responsible positions at smaller market stations or move up at their own stations by becoming Engineering Supervisors in charge of equipment maintenance. In addition, there are opportunities for advancement to engineering jobs in cable, MDS and STV firms, or in educational, health, industrial, and other non-profit TV settings.

Education

A high school diploma and some training in electronics at a vocational or technical school are required. Nearly all Maintenance Engineers have a Second Class Radiotelephone License from the Federal Communications Commission (FCC) and the majority have passed the examination for the First Class License.

Experience/Skills

A minimum of one year of television maintenance experience is expected in individuals promoted to the position of Maintenance Engineer. A solid knowledge of engineering test equipment and its functions is required. An ability to read and interpret schematic diagrams is also a must.

An inquisitive mind is most useful, as is the ability to analyze electronic problems. Most good Maintenance Engineers are excellent problem-solvers who combine experience and intuition when troubleshooting malfunctions in complex electronic gear.

Minority/Women's Opportunities

The majority of Maintenance Engineers in commercial television are white males. In public television, women held less than 1 percent of the positions in 1980, while minority-group members held 8 percent, according to the Corporation for Public Broadcasting (CPB).

Unions/Associations

Almost all Maintenance Engineers in commercial television are members of the National Association of Broadcast Employees and Technicians AFL-CIO (NABET) or the International Brotherhood of Electrical Workers (IBEW). In public TV most are not represented by a union.

TRANSMITTER ENGINEER

CAREER PROFILE

Duties: Operating and maintaining a television transmitter

Alternate Title(s): None

Salary Range: $7,600 to $30,000 +

Employment Prospects: Good

Advancement Prospects: Good

Prerequisites:

Education—High school diploma; some technical training; First Class FCC License

Experience—Two years as an Engineering Technician

Special Skills—Technical aptitude; mechanical skills; diligence

CAREER LADDER

Engineering Supervisor

↑

Transmitter Engineer

↑

Engineering Technician

Position Description

A Transmitter Engineer is involved in all phases of the direct transmission of a television signal to the viewing audience. He or she is responsible for operating a TV transmitter and antenna system in accordance with Federal Communications Commission (FCC) regulations. In co-owned radio-TV stations, this person often operates the AM or FM transmitter as well.

In some instances, the transmitter is located with the station's production, administrative, and other engineering facilities in a metropolitan area or suburb. In other circumstances, the transmitter tower and antenna are located in a remote spot (usually on high terrain) away from the main studio. A Transmitter Engineer works at either location, under the direction of the Engineering Supervisor in charge of the transmitter.

A Transmitter Engineer is responsible for the actual broadcasting of a TV signal. He or she also handles the day-to-day maintenance of the transmitters, antennas, and associated equipment.

A Transmitter Engineer is assigned to various shifts during the broadcast day. During a typical eight-hour period, he or she makes the immediate technical adjustments necessary to keep the station on the air, and to ensure uninterrupted operation.

Additionally, the Transmitter Engineer:

- makes proof of performance tests, measures transmitter and antenna output, and performs other tests in accordance with FCC regulations;
- conducts daily inspections of building, tower, and antenna lights to ensure continued operation;
- adjusts and repositions the equipment associated with microwave receiving and transmitting parabolas located on towers;
- adjusts and repositions a satellite-transmission receiving unit and associated equipment;
- designs, constructs, and installs prototype or experimental components and equipment, and makes test measurements and reports;
- maintains hour-by-hour and program-by-program transmission records, according to FCC rules;
- continually monitors all incoming and outgoing transmissions, including satellite, studio, network, or regional broadcasts;
- keeps an inventory of supplies and parts for all electronic components and systems associated with television transmission.

Salaries

Salaries for Transmitter Engineers in commercial television varied from $12,000 for beginning personnel in small market stations to more than $22,000 in 1980, according to industry sources. In a few major market network-owned stations with union representation, the minimum amount paid for the position in 1980 was $30,000.

In public TV, the salaries for any engineer with a First Class Radiotelephone License from the FCC ranged from $7,600 at small college- or school-owned stations to $29,400 for people with seniority at larger stations, according to the Corporation for Public Broadcasting (CPB). The average salary in 1980 was $16,600, an increase of 8 percent over the previous year.

Employment Prospects

The possibilities for employment are good, since there is generally a scarcity of qualified applicants. Most stations employ four or more full-time Transmitter Engineers for various shifts and specific duties, and opportunities exist for an alert Engineering Technician or other individual with some broadcast experience. In addition, there are new opportunities for Transmitter Engineers at subscription television (STV) and multipoint distribution service (MDS) companies.

Advancement Prospects

The opportunities for qualified and ambitious Transmitter Engineers to move to more responsible positions within an engineering department are good but the competition is heavy. All Engineering Supervisors in charge of transmitters, Assistant Chief Engineers, and Chief Engineers have worked as Transmitter Engineers at some time in their careers. Because of their experience in broadcast technology, many Transmitter Engineers are in demand at STV and MDS operations.

Education

Most Transmitter Engineers have a high school diploma and some training in broadcast engineering at a trade or vocational school. Some course work in electrical engineering and physics is also helpful. All have passed an examination for a First Class FCC License.

Experience/Skills

A Transmitter Engineer usually has at least two years of experience as an Engineering Technician with some of it at a transmitter. He or she must be familiar with the specialized equipment used to keep a TV station on the air. Considerable experience with standard engineering test equipment is also required, as well as some ability at reading schematic diagrams.

Most employers seek conscientious, technically-oriented individuals who are good with details and at solving problems.

Minority/Women's Opportunities

The vast majority of Transmitter Engineers were white males in 1980, according to industry sources. However, employment of women and minority-group members in the position has increased somewhat in the past ten years.

Unions/Associations

Most Transmitter Engineers at commercial television stations are members of and are represented by the National Association of Broadcast Employees and Technicians AFL-CIO (NABET) or by the International Brotherhood of Electrical Workers (IBEW). While the majority of such workers in public TV are members of neither union, some engineering employees at major market community-owned public television stations do belong to one of them.

AUDIO/VIDEO ENGINEER

CAREER PROFILE

Duties: Operating electronic audio and video equipment at a television station

Alternate Title(s): Audio Technician; Video Technician

Salary Range: $7,100 to $27,800

Employment Prospects: Good

Advancement Prospects: Good

Prerequisites:

Education—High school diploma; technical school training; Second Class FCC License

Experience—Minimum of one year as an Engineering Technician

Special Skills—Technical aptitude; quick reflexes; versatility

CAREER LADDER

Engineering Supervisor; Technical Director; Technician

↑

Audio/Video Engineer

↑

Engineering Technician

Position Description

An Audio/Video Engineer is responsible for operating all electronic controls of audio and video equipment used in a TV station's studio and on-location productions. Most engineering departments cross-train personnel, but Audio/Video Engineers usually concentrate on and are more experienced in one of the two functions.

The audio specialist is responsible for the sound portion of a production which includes voices, music, and special effects. He or she sets and places microphones, pre-records all necessary material, and, during production, monitors all sound levels and cues records and tapes. During videotape editing of a major show, he or she may also add to the tape pre-recorded elements such as music, and modify certain portions to improve overall sound quality.

The video specialist is in charge of the quality of the picture that is created. He or she sets up and aligns cameras and, during production, controls their brightness and color levels. Because of the sensitivity of video equipment, it must be monitored continuously to ensure that the best possible image is being broadcast.

The Audio/Video Engineer usually reports to an Engineering Supervisor who makes shift and project assignments. During rehearsals and actual production, he or she is responsible to the Technical Director.

During a studio production, the Audio/Video Engineer works in the control room. For on-location programs he or she is in a remote truck that contains all of the necessary equipment.

Additionally, the Audio/Video Engineer:

- maintains and makes minor repairs on audio or video equipment;
- assists the Director in achieving special sound or visual effects.

Salaries

Salaries for Audio/Video Engineers in commercial TV in 1980 ranged from $13,000 at small market stations for employees with one or two years of experience to $22,000 for experienced people with seniority at major market stations, according to industry sources. In major market union shops in 1980, salaries ranged from $15,900 for those with less than one year of experience to $27,800 for people with more than four years.

In public television, the 1980 salaries for Audio/Video Engineers ranged from $7,100 for newer employees to $26,800 for experienced personnel with seniority, according to the Corporation for Public Broadcasting (CPB). The average salary was $17,000, an increase of 27 percent over the previous year.

Employment Prospects

Most stations employ six or more Audio/Video Engineers. At least two are assigned to studio productions, one specializing in audio and the other in video. Complex studio or on-location productions require more individuals, particularly in the video area. In addition, the majority of government, health, education, and independent production operations require Audio/Video Engineers. Cable TV systems that produce programs also employ them. As a result of all of these available positions, employment opportunities are good.

Advancement Prospects

Qualified Audio/Video Engineers are in demand, and those with experience have good opportunities to obtain more responsible engineering positions. Many alert individuals are promoted to Engineering Supervisors at their own stations, while some move to larger stations as Technical Directors. The opportunities for employment in middle management engineering positions at independent production companies and health, government, or educational production studios are also good for those who exhibit leadership abilities. Competent Audio/Video Engineers are also sought as Technicians by cable TV, subscription TV (STV), and multipoint distribution service (MDS) operations because of their general broadcasting experience.

Education

A high school diploma and some training at a technical or vocational school are usually necessary to obtain the position of Engineering Technician and to be promoted to Audio/Video Engineer. All prospects must have a minimum of a Second Class Radiotelephone License from the Federal Communications Commission (FCC). Many employers require a First Class License.

Experience/Skills

A minimum of one year's experience as an Engineering Technician is usually required for promotion to the position. The individual must be thoroughly familiar with the operation of all audio and video equipment used in a television production and must be capable of routine maintenance.

In promoting personnel to the position, most Technical Directors and Engineering Supervisors look for young people who are versatile but display a particular skill in the operation of either audio or video equipment. Audio personnel should possess a discriminating "ear" and for both positions quick reflexes and an understanding and feel for television production, are required.

Minority/Women's Opportunities

While employment of women and minority-group members has increased in the position, the majority of Audio/Video Engineers in commercial television were white males in 1980. In public TV, 9.3 percent of the Audio/Video Engineer (and related) positions were held by women in 1980, and 17.9 percent by members of minorities, according to the CPB.

Unions/Associations

Most Audio/Video Engineers in commercial television are members of the National Association of Broadcast Employees and Technicians AFL-CIO (NABET), or of the International Brotherhood of Electrical Workers (IBEW). In public TV, some employees of major market stations also belong to one of these unions.

VIDEOTAPE ENGINEER

CAREER PROFILE

Duties: Operating videotape machines at a television station

Alternate Title(s): Videotape Operator; Videotape Editor

Salary Range: $7,100 to $27,800

Employment Prospects: Good

Advancement Prospects: Fair

Prerequisites:

Education—High school diploma; some technical training; Second Class FCC License

Experience—Minimum of one year as an Engineering Technician

Special Skills—Technical aptitude; creativity; ability to work with a variety of people

CAREER LADDER

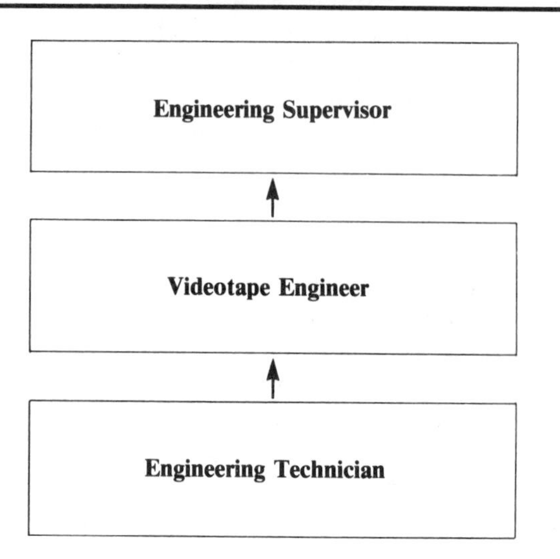

Position Description

A Videotape Engineer is responsible for the setup and operation of all types of videotape machines, and for recording, playback, and editing at a television station. He or she is also in charge of the evaluation of videotapes, the duplication (dubbing) of taped material, and the assembling of videotapes for production use and broadcast.

Videotape Engineers record programs being produced in studios at their stations or those from networks and other outside sources. They also assemble and edit commercials, promotional spots, and station breaks for each day's broadcasting. They work with Producers, Associate Producers, Directors, Assistant Directors, and Reporters in editing various pre-taped segments into a finished program. They also provide dubbing services for the programming and sales departments and for the promotion and publicity departments.

At most television stations, the Videotape Engineer works in a clean and secure environment adjacent to the master control room. Two or more Videotape Engineers are assigned by an Engineering Supervisor to each shift. Depending on the number of machines and the production and dubbing schedule, the Videotape Engineers work on various assignments during each shift.

The Videotape Engineer is one of the busiest individuals in the engineering department during a broadcast day. It is his or her responsibility to juggle various videotape requirements among machines and to determine priorities necessary for a smooth and professional videotape operation.

Additionally, the Videotape Engineer:

- sets up, aligns, adjusts, cleans, and monitors videotape machines prior to operation;
- files technical and malfunction reports;
- works on remote productions (outside the studio).

Salaries

Salaries for Videotape Engineers in commercial television in 1980 ranged from $12,000 at small market stations to more than $22,000 for people with seniority and experience in major market operations. In union shops at major market network-owned stations, the salaries ranged from $15,800 to $27,800 for employees with more than four years of experience.

In public television, 1980 salaries for the position ranged from $7,100 at small stations to $26,800 for those with seniority, according to the Corporation for Public Broadcasting (CPB). The average was $17,000.

Employment Prospects

A station usually employs six or more Videotape

Engineeers to cover all shifts and contingencies. Some Videotape Engineers are recruited from other stations. Opportunities for employment in this position are also available at government, health, education, or corporate organizations that specialize in production. Some cable TV, multipoint distribution service (MDS), and subscription television (STV) operations that originate programming on videotape also require Videotape Engineers. Thus, the opportunities are good for qualified individuals.

Advancement Prospects

The chances for advancement from the position are generally only fair. Talented Videotape Engineers may seek advancement to the position of Engineering Supervisor. Others move to smaller market stations to advance their careers or to independent production companies. Some use their skills and experience to move to more responsible positions at large advertising agencies that operate their own production or engineering units. There are also some chances for moving to a network or to more responsible engineering positions at STV, MDS, and cable TV operations.

Education

Most employers require a minimum of a high school education and some post-high school technical or vocational training in employing or promoting personnel to the position. While a First Class Radiotelephone License from the Federal Communications Commission (FCC) is not required for the position, evidence of study toward the exam is helpful. Most employers do require a Second Class License for the position.

Experience/Skills

The position is usually not an entry-level one. Most Videotape Engineers are promoted to the spot after a year or two as Engineering Technicians or in some other engineering position at the station. Sound knowledge of and experience with all types of videotape machines is required, as well as basic engineering knowledge of other types of technical equipment.

The Videotape Engineer must have technical aptitude and feel comfortable in working with broadcast equipment. He or she must also be capable of working well with a variety of non-engineering personnel in editing productions and must possess some creative instincts in assembling pre-recorded segments into a polished program.

Minority/Women's Opportunities

An increasing number of minority-group members and women have assumed the position of Videotape Engineer in commercial television during the past ten years. In public TV, 6 percent of the people in this job were women in 1980 and 11.9 percent were minorities, according to the CPB.

Unions/Associations

The majority of Videotape Engineers in commercial television are members of the International Brotherhood of Electrical Workers (IBEW) or of the National Association of Broadcast Employees and Technicians AFL-CIO (NABET). In public broadcasting, some individuals at large community-owned stations are members of one of these unions, but most are not represented by any group.

MASTER CONTROL ENGINEER

CAREER PROFILE

Duties: Coordinating all audio and video inputs for broadcasting

Alternate Title(s): Master Control Operator; Air Operator

Salary Range: $6,400 to $30,000

Employment Prospects: Fair

Advancement Prospects: Good

Prerequisites:

Education—High school diploma; some technical training; First Class FCC License

Experience—Minimum of one year as an Engineering Technician

Special Skills—Knowledge of broadcast equipment; decision-making ability; quick reflexes; sound judgment

CAREER LADDER

Operations Manager; Engineering Supervisor; Technical Director; Audio/Video Engineer

↑

Master Control Engineer

↑

Engineering Technician

Position Description

The Master Control Engineer is charged with the successful coordination of all audio and video input from various sources at a television station and with delivering their signals to the transmitter for broadcasting. As the chief operator in the nerve center of a TV station, he or she is directly responsible for ensuring that all transmissions are made in accordance with Federal Communications Commission (FCC) regulations.

The Master Control Engineer combines videotape, studio and on-location feeds, network programs, and pre-recorded audio announcements. He or she also cues and rolls film and videotape programs and switches video and audio inputs from different sources. Another important function is to ensure the smooth transition from one program to another and to control the switching to and from station breaks. This includes coordinating the broadcasting of commercials, public service messages, slides, and promotional tapes as well as cueing Announcers.

The Master Control Engineer is also responsible for handling all emergency situations, such as the loss of picture, program overruns and underruns, and equipment malfunctions. He or she has to make immediate corrections so that broadcast interruptions are kept to a minimum.

In assuring the smooth transmission of all of the station's programming, the Master Control Engineer works with the Operations Manager, Videotape Engineers, the Film Manager, Announcers, and the Traffic/Continuity Supervisor.

Additionally, the Master Control Engineer:

- ensures that all scheduled program elements are ready prior to broadcast;
- operates the master control switcher and audio console during station breaks;
- maintains the official station log and discrepancy records.

Salaries

Master Control Engineers are not as well rewarded financially as many of their peers. Salaries ranged from $11,000 to $18,000 in commercial television in 1980, according to industry sources. In unionized major market stations owned and operated by the networks, the minimum salary for the position was $30,000 in 1980. In public TV, salaries went from $6,400 at small market stations to $28,000 for people with experience and seniority at major market community stations in 1980, according to the Corporation for Public Broadcasting (CPB). The 1980 average was $14,100, an increase of 2 percent over the previous year.

Employment Prospects

Most stations employ three or more Master Control Engineers who work eight-hour day, night, and weekend shifts. In complex major market station operations, two or more Engineers are usually assigned to each shift. The opportunities are fair for bright, alert Engineering Technicians who demonstrate good judgment in the operation of television equipment to become Master Control Engineers.

Advancement Prospects

Opportunities for advancement are good. Many Master Control Engineers move to smaller market stations as Operations Managers or become Engineering Supervisors at their own stations. Others advance to positions as Technical Directors or Audio/Video Engineers, usually at higher salaries. A few who have the requisite skills become Maintenance Engineers.

Education

Most Engineering Supervisors require a minimum of a high school education and some technical training in promoting employees to the position of Master Control Engineer. A First Class Radiotelephone License from the FCC is also required.

Experience/Skills

A minimum of one year as an Engineering Technician is usually required for promotion to the position. Some major market stations require more engineering experience.

A Master Control Engineer must have a working knowledge of the limitations and operating characteristics of a wide variety of audio and video equipment. He or she must be alert and capable of making decisions quickly, must have good reflexes, and must display sound judgment.

Minority/Women's Opportunities

The opportunities for women and minority-group members in obtaining the position in commercial TV have increased in the past ten years, according to industry sources. The position, however, is still largely occupied by white males. In public television, 10 percent of the Master Control Engineers were women, and 12.6 percent were members of minorities in 1980, according to the CPB.

Unions/Associations

The majority of Master Control Engineers in commercial television were members of the National Association of Broadcast Employees and Technicians AFL-CIO (NABET) or the International Brotherhood of Electrical Workers (IBEW) in 1980. In public television, the Master Control Engineers at major market community stations are represented by the IBEW.

ENGINEERING TECHNICIAN

CAREER PROFILE

Duties: Carrying out various engineering duties at a television station

Alternate Title(s): Engineer; Technician; Operating Engineer

Salary Range: $10,000 to $22,400

Employment Prospects: Excellent

Advancement Prospects: Good

Prerequisites:

Education—High school diploma; some technical school; Second or Third Class FCC License helpful

Experience—Familiarity with electronic and technical equipment

Special Skills—Mechanical aptitude

CAREER LADDER

Audio/Video, Maintenance, Master Control, Transmitter, or Videotape Engineer

↑

Engineering Technician

↑

High School/Technical School

Position Description

Engineering Technician is usually the entry-level position in the engineering department at a commercial or public television station. The person in this job is responsible for performing various technical tasks in one or more units of the department. The Engineering Technician is responsible for the operation, maintenance, set-up and construction of equipment. He or she installs components and systems and collates and summarizes test data. He or she also works on various pieces of equipment, including cameras, video and audio tape recorders, microphones, audio switchers and mixers, video switchers and special effects boards, transmission equipment, audio and video testing devices, lighting equipment, and slide and film projectors.

The average TV station employs several Engineering Technicians in various capacities. It is usually the goal of the engineering department to cross-train them so that each person can gain experience in as many areas as possible.

The Engineering Technician usually reports to a Technical Director or to an Engineering Supervisor. After showing talent in a particular area, he or she is often trained to become an Audio/Video, Maintenance, Master Control, Transmitter, or Videotape Engineer. In some stations where Camera Operators are considered part of the engineering department, an Engineering Technician can also be trained for that position.

The responsibilities of the engineering department are diverse. The Engineering Technician can, therefore, work in almost any area of the station. He or she may be assigned to a studio, the videotape room, the transmitter, the master control room, the maintenance shop, or an on-location spot outside the station.

Additionally, the Engineering Technician:

- collects engineering records, including Federal Communications Commission (FCC) program and transmitter logs, facilities utilization forms, and testing and monitoring reports;
- maintains an inventory of engineering supplies and replacement parts.

Salaries

Salaries for Engineering Technicians are fairly good. At commercial stations in 1980 they ranged from $10,000 for entry-level personnel at small market stations to more than $20,000 for individuals with experience and seniority. The 1980 average was $15,300, according to industry sources. In union shops, the range was between $15,800 for employees with less than a year of experience and $21,200 for those with more than three years of experience.

Salaries in public television in 1980 ranged from

$5,800 for part-time personnel to $22,400 for full-time Engineering Technicians with experience, according to the Corporation for Public Broadcasting (CPB). The average was $12,600.

Employment Prospects

There are usually more beginning engineering positions open than there are qualified candidates to fill them. The average commercial or public television station employs more than 14 Engineering Technicians, and many major market stations have more than 60 on the payroll. The opportunities for bright, capable, and mechanically- and electronically-oriented young people are excellent.

Advancement Prospects

The opportunities for advancement to other Engineering positions are good. With application and diligence, an Engineering Technician can expect to be promoted to Audio/Video, Maintenance, Master Control, Transmitter, or Videotape Engineer, after a year or two of experience.

Many Engineering Technicians move further up the career ladder, in time, to Engineering Supervisors or Assistant Chief Engineers. Some are promoted to more responsible positions within their stations or move to better paying positions at major market stations. Advancement and promotion, however, usually require the acquisition of a First Class Radiotelephone License from the FCC, and the competition is heavy. Some people seek opportunities in cable TV, subscription TV (STV), or multipoint distribution service (MDS) operations.

Education

A high school diploma and some electronics training in a technical or vocational school are required by most employers. Some stations operate intern programs to help train prospective employees. Individuals who have studied for and passed the examination for a Third or Second Class FCC License are particularly sought after by employers.

Experience/Skills

Most stations do not require extensive experience in employing personnel for the position of Engineering Technician. Some interest in and understanding of electronics and the repair of mechanical and audio-visual equipment is helpful. A basic grasp of physics and the ability to read schematic diagrams are also useful. Employers seek diligent, alert, and ambitious people who have a love of electronics and who enjoy working with equipment.

Minority/Women's Opportunities

The opportunities for members of minorities and women are excellent. An increasing number of minority-group members has entered the field in the last ten years and, according to the FCC, they held 15.9 percent of the Engineering Technician positions in 1980. While the increase in the employment of women has not been as rapid, more are being hired. Women occupied 9.7 percent of the technical positions (including Engineering Technician) in 1980, according to the FCC.

In public TV, 10 percent of the positions were held by women and 13 percent by minority-group members in 1980, according to the CPB. These figures represent increases of 3.5 and 2 percent, respectively, over the previous year.

Unions/Associations

The majority of Engineering Technicians in commercial television are represented by the National Association of Broadcast Employees and Technicians AFL-CIO (NABET) or the International Brotherhood of Electrical Workers (IBEW). Some public television employees are represented by these unions, but most work in non-union stations.

TECHNICAL DIRECTOR

CAREER PROFILE

Duties: Overseeing technical quality of a television production; operating production switcher

Alternate Title(s): Switcher

Salary Range: $6,400 to $30,900

Employment Prospects: Fair

Advancement Prospects: Good

Prerequisites:

Education—High school diploma; some technical training; Second Class FCC License

Experience—Minimum of one year as an Audio/Video Engineer, Lighting Director, or Camera Operator

Special Skills—Knowledge of TV production and engineering; quick reflexes; leadership qualities; cooperativeness

CAREER LADDER

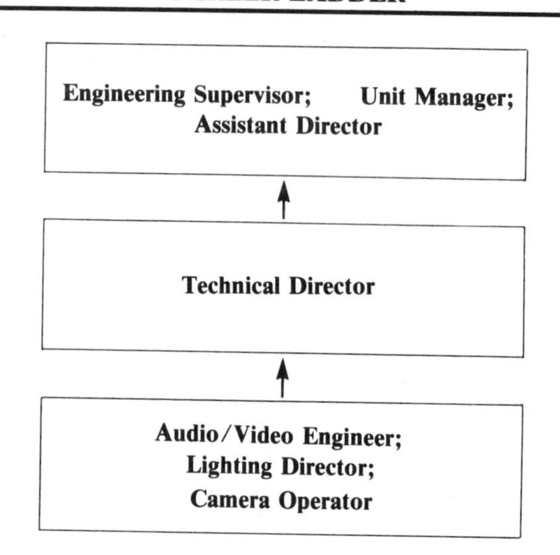

Position Description

The Technical Director is an important element in the production of a television show. He or she oversees the technical quality of the program and operates the production switcher, the unit that controls which camera images and special effects are broadcast or transmitted to videotape.

The Technical Director is the principal link between the Director and the technical crew assigned to the production. He or she usually supervises from three to nine engineers and assistants. At stations that do not have a Lighting Director, the Technical Director is often responsible for fulfilling that role.

During the pre-production stage, the Technical Director analyzes specific production requirements and makes technical recommendations to the Director. During rehearsals and actual production, the Technical Director usually sits on the right of the Director in the control room and transmits instructions (via headphones) to Camera Operators about the positioning of cameras and directs all other studio and control room technical personnel. He or she also operates the production switcher and follows the Director's instructions about changing from one camera shot to another and about selecting other picture sources, such as film and videotape. He or she presets and controls the switcher for special electronic effects and transitions called for by the Director.

The Technical Director normally reports to an Engineering Supervisor and is assigned by him or her to specific productions. As indicated above, the Technical Director reports to the program's Director during actual production. In most smaller stations where there is no staff Technical Director, the responsibilities are often shared by an Engineering Supervisor and Assistant Director. In very small stations, Directors frequently do their own switching.

Additionally, the Technical Director:

- provides technical leadership during production emergencies;
- assigns and trains studio and control room crew members and monitors their performance;
- prepares reports on the facilities used in a production for the Engineering Supervisor or Assistant Chief Engineer.

Salaries

Salaries for the the position in commercial TV range from $12,000 in small market stations to more than $25,000 in major market operations with heavy production schedules. The average 1980 salary for a Technical Director was $12,700, according to industry sources. In stations where Technical Directors are represented by a union, the minimum yearly salary was $30,900 in 1980.

Salaries for Technical Directors in public television ranged from $6,400 to $28,200 in 1980, according to the Corporation for Public Broadcasting (CPB). The average was $14,100.

Employment Prospects

The position of Technical Director is not an entry-level job. Some experience in TV engineering, and specifically in studio engineering, is required. Although some major market commercial and public television stations with heavy production schedules employ four more Technical Directors, most large stations require only two on staff. In many smaller stations, there are none. Technical Directors are also employed by networks and major independent production companies, but they tend to hire experienced people. The opportunities for employment are generally only fair.

Advancement Prospects

Technical Directors usually seek advancement to the position of Engineering Supervisor or, in some major production organizations, they are promoted to Unit Manager. Since a good part of the job involves the positioning of cameras and the preparation of shots, many individuals use their experience to move into Director positions, usually by first becoming an Assistant Director. Some Technical Directors move to smaller market stations at higher salaries and in more responsible technical or production positions. Those occupying the position usually view it as a transitional job that will help them to move to more responsible positions in engineering or production. The opportunities are good for ambitious individuals with initiative.

Education

The minimum requirements for the position are a high school diploma and some education at a vocational or technical school. Courses in TV production techniques can be helpful. Most employers require a Second Class Radiotelephone License from the Federal Communications Commission (FCC).

Experience/Skills

Most Engineering Supervisors require a minimum of one or two years of experience as an Audio/Video Engineer, Lighting Director, or Camera Operator when promoting individuals to the position of Technical Director. Experience in professional studio and remote TV production is a must.

To be effective, the Technical Director must have a thorough understanding of all production and engineering aspects of putting together a television program. In addition, he or she must be able to react to directions quickly and to make decisions rapidly in pressured situations. An ability to lead and to cooperate with other crew members is also important.

Minority/Women's Opportunities

There were only a handful of women employed as Technical Directors in commercial television in 1980. Male minority-group members, however, are increasingly represented in the position, according to industry sources. In public TV, 8.3 percent of the Technical Directors were women, and 25 percent were minority males in 1980, according to CPB surveys.

Unions/Associations

Technical Directors in commercial TV are considered part of the engineering staff and are usually represented by the International Brotherhood of Electrical Workers (IBEW) or the National Association of Broadcast Employees and Technicians AFL-CIO (NABET). In public television, they are usually not represented by a union.

CAMERA OPERATOR

CAREER PROFILE

Duties: Setup and operation of television cameras on location or in a studio

Alternate Title(s): Cameraman

Salary Range: $6,700 to $27,800

Employment Prospects: Fair

Advancement Prospects: Good

Prerequisites:

Education—High school diploma; training in photography or audio-visual equipment helpful

Experience—Some TV production, usually as a Production Assistant

Special Skills—Creativity; versatility; fast reflexes

CAREER LADDER

Floor Manager; Lighting Director; Audio/Video Engineer

↑

Camera Operator

↑

Production Assistant; Engineering Technician

Position Description

A Camera Operator is responsible for operating a television camera during the rehearsals for, and the actual production of, a TV program. He or she may operate full-size cameras mounted on tripods or dollies, or the smaller "mini-cams" or Electronic News Gathering (ENG) cameras that are used for on-location shooting.

Most stations have a minimum of two Camera Operators on staff. At some stations, more than eight are employed for various shifts and assignments. For some simple productions, only one or two cameras are used. For more complex programs, five or more cover various elements of the program.

Camera Operators are sometimes considered part of the engineering department and report to an Engineering Supervisor. At other stations, they are members of the production department and report to the Production Manager. Many commercial stations assign some Camera Operators to the news department where they are given specific assignments by the Assistant News Director.

In most situations, the Camera Operator receives directions (via an intercom system and headset) from the Director or Technical Director during rehearsal and actual production of a program. In news gathering operations, the Camera Operator usually follows the instructions of a Reporter when covering an on-location story. As an integral part of the production team, the Camera Operator is important for the continuing flow and appearance of any show or newscast. In addition to responding immediately to the Director or Reporter, a good Camera Operator finds new, interesting, and imaginative angles and shots for use during a production.

Additionally, the Camera Operator:

- assists the Audio/Video Engineer in the setup, technical check, and simple maintenance of television cameras;
- assists the Lighting Director in the setup of scenery, lighting equipment, and other properties prior to a production;
- assists the Floor Manager or Unit Manager in the strike (dismantling) of a production.

Salaries

Salaries in commercial television ranged from $10,000 at smaller market stations to $18,000 for experienced personnel in 1980, according to industry sources. In some major market union stations, the earnings ranged from $15,900 to $27,800 for Camera Operators with many years of seniority and experience.

In public television, Camera Operators were paid between $6,700 and $26,300 at major market community stations in 1980, according to the Corporation for Public Broadcasting (CPB). The 1980 average was $12,700, an increase of 21.7 percent over the previous year.

Employment Prospects

Camera Operator is considered an entry-level job at some stations but, more often, an experienced Production Assistant is promoted to the position. Sometimes an Engineering Technician is assigned to the job after some experience at a station. The number of openings depends on the amount of production done. Camera Operator positions are also available at education, government, and health television studios, and at many independent production firms. A few cable systems also employ full-time camera people. Because of competition, however, employment prospects are considered to be fair.

Advancement Prospects

Chances for moving up in television are good. Most Camera Operators seek a career in production. The usual advancement path for production-oriented individuals is to Floor Manager or Lighting Director. In stations where a Camera Operator starts as an Engineering Technician and is considered part of the engineering department, he or she often seeks advancement to Audio/Video Engineer.

Some advancement opportunities are also available in production or engineering roles at government, health, or educational organizations.

Education

A high school diploma is required by most employers. Additionally, some training in photography (still or motion) or audio-visual equipment is helpful. Some employers seek evidence of study toward a Second Class Radiotelephone License from the Federal Communications Commission (FCC).

Experience/Skills

Most Camera Operators have some experience in television production, usually as Production Assistants, prior to assuming the position. They need to have a creative sense of composition and an ability to initiate imaginative angles when operating a TV camera. Camera Operators should also have the ability to react rapidly to Directors during a production.

Minority/Women's Opportunities

Although there has been some increase in the number of women holding the job in commercial television during the past ten years, the majority of Camera Operators in 1980 were male, according to industry sources. Minority male employment in the position has increased greatly in the past few years. In public television, 13.4 percent of the positions were occupied by women, and 17.6 percent by minority-group members in 1980, according to the Corporation for Public Broadcasting (CPB).

Unions/Associations

At many commercial television stations the position of Camera Operator is represented by the International Brotherhood of Electrical Workers (IBEW) or by the National Association of Broadcast Employees and Technicians AFL-CIO (NABET). In public TV, the majority of Camera Operators are not represented by a union. Some individuals in both commercial and public television belong to the American Society of TV Cameramen (ASTVC) in order to share professional concerns and advance their careers.

GENERAL SALES MANAGER

CAREER PROFILE

Duties: Responsibility for producing all advertising revenue for a television station

Alternate Title(s): Sales Director; Advertising Sales Manager

Salary Range: $25,000 to $90,000+

Employment Prospects: Poor

Advancement Prospects: Good

Prerequisites:

Education—Undergraduate degree in marketing, advertising, or business administration preferable

Experience—Minimum of five years of experience in broadcast advertising sales

Special Skills—Leadership qualities; aggressiveness; poise; perseverence; organizational ability; interpersonal skills

CAREER LADDER

General Manager

↑

General Sales Manager

↑

Assistant Sales Manager

Position Description

The General Sales Manager of a commercial television station is responsible for generating the advertising revenue that enables a station to pay for programming, salaries, technical equipment, and operations. He or she deals with a limited inventory of commercial air time availabilities for advertising sales. The General Sales Manager works closely with the Program Manager and the General Manager in selecting programs during and around which such advertising can be sold.

A General Sales Manager is in day-to-day charge of all advertising sales activities and of the station's sales staff, usually from two to fifteen or more employees. Duties entail administering and coordinating all local, regional, and national programming and spot commercial sales accounts with clients, advertising agencies, and the station's national representative (rep). This includes developing the station's overall advertising sales plan and targets, previewing programs, and directly supervising the effectiveness of all sales efforts.

The General Sales Manager reports to the General Manager of the station and, with that person, establishes the station's advertising policies. A General Sales Manager must know the local market and the competition in order to establish advertising rates that are competitive. In addition, he or she develops audio-visual and statistical sales tools to assist Salespersons in their work.

A General Sales Manager must be aware of economic factors, retailing, merchandising, barter and trade arrangements, programming trends, and audience demographics and psychographics at both the local and national levels. He or she works closely with advertising agencies, reps, and local clients to ensure a continuing and profitable relationship.

Additionally, the General Sales Manager:

- monitors ARB and Nielsen ratings and research studies of the station's audience;
- supervises all commercial scheduling and copy writing;
- services top agency accounts, in order to keep close to the selling scene;
- ensures that the commercial time sold by the station conforms with National Association of Broadcasters (NAB) codes, Federal Trade Commission (FTC) regulations, and station policies.

Salaries

The General Sales Manager in a local commercial television station is usually the second highest paid employee. The compensation has increased steadily since 1970 and is expected to increase even more in the future. According to a Broadcast Information Bureau (BIB) survey of commercial station management, the average salary was $55,500 in 1980, an increase of more than 31 percent from the 1978 BIB figure.

Salaries ranged from $25,000 in small markets to more than $90,000 in major market stations in 1980. According to the BIB survey, 46.3 percent of the General Sales Managers made more than $50,000 in 1980. They are often paid a base salary and a bonus, but most receive a salary plus a commission.

Employment Prospects

The opportunity to acquire a position as a General Sales Manager in a commercial TV station is more limited than for other management personnel, and the competition is heavy. There is little turnover in the position, and the majority of respondents to the BIB survey had been with their employers for more than ten years.

Most General Sales Managers are promoted from within a television station organization, having first served as an Assistant Sales Manager. Some are recruited from the sales staffs of larger market stations or even from competitors in the local community. Others are recruited from advertising sales forces in print or other media.

Advancement Prospects

The opportunities for promotion for General Sales Managers are good. The typical move is to the position of General Manager, although some individuals transfer to larger markets, rep firms, group-owned stations, or even into advertising sales management at a network.

The skills of a General Sales Manager are easily transferable to cable TV systems and to subscription TV (STV) or multipoint distribution service (MDS) operations. Although opportunities in these areas are also good, the salary levels are often not as high as in commercial television.

Education

A degree is less important in obtaining the job of General Sales Manager than is the ability to sell effectively and consistently. An applicant with a degree, however, is usually chosen over someone without a degree.

An undergraduate degree in marketing, advertising, or business administration provides an excellent background for entering into sales positions and for eventual advancement to a higher position. In larger market stations, some General Sales Managers have a master's degree in business administration.

Experience/Skills

Most General Sales Managers are expected to be profit-oriented administrators with proven track records. Successful experience in TV advertising sales is a prime requisite, as is the ability to organize and motivate a sales staff and to develop the station's sales efforts. The General Sales Manager usually has at least five years of successful TV sales experience as an Assistant Sales Manager, a good feel for organization and logistics, and an outgoing and positive personality.

Most successful General Sales Managers are able to analyze research data easily and are experienced in marketing, merchandising, and retailing. They are capable of and enthusiastically interested in actual sales. As a group, they are aggressive, proficient at controlling expenses, poised, and persistent.

Minority/Women's Opportunities

There has been considerable improvement in the opportunities for women and minority-group members in the television sales field in the past four years. The majority of General Sales Managers in 1980, however, were white males. According to Federal Communications Commission (FCC) statistics, the percentage of women in all sales positions in commercial TV increased from 15.9 in 1976 to 28.6 in 1980. The percentage of minority-group members in all commercial TV sales positions rose from 7.3 in 1976 to 8.5 in 1980. According to some industry estimates, however, less than 5 percent of the General Sales Manager positions are currently filled by women or members of minorities.

Unions/Associations

There are no unions or professional organizations that represent General Sales Managers.

ASSISTANT SALES MANAGER

CAREER PROFILE

Duties: Assisting the General Sales Manager in local and national advertising sales

Alternate Title(s): Local Sales Manager; National Sales Manager

Salary Range: $20,000 to $50,000

Employment Prospects: Fair

Advancement Prospects: Good

Prerequisites:

Education—High school diploma and some college; undergraduate degree in marketing or advertising preferable

Experience—Minimum of three years in television advertising sales

Special Skills—Analytical mind; motivational talent; leadership and supervisory abilities

CAREER LADDER

General Sales Manager

↑

Assistant Sales Manager

↑

Advertising Salesperson

Position Description

An Assistant Sales Manager at a commercial television station assists the General Sales Manager in all aspects of the station's advertising sales activities. He or she relieves the General Sales Manager of some burdensome details and personnel management, so that the senior person can concentrate on long range sales planning. He or she helps the General Sales Manager establish and monitor quotas for each Advertising Salesperson, and assigns each to particular accounts. An Assistant Sales Manager is usually an experienced professional who understands both the advertising business and the local market.

As a rule, an Assistant Sales Manager can be responsible for one particular account area, such as local, national, or spot sales, in addition to more general duties. In many large market stations, there are two or more Assistant Sales Managers, often with titles reflecting their area of responsibility, such as Local Sales Manager or National Sales Manager. When an Assistant Sales Manager is responsible for local sales activities, he or she oversees local Advertising Salespersons and accounts. In developing, selling, and servicing such accounts, he or she supervises from three to eight Salespersons and keeps them informed about changing station policies and sales procedures. On a daily basis, he or she meets with the station's sales staff.

An Assistant Sales Manager may specialize in air time for specific types of programs, such as news, special events, or sports. In some stations, he or she serves as a troubleshooter in dealing with difficult accounts or in soliciting new business. He or she is often assigned direct consumer response sales, or concentrates on persuading advertisers who are traditionally print-oriented to buy ads on television. In some of the larger stations, an Assistant Sales Manager is in charge of supervising the servicing of all accounts after they are sold.

Although the position is generally a supervisory one, an Assistant Sales Manager also actively sells air time on a day-to-day basis, and is responsible for the continued sales to his or her own accounts. He or she is usually in charge of all sales efforts in the absence of the General Sales Manager.

Additionally, the Assistant Sales Manager:

- oversees all advertising proposals and contracts;
- maintains advertiser, agency, and prospect files;
- analyzes ARB and Nielsen ratings statistics;
- prepares competitive product studies, internal sales analyses, and reports for specific clients;

• monitors all available advertising air time.

Salaries

Most Assistant Sales Managers are paid on a salary-plus-commission basis. Commissions range from 10 to 25 percent, with the majority at the rate of 15 percent of the advertising income received by the station for a particular account. In almost all cases, an Advertising Salesperson is allowed to retain his or her sales accounts upon promotion to Assistant Sales Manager and receives commissions accordingly. Often with the assumption of administrative duties, the initial employment contract is renegotiated to provide for a larger base salary and smaller commissions. Sometimes the commission rate is raised appreciably and the Assistant Sales Manager is given a modest salary increase.

Earnings for Assistant Sales Managers in 1980 ranged from $20,000 to $50,000, averaging $35,300, according to industry sources. Most individuals have expense accounts, and many have company-paid life insurance and group health plans. The majority participate in company pension plans.

Employment Prospects

The possibilities are only fair for securing this position. Nearly 10 percent of the staffs of local TV stations are in the sales department. In general, only the brightest, most successful, and experienced people within sales are promoted to Assistant Sales Manager. The relative scarcity of good sales people and the high turnover rate in this area create openings, but competition is strong.

Assistant Sales Managers are usually chosen from those Advertising Salespersons within the station who have good sales records and experience with the department. Some are recruited from other stations where they have been successful in specific sales areas.

Advancement Prospects

Advancement opportunities are good. Most Assistant Sales Managers actively seek the position of General Sales Manager, and, ultimately, of General Manager. The chances at a commercial station, however, are limited to the number of TV stations on the air, and only the very able make the transition. Some Assistant Sales Managers move to smaller market stations as General Sales Managers or to network or syndicated program sales positions.

The skills and experience acquired in television time sales are readily transferable to other media-related industries. Cable television systems and subscription television (STV) and multipoint distribution service (MDS) companies are increasingly moving into advertising sales. They often recruit people who are experienced in TV advertising and an Assistant Sales Manager's background is, therefore, applicable.

Education

A minimum of a high school diploma is usually required, along with some college work. Many employers prefer to hire, and promote, people who have an undergraduate degree in marketing or advertising. Some courses in business administration and communications are also helpful. In general, however, a degree is less important than the ability to generate sales effectively and consistently.

Experience/Skills

For promotion to Assistant Sales Manager, the candidate should have at least three years of experience with a consistent and successful track record in TV advertising sales, and a special talent in a particular sales area.

An ability to analyze data quickly and to motivate other sales employees are prime requisites, as are competitive drive and demonstrated initiative. Supervisory and leadership skills are also important since most Assistant Sales Managers oversee less experienced sales people.

Minority/Women's Opportunities

Opportunities for women and minority-group members in TV sales are increasing yearly. Women held 28.6 percent of all broadcast sales positions in 1980, according to the Federal Communications Commission (FCC). They are increasingly being employed in television sales, although promotions to Assistant Sales Manager jobs appear to be slower than for their male colleagues.

The opportunities for members of minorities are not as good. Unless the station is minority-owned and serves a predominantly minority-member audience, the chances of such promotion are limited. In general, the position of Assistant Sales Manager is occupied by white males between the ages of 25 and 35.

Unions/Associations

There are no unions or professional organizations that represent Assistant Sales Managers.

ADVERTISING SALESPERSON

CAREER PROFILE

Duties: Selling advertising time locally, regionally, or nationally for a television station

Alternate Title(s): Time Salesman; Broadcast Salesman

Salary Range: $15,000 to $35,000

Employment Prospects: Good

Advancement Prospects: Good

Prerequisites:

Education—High school diploma; some college or undergraduate degree in marketing or advertising preferable

Experience—Minimum of one year in retail or print advertising sales helpful

Special Skills—Competitive drive; initiative; persuasiveness; persistence; gregariousness

CAREER LADDER

Assistant Sales Manager

↑

Advertising Salesperson

↑

Salesperson; Advertising Salesperson (Print Medium)

Position Description

An Advertising Salesperson at a commercial television station is responsible for selling advertising air time to businesses or to advertising agencies acting on behalf of businesses thus generating income for the station.

At most stations, Advertising Salespersons are assigned specific territories or responsibilities. Some concentrate on local sales, some on national sales, and others on obtaining sponsors for particular programs.

TV advertising time usually consists of 10- or 30-second commercials, or the partial or full sponsorship of a television program. The Advertising Salesperson matches the station's available advertising air time with a client's need, selling a time slot or program that will improve the client's sales. He or she analyzes a client's products or services, obtains detailed information about its business, and explains the size and type of audience that can be reached with particular advertising formats. To be effective, the individual must understand many kinds of products and merchandising techniques.

The Advertising Salesperson must also be familiar with the station's programming (syndicated, network, and local) and specific audience shares and ratings. He or she analyzes all research data and devises sales presentations, including charts, graphs, ratings records, and other audience research data, and often screens programs for a client.

At larger stations, the Advertising Salesperson works with and sells to the advertising agency representing the client. The agency actually prepares and produces the television commercial and delivers it to the station. In other instances, the Advertising Salesperson works directly with the advertiser in creating and producing commercial spots and announcements, using the station's production facilities.

Additionally, the Advertising Salesperson:

- serves as the continuing liaison between the client and the TV station in all advertising and program matters;
- assists in or supervises the copywriting and production of spot commercials for the client;
- works with the Sales Coordinator and Traffic/Continuity Supervisor to ensure proper scheduling for advertising;
- attends regular sales staff meetings to exchange ideas, submit reports, and receive assignments.

Salaries

The remuneration for Advertising Salespersons varies considerably, according to market size and the individual's experience. The top 100 markets generally offer a chance for a better income, although some stations in smaller markets pay more because of special competitive or geographic circumstances.

Nearly 50 percent of the TV and radio stations in the United States pay by a combination of salaries and commissions, 14 percent pay straight commissions, 8 percent pay straight salaries, and 28 percent pay a draw against commissions, according to the *Chronicle Occupational Brief for Radio-TV Time Salespeople* of January 1980 (Chronicle Publications Inc., Moravia, NY). Most commissions are paid after the client is billed, but some are not paid until after the income is actually received by the station.

The annual salary for Advertising Salespersons in television in 1980 ranged from $15,000 for beginning employees to more than $35,000 for more experienced people in middle or large market stations. The average was $25,200, according to industry sources. Because of commissions, the income for an individual may vary from year to year, and often from month to month. Most stations offer Advertising Salespersons the opportunity to contribute to life and medical insurance plans, and almost all provide them with expense accounts.

Employment Prospects

The chances of employment and eventual promotion in the field are good. Most small stations employ at least two Salespersons, and a sales staff of 15 or more is not unusual in a major market station.

There is a constant need in commercial television for effective Advertising Salespersons. It is often more difficult for station management to find qualified sales people than it is to find other TV and media specialists.

In addition, there is a rather rapid turnover of sales personnel and stations usually have openings, particularly in the Fall. As a result, the general outlook for employment as an Advertising Salesperson in TV is better than that of most other occupations in the medium.

Advancement Prospects

The majority of Advertising Salespersons actively seek advancement to Assistant Sales Manager or to General Sales Manager. The more aggressive and successful individuals move to higher administrative sales positions within their own stations, or assume such positions at larger market stations. Some become General Sales Managers at smaller market stations. The skills necessary for a good Advertising Salesperson in broadcast television are readily transferable to other media firms such as subscription TV (STV) and multipoint distribution service (MDS) companies or cable TV systems, where advertising time is increasingly being sold. In most instances, the personality of a successful Salesperson demands further advancement and many easily obtain higher positions and salaries during their careers.

Education

Although a minimum of a high school education is usually required, the ability to sell is often more important than formal schooling in obtaining a position as an Advertising Salesperson. Having some college education, however, indicates that an individual is serious and motivated in pursuing a career. Employers at some major market stations look for candidates who have an undergraduate degree in advertising, marketing, or business administration. This is particularly true if the position to be filled is in national sales, where more complicated and sophisticated sales environments are the rule.

Experience/Skills

At least one year of sales experience in the retail field or in newspaper or other print advertising is extremely helpful for aspiring TV Advertising Salespersons. Some background in television production and programming can be very useful.

Most Advertising Salespersons have an extroverted personality, an ability to relate to others, and initiative and perseverance. They should also have a good appearance, correct manners, and competitive drive. It is also important that they demonstrate confidence and imagination when making sales presentations and be able to express ideas well when speaking and writing. Above all, an Advertising Salesperson must like to sell and be interested in making money.

Minority/Women's Opportunities

The opportunities for women and minority-group members in the television sales area are increasing yearly. In commercial TV, 28.6 percent of all sales employees were women in 1980, and 8.5 percent were members of minorities, according to Federal Communications Commission (FCC) studies. The percentage of women employees in this field has nearly doubled in the past four years. While employment has not risen as fast for members of minorities, opportunities for them in the sales area are increasing as more stations seek wider audiences and advertising clients, and a sales staff to reach them.

Unions/Associations

There are no unions or professional organizations that represent Advertising Salespersons.

SALES COORDINATOR

CAREER PROFILE

Duties: Coordinating all advertising activities for a television station

Alternate Title(s): Traffic/Sales Assistant; Order Processor

Salary Range: $8,000 to $13,000

Employment Prospects: Good

Advancement Prospects: Good

Prerequisites:

Education—Minimum of a high school diploma; some business or secretarial training preferable

Experience—Some general office and retail sales work

Special Skills—Typing ability; good math skills; sales talent; organizational ability; detail orientation

CAREER LADDER

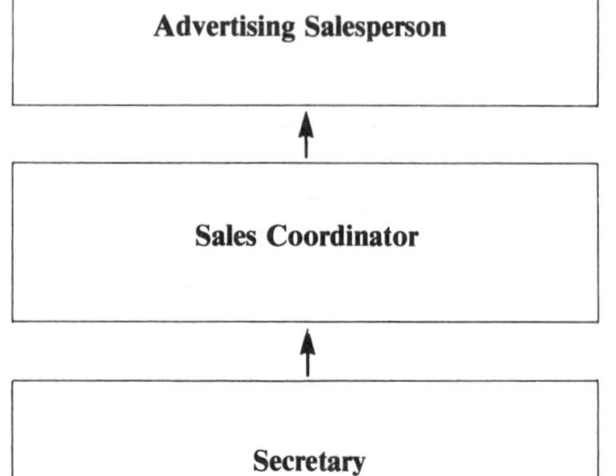

Position Description

At most middle and major market commercial television stations, the Sales Coordinator is in charge of all traffic within the advertising sales department and he or she monitors the activities of the Advertising Salespersons. His or her responsibilities include writing advertising orders, maintaining a schedule of available air time, and keeping track of all commercial advertising matters. In addition, he or she serves as a general assistant to the people in the sales department.

The Sales Coordinator must keep abreast of the changing status of all actual and potential sales, scheduled commercials, make-goods, exceptions, and cancellations. He or she maintains a master scheduling board or computer terminal giving the up-to-date line-up of all sales traffic for use by the sales department staff members. It is the responsibility of the Sales Coordinator to keep the traffic/continuity department informed of all sales activities. He or she also coordinates with the production department all dubbing of commercials for clients, commercial production schedules, and other client production services.

In addition, the Sales Coordinator maintains the master sales files and schedules, order control sheets, and contracts for the Advertising Salespersons. In many stations, he or she also serves as a telephone solicitor of advertising sales, and as a backup for the Advertising Salespersons when they are absent from the office.

As a rule, the Sales Coordinator reports to the Assistant Sales Manager or directly to the General Sales Manager.

Additionally, the Sales Coordinator:

- checks all sales contracts for accuracy against work orders, control sheets, and clients' time orders;
- researches competitive product and media reports for use by the sales department;
- rearranges all pre-log schedules to assist in clearing and changing commercial air time availabilities.

Salaries

According to industry sources, the salaries for Sales Coordinators are relatively low. Earnings in 1980 ranged from $8,000 at smaller market stations for beginning employees with no experience to $13,000 at major market stations for individuals with some experience and seniority. The average was $9,400. Sales Coordinators are often promoted to their jobs from within the station, and carry with them their previous seniority and a modest increase in salary.

Employment Prospects

The position of Sales Coordinator is often the first promotion for a secretary in the Sales Department at a commercial station. In some settings it is considered and entry-level slot.

TV sales people are generally quite mobile, with some 30 percent moving to other positions each year, according to a study by the Television Bureau of Advertising (TBA). In addition, many major market stations often employ two or more Sales Coordinators in specific sales areas (such as national and local) who perform approximately the same duties. The opportunity for employment or promotion to the position at commercial stations is good.

Advancement Prospects

Experience gained as a Sales Coordinator often leads to advancement in a television sales career.

The opportunities for a bright and alert Sales Coordinator to move into a more responsible position in the sales department are good, even though the competition is heavy. Individuals who work well with clients and assist in the servicing of accounts are often promoted to direct selling as an Advertising Salesperson. The promotion to actual day-to-day sales, however, often depends entirely on an individual's aggressiveness and persistence, qualities needed by an Advertising Salesperson.

Education

A minimum of a high school diploma is required for the position. Many employers look for some further education at a business or secretarial school. Some require training in computer terminal operations or in word processing equipment. A few seek a person with an undergraduate degree in business administration, marketing, or mass communications.

Experience/Skills

Some experience in general office work is helpful in getting the job, as is some background in sales at the retail level. Many employers promote the brightest and most able secretaries to the position.

Typing ability of at least 50 words per minute and good business math skills are required. Many employers look for aptitude in the operation of computer terminals, while others are impressed by the candidate's interest in and talent for actual sales work. All stations require that the Sales Coordinator be precise, well-organized, and have a penchant for detail.

Minority/Women's Opportunities

The vast majority of Sales Coordinators at smaller local television stations are women, according to most industry estimates. In some major markets, the position is held by men who are in training for other positions within the sales department.

The opportunities for minority-group members as Sales Coordinators are somewhat better than in other sales positions in commercial television, especially at major market stations.

Unions/Associations

There are no unions or professional organizations that represent Sales Coordinators.

TRAFFIC/CONTINUITY SUPERVISOR

CAREER PROFILE

Duties: Scheduling programs and commercials, developing logs and writing station identifications and announcements

Alternate Title(s): None

Salary Range: $6,500 to $20,200

Employment Prospects: Good

Advancement Prospects: Good

Prerequisites:

Education—High school diploma; some college preferable

Experience—Writing; computer operation

Special Skills—Organizational ability; detail-orientation; precision

CAREER LADDER

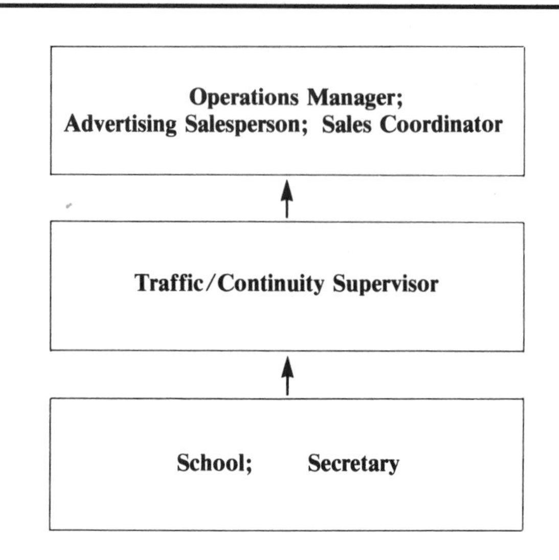

Position Description

The Traffic/Continuity Supervisor is one of the most indispensible employees at a television station. Although the duties vary from large to small markets and between public and commercial stations, the position is vital to day-to-day operations.

The Traffic/Continuity Supervisor is in charge of the detailed scheduling of all programming for the station, including programs, station breaks, commercials, and public service announcements. He or she prepares the daily operational Federal Communications Commission (FCC) log, which details the minute-by-minute broadcasting operation from sign-on to sign-off. In many circumstances, the Traffic/Continuity Supervisor writes on-air promotional copy for the station announcers to use during station breaks and as transitions between programs.

At commercial stations he or she usually reports to the General Sales Manager but may report to the Program Manager instead. The Traffic/Continuity Supervisor keeps a continuous record of all available commercial air time and informs the sales department when slots are open and when they are sold and scheduled. He or she also ensures that there is a time separation between advertised products of a similar nature, and that commercials adhere to FCC and Federal Trade Commission (FTC) regulations and to the station's broadcast standards.

At a public TV station, the Traffic/Continuity Supervisor usually reports to the Program Manager and does not deal with the availability of commercial air time. He or she compiles all program information, creates and distributes the FCC log to engineering and operations personnel, writes the station identifications (ID's), and often writes and schedules all public service announcements and station promotional spots.

Additionally, the Traffic/Continuity Supervisor:

- assembles and processes all audio and video information needed to develop the station's operating log;
- creates, in cooperation with the programming or production department, slide, film, or videotape spots to accompany the audio for local station ID's;
- schedules announcers for recording station breaks and promotional or live announcements.

Salaries

Traffic/Continuity Supervisors are often classified as support personnel and thus, in spite of the responsibility, their salaries are rather low. In commercial TV, the range is from $9,000 for a beginner in a small market to $18,000 for a Supervisor in a major market station. The average was $14,100 in 1980, according to industry sources.

In public television, salaries for the position ranged from $6,500 to $20,200 in 1980, according to Corporation for Public Broadcasting (CPB) surveys. The average amount paid was $10,200 in 1980, an increase of 8.2 percent over the previous year.

Employment Prospects

Traffic/Continuity Supervisor is often an entry-level job at both commercial and public television stations. Attrition and promotion rates are high, and many opportunities are available for beginners. Most cable TV systems and subscription TV (STV) and multipoint distribution service (MDS) operations also have similar positions.

Many large market stations employ three or more people in their traffic/continuity departments, with each assigned specific duties and responsibilities. In many instances, secretaries and stenographers in the programming department are promoted to the position. The opportunities for employment in this job are good.

Advancement Prospects

Opportunities for advancement to other positions at a station are also good. In this spot, an individual has the chance to learn some of the most fundamental aspects of broadcasting.

Possibilities often exist for promotion to the more responsible job of Operations Manager. In addition, the alert and bright Traffic/Continuity Supervisor can move into the sales department as an Advertising Salesperson or Sales Coordinator.

Education

The minimum educational requirement for a position as Traffic/Continuity Supervisor is a high school diploma. Some employers in major market stations prefer an undergraduate degree in mass communications or advertising. Courses in broadcasting or marketing and in copy writing are also valuable.

Experience/Skills

Since the position is very often an entry-level one, employers do not require extensive television experience.

Most stations look for alert, bright young people who are extremely detail-oriented and precise. Experience in computer operations or word processing can be very helpful since many stations are using such equipment for traffic and continuity. A candidate should have some background in writing to be able to create good copy for promotional announcements and station breaks. In addition, a facility with numbers is important since a Traffic/Continuity Supervisor works with specific segments of a broadcast day.

Minority/Women's Opportunities

In commercial TV, well over 60 percent of the Traffic/Continuity Supervisors are women, according to most industry estimates. In public television, 86.4 percent of the positions were held by women in 1980, according to CPB surveys. This was an increase of 8.4 percent over 1976.

Minority-group employment in the job is not as high at either commercial or public stations. According to the CPB, 18.2 percent of the positions were held by members of minorities in 1980.

Unions/Associations

There are no unions that serve as collective bargaining agents or representatives for Traffic/Continuity Supervisors.

MEDIA DIRECTOR

CAREER PROFILE

Duties: Selection of different types of media; identification of target audiences for advertising campaigns

Alternate Title(s): Senior Vice President; Director of Media Services

Salary Range: $60,000 to $85,000

Employment Prospects: Poor

Advancement Prospects: Poor

Prerequisites:

Education—Undergraduate degree in advertising, journalism, or mass communications; graduate degree preferable

Experience—Minimum of five years of advertising agency work

Special Skills—Leadership qualities; knowledge of all media; analytical mind

CAREER LADDER

Marketing VP or President (Ad Agency)

↑

Media Director

↑

Media Buyer

Position Description

A Media Director in an advertising agency determines the specific market a client is trying to reach, and the media that will most effectively reach that market. He or she also decides how a commercial message should be positioned in various media in order to achieve maximum results.

Media are the channels through which products or services may be advertised. The Media Director determines the placement of advertising in broadcast media (television or radio) as well as in print media (magazines, trade publications, newspapers, displays and posters, direct mail). He or she is a creative executive who relies on extensive research as well as highly developed marketing instincts.

The selection of the media is arrived at after an in-depth study of television ratings (ARB, Nielsen), surveys, demographic studies, and other analytical reports. Market surveys are carefully analyzed to define the type of consumers and the target audience. In considering television as a vehicle for the advertising message, the Media Director deals with the advertising sales departments at television stations. He or she computes the costs of placing the television commercial in various positions during the broadcast day, as well as the cost of creating the commercial itself. After these costs are combined, the Media Director is involved in developing an overall media campaign that best fits the marketing goals and the budget of the client.

The Media Director in a large advertising agency supervises a media staff that handles most of the detail work. In order to perform effectively, the Media Director must communicate regularly with other departments of the agency. He or she culls information from the research staff, and also works closely with the creative department (particularly artists and writers) in the development of a television commercial message.

Additionally, the Media Director:

- oversees the selection of the specific type of advertising (spot announcement, partial program sponsorship, etc.);
- supervises the actual buying of all television time.

Salaries

According to a 1980 survey of the industry, Media Directors often carry the rank of senior vice president, and salaries range from $60,000 to $85,000, with a generous expense account. The senior vice president position often includes additional benefits, such as bonuses, stock options, and profit-sharing plans. Other perquisites may include a new company car every two years, club dues and assessments, and fully-paid medical and dental benefits, as well as some company-paid financial services.

Employment Prospects

The Media Director's position is among the higher-ranking and better-paid jobs in an advertising agency. Competition is very heavy, and only the brightest and most capable people achieve this position. When openings do occur, they are usually filled by the promotion of an agency's Media Buyer or (in larger companies) assistant media director, from a smaller agency, or from the advertising sales department of a TV station. In a few instances, an Account Executive can move up to the Media Director position.

Advancement Prospects

Opportunities for promotion from the position of Media Director to other advertising jobs are poor, except for an extremely talented person, whose services may be in demand at a larger, more prestigious agency. Occasionally, a Media Director will become marketing vice president, and sometimes president of an advertising agency.

Education

An undergraduate degree in advertising, journalism, or mass communications is essential, and a graduate degree is preferable. Studies should include all of these disciplines as well as business administration and statistical courses.

Experience/Skills

Media Directors should be knowledgeable about or have worked in all aspects of advertising. They usually have five to ten years of agency experience. A background in time selling at a television station is also helpful. Previous experience in data analysis is essential, as is an understanding of marketing philosophies and merchandising techniques.

Media Directors should be strong leaders and administrators. They must have a thorough understanding of all advertising media and how best to use them for successful product promotion.

Minority/Women's Opportunities

Employment in the advertising industry is based largely on personal talent and ability, and historically there is perhaps less discrimination within it than in some other businesses. According to the American Association of Advertising Agencies (4As), women hold some of the most important jobs in the industry and there are generous opportunities for advancement, as well, for members of minority groups.

The increasing influence of women in the advertising industry is reflected in the 1981 election of Ms. Patricia Martin of the Warner-Lambert Company to the chairmanship of the American Advertising Federation (AAF), the first woman to chair that respected 76-year-old organization.

Unions/Associations

As part of management, Media Directors are not represented by a union. They may, however, be members of professional associations such as the 4As and the AAF for professional guidance and support.

MEDIA BUYER

CAREER PROFILE

Duties: Purchasing broadcast air time on television stations for an advertising agency

Alternate Title(s): Time Buyer

Salary Range: $14,000 to $23,000

Employment Prospects: Fair

Advancement Prospects: Poor

Prerequisites:

Education—Undergraduate degree in marketing, communications, or advertising; graduate degree desirable

Experience—Minimum of two years in advertising; television sales background useful

Special Skills—Objectivity; good judgment; ability to interpret statistics; intuition

CAREER LADDER

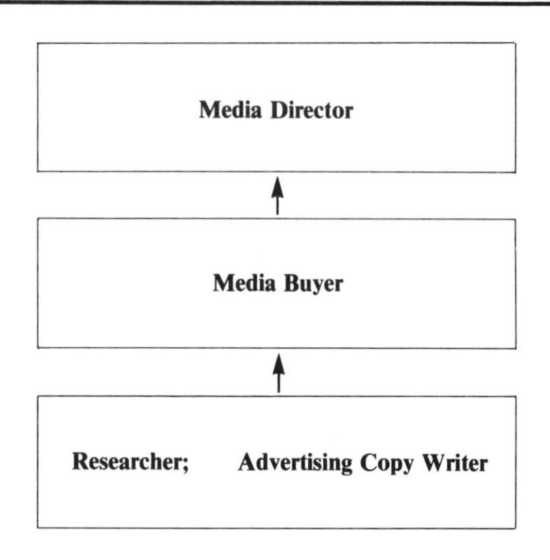

Position Description

A Media Buyer in an advertising agency selects the proper media for the placement of advertising. In some middle market and large market areas, the Media Buyer specializes in the purchase of time for television advertising. He or she selects, negotiates, and purchases air time on TV stations and networks for spot announcements and program sponsorship for the agency's clients.

The most important asset of a Media Buyer is a comprehensive knowledge of television stations and networks—their audience ratings, rate cards, influence, image, types of programming, coverage, audience demographics, and general track records.

The Media Buyer must also have a thorough understanding of each client's product and marketing goals. He or she must consider particular emphasis of the product to be advertised when selecting the appropriate TV air time, as well as the ages and income groups of the target audience. Television market research and statistics are valuable tools in this regard, and the Media Buyer must be able to study and interpret them, relating that information to the client's product and to the overall advertising campaign that has been devised for that product.

With the product and market factors clearly in mind, the Media Buyer decides which of the television possibilities will allow the commercials to reach the maximum number of potential customers at the best cost. The Media Buyer's knowledge of specific television rates and of the demographics and psychographics of each station's viewers are important factors in arriving at the decision.

Once the Media Buyer has determined what station(s) or network(s) will carry the client's message, he or she considers what position within the broadcast day will be most effective for spot announcements or, if the advertising will consist of program participation, what particular program will best serve the client's needs.

Additionally, the Media Buyer:

- compares prevailing advertising rates and program sponsorship costs in order to stay within the allocated budget;
- studies and interprets television market research and statistics.

Salaries

A 1980 salary survey of the advertising industry places Media Buyers in the $14,000 to $23,000 range. An individual's income depends on the size, billings, and importance of the agency, as well as its geographical location. The competitive climate between agencies in a particular market also has an influence on salaries.

Employment Prospects

Many larger agencies have more than one Media Buyer, each specializing in a particular medium (television, print, radio, etc.), as well as assistant positions in media buying and planning. Smaller agencies often have only one Media Buyer who purchases advertising on other media as well as on television. In very small agencies, the duties are often handled by another staff position. There are, therefore, relatively few Media Buyer positions available, and employment opportunities are only fair. A Researcher or Advertising Copy Writer who has a good understanding of marketing, merchandising, and television time sales should, however, be able to attain the position.

Advancement Prospects

While the job of Media Buyer is sometimes a terminal career position, the brighter and more diligent individual may, after a number of years, be promoted to Media Director, and possibly to a higher position in an advertising agency. Good Media Buyers may also find themselves in demand at the more prestigious agencies, at higher salaries and with more responsibilities. Nevertheless, the opportunities for advancement are poor.

Education

At most agencies, an undergraduate degree in marketing, communications, or advertising is preferred, as well as some training in statistical research. Some employers in large agencies require a graduate degree in marketing.

Experience/Skills

Media sales and media research positions offer excellent training for Media Buyers. A Media Buyer usually has two years of experience in various positions in an advertising agency. Additional experience in the sales department of a television station is also useful.

A Media Buyer should be objective to be able to make sound decisions for a client. He or she also needs good judgment, skill at interpreting statistics, and an intuitive sense.

Minority/Women's Opportunities

Opportunities in advertising are predicated more on ability, talent, and performance than on sex or race. Women and minority group members should not find discrimination to be a problem when seeking Media Buyer positions.

Unions/Associations

No union serves as a bargaining agent for Media Buyers, but many individuals belong to organizations such as the American Association of Advertising Agencies (4As) and the American Advertising Federation (AAF) to share mutual professional concerns.

ACCOUNT EXECUTIVE

CAREER PROFILE

Duties: Overall supervision of the advertising campaign for an account

Alternate Title(s): None

Salary Range: $25,000 to $40,000

Employment Prospects: Good

Advancement Prospects: Fair

Prerequisites:

Education—Minimum of undergraduate degree in advertising or business administration

Experience—Minimum of five years with all departments in an advertising agency

Special Skills—Organizational ability; leadership qualities; creative mind

CAREER LADDER

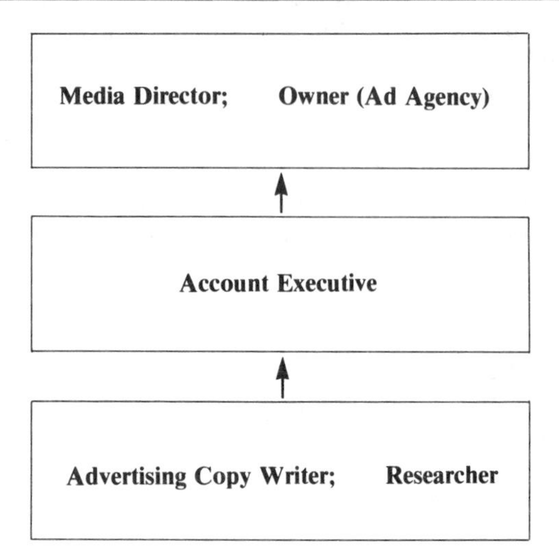

Position Description

An Account Executive is the advertising agency's representative to a particular client and is one of the most capable professionals in the organization. He or she looks after the client's interests, coordinates all of the creative and production work, meets deadlines, and controls costs. A thorough knowledge of the client's advertising and marketing problems is essential, as well as an understanding of how the agency can respond to and solve them. The creative staff relies on the Account Executive for product information, market analysis, and the timetable for the various stages and completion of the advertising campaign. He or she determines the basic facts about a product, analyzes them, and leads the creative team in developing the overall advertising strategy.

It is essential that the Account Executive have a good understanding of all advertising media, including television, radio, newspapers, magazines, direct mail, etc. When television advertising is called for, he or she uses the services of the Media Director and Researcher to help determine the markets, time slots, stations, networks, audiences, shows, and frequency suitable to the client's product and budget. The Account Executive is also involved in overseeing the creation of the commercial, from concept to story board (a rough representation of the frame-by-frame action) through production, and makes certain that the television advertising complements the full campaign. As the client's representative, he or she must be satisfied with all elements before a presentation is made to the client.

While much of the detail work for an advertising campaign is delegated within the agency, an Account Executive must have a firm grasp of the entire operation. He or she must be able to develop concepts that fulfill the client's advertising needs.

Additionally, the Account Executive:

• develops the overall advertising strategy and sales message;

• analyzes the marketing, advertising, and quality of competing products;

• manages the allocated budget to the best advantage for the client, and to ensure profitability for the agency;

• monitors the effectiveness of the campaign strategy and recommends alternatives as necessary;

• promotes agency relationships with clients and the media.

Salaries

A 1980 survey of the industry showed that Account Executives earned from $25,000 to $40,000. Expense accounts are generally healthy, and other benefits may include bonuses, stock options, profit sharing, a company car, club dues, and medical and dental insurance, depending on the individual agency. While higher incomes are often associated with larger, more prestigious agencies, many small firms also maintain a respectable salary schedule. The person servicing high-budget accounts is usually compensated more than colleagues who work on those with lower budgets, whatever the size of the agency.

Employment Prospects

For the qualified, experienced individual, prospects for becoming an Account Executive are good. Openings are usually filled from within the ranks of an advertising agency, although sometimes by people in similar positions at other agencies. Some agencies employ several Account Executives. In addition, some multi-product large corporations have their own in-house ad agencies that assign Account Executives to specific products or divisions.

Practical knowledge can be gained working as a Researcher or as an Advertising Copy Writer, but an Account Executive must also be well-grounded in all aspects of advertising media. Even with years of experience, only the most capable and versatile people who have strong instincts for business, marketing, and merchandising are qualified for this position. Because of the number of accounts to be serviced, and the considerable mobility within this position, many Account Executives are needed.

Advancement Prospects

Although turnover among Account Executives is high, people tend to move from one agency to another at approximately the same level. Advancement opportunities, in general, are only fair. Some Account Executives advance to higher-paying and more responsible positions within the agency, either to Media Director or in broadcast operations. Advancement to vice president, or even senior vice president in charge of media operations, media services, research, and broadcast management, while possible, occurs infrequently.

Because of the nature of their work, Account Executives are extremely knowledgeable about all phases of agency operations and are, therefore, capable of managing their own agencies. Many do exactly that. Leaving the large corporate environment, they open their own companies, usually starting with a small staff and a few reliable and stable clients.

Education

Within the industry, it is generally recognized that a top-notch Account Executive must have an undergraduate degree in advertising or business administration. Further graduate study or a master's degree is helpful. Courses in journalism, broadcasting, and marketing are also useful.

Experience/Skills

There is no one training area that provides all of the knowledge an Account Executive must have; rather, the position demands at least five years of advertising agency experience, including copy writing, art, design, layout, and television or film production.

A successful Account Executive must be able to organize many elements, from creative to financial, into an effective advertising approach to satisfy a client's needs. He or she must have leadership qualities in order to coordinate the efforts of the agency's departments, and must feel comfortable with different types of clients in a variety of situations. The Account Executive has business acumen, marketing expertise, and a creative, facile mind.

Minority/Women's Opportunities

There is little statistical information on the status of minority-group members and women as Account Executives. The American Association of Advertising Agencies (4As) reports, however, that some of the most important jobs in the industry are held by women, and that opportunities for advancement are also available to members of minorities.

Unions/Associations

There are no unions that represent or serve as bargaining agents for Account Executives. There are, however, many groups that give support to, and provide a professional forum for, advertising agencies and executives. Most prominent are the 4As and the American Advertising Federation (AAF).

RESEARCHER

CAREER PROFILE

Duties: Obtaining and analyzing information essential to marketing and advertising decisions

Alternate Title(s): Researcher/Planner; Research Analyst; Research Specialist; Market Researcher

Salary Range: $12,000 to $18,000

Employment Prospects: Fair

Advancement Prospects: Poor

Prerequisites:

Education—Undergraduate degree in business administration, advertising, or mass communications

Experience—Some experience in a research firm or research department outside the advertising industry helpful

Special Skills—Analytical aptitude; statistical skill; orientation to detail; patience; writing ability

CAREER LADDER

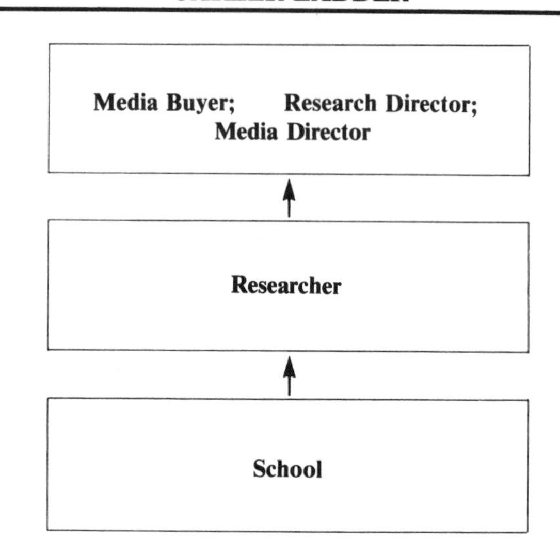

Position Description

A Researcher in an advertising agency acquires and analyzes the information on which profitable marketing and advertising decisions are made. Television is a highly effective advertising medium. Many of the ads seen on TV were purchased for the advertiser by an agency, but only after a considerable amount of research had been done.

Research data may come from interviews, questionnaires, library study, government agencies, or trade magazines and associations. The goal of advertising research is to determine who will buy a product and why. Research also explores the potential market, and helps to determine how a new product can best be presented or how an existing product can be presented more attractively.

Research at a large advertising agency may be conducted by a research department, with a director supervising several Researchers. At a small agency, such responsibility may be handled by one person. Research duties at very small agencies may be added to those of another staff member, and on occasion all research is sub-contracted by the agency to an outside independent research organization.

There are usually three areas of advertising research, and a Researcher may work in all of them or, in a larger agency, specialize in a single area. One area centers on the advertising markets. Research must uncover who the consumers are, how they think, how they are most likely to respond to a televised commercial message, and what advertising techniques will appeal to them most strongly.

Another area concentrates on the product to be advertised. It is the responsibility of the Researcher to determine what the product does, how it is used, whether the packaging and pricing would best be presented on television or in other media, how to attract buyers, why consumers will (or will not) buy it, and how the competition treats these questions.

A third area of research analyzes the effectiveness of the advertising. This follow-up work, done after television commercials have appeared, evaluates the specific ad campaign and provides guidelines for future advertising of the client's products. This aspect of research examines the viewers' degree of product awareness, their buying motivations, and the direct results, in sales, of the advertising.

The good Researcher must be able to think and write clearly, to design a research methodology, to interpret data from all areas of study and to present results in clear, concise, sound recommendations that address specific marketing problems of the client and the product. Equally important is an understanding of human behavior, particularly consumer motivations, and the ability to deal with people in a manner most likely to elicit such information.

Additionally, the Researcher:

- determines what research information should be secured, and the most efficient manner of obtaining it;
- analyzes already published research as well as statistical data relating to products, markets, consumer preferences, and buying trends;
- designs and prepares questionnaires and conducts field or telephone research;
- makes recommendations to the media and creative departments.

Salaries

Salaries for Researchers at advertising agencies range from a beginning level of $12,000 per year to a high of $18,000 for personnel with some experience and seniority.

Employment Prospects

There is a reasonable amount of mobility among advertising research people, which creates some openings for an entry-level candidate. While some smaller agencies contract for all research services with outside independent research firms, many agencies employ Researchers on staff. In addition, many major market stations employ Researchers to assist their advertising sales staff.

Advancement Prospects

Opportunities for promotion are generally poor because of heavy competition. The bright and dedicated Researcher, however, can sometimes move up to the position of research director at a larger advertising agency. At smaller companies, promotion to more responsible positions in other areas of the agency is possible, and many Media Buyers and Media Directors have had some experience in research.

Some Researchers join, or return to, independent research organizations at higher salaries to advance their careers. Still others utilize their skills and experience to obtain similar positions in the sales departments of major market television stations.

Education

An undergraduate degree from an accredited college or university is almost mandatory for individuals seeking the position of Researcher at an advertising agency. A degree in business administration with a solid liberal arts background, or a degree in advertising, or mass communications with some marketing and statistical analysis course work is desirable. Courses in sociology are helpful.

Experience/Skills

Some analytical experience in research methodology and design is helpful. Most useful is some actual experience, either with an independent research firm or in the research department of a company in a non-related field. Because the position is often an entry-level job, part-time work in college, at an agency, or at a research organization can provide valuable experience in advertising, marketing, and research techniques. A Researcher should be skilled at statistical interpretation and have the patience to carry out detailed projects. The ability to write clearly and logically is essential in this job.

Minority/Women's Opportunities

As is usual in advertising, opportunities are predicated more on ability, talent, and performance than on sex or race. Women and minority group members, therefore, should find more opportunities in advertising (including research) than in some other areas of the television marketplace.

Unions/Associations

No union serves as a representative or bargaining agent for Researchers, but membership groups such as the American Association of Advertising Agencies (4As) and the American Advertising Federation (AAF) provide professional guidance and support.

ADVERTISING COPY WRITER

CAREER PROFILE

Duties: Writing copy for television commercials

Alternate Title(s): None

Salary Range: $10,000 to $20,000

Employment Prospects: Good

Advancement Prospects: Good

Prerequisites:

Education—Undergraduate degree in journalism, advertising, English, or mass communications

Experience—Amateur or professional advertising or writing background helpful

Special Skills—Excellent writing ability; originality; understanding of human motivation

CAREER LADDER

Media Buyer; Copy Chief

↑

Advertising Copy Writer

↑

College; Secretary; Clerk

Position Description

An Advertising Copy Writer (Copy Writer, for short) works in an advertising agency and conceives of and writes the announcements, voice-overs, dialogue, special effects, and expository material heard on a television commerical. This work requires more than a mastery of the language. It demands an understanding of the total advertising campaign, a clear perception of the potential audience, and a knowledge of the client's marketing philosophy about the product being advertised.

The Copy Writer is not usually involved until after the initial development of an advertising plan or campaign, the study of the product or service, an analysis of the market, the determination of the media and marketing approach to be used, and the completion of market research. The Copy Writer does become directly involved at this point, drawing on all this information in his or her creation of the advertising copy.

The Copy Writer works closely with the Researcher or Media Director to determine the overall thrust of the commercial message; with the research department to become knowledgeable about the product, the competition, and the intended audience; and with the other members of the creative team, especially the Art Director, to ensure an imaginative, integrated, informative commercial that attracts the attention of the potential consumers and motivates them to respond.

In addition to preparing final copy, the Copy Writer also writes the material for the story board (a preliminary treatment that is a rough representation of the frame-by-frame action of a television commercial).

Good Copy Writers use the English language not only to meet the needs of clients, but also to stimulate buying impulses and product acceptance by the consumer. In addition to being writers, they are part sales people, part word inventors, and part artists.

Additionally, the Advertising Copy Writer:

- creates slogans and catchy phrases for products to help ensure viewer recognition and acceptance;
- writes copy for brochures and promotional pieces used in advertising campaigns.

Salaries

Salaries for Advertising Copy Writers vary considerably, depending on the geographical location and the local competition, the size and billings of the agency, the clients' commitments to advertising, and the experience of the writer. According to the American Advertising Federation (AAF), college-trained beginners can expect to start at about $10,000 to $12,000 per year and advance to about $20,000 within a few years. Some advertising firms have introductory training programs that provide valuable basic education and give the trainee a head start on the salary scale.

Employment Prospects

With 6,000 advertising agencies in the U.S., many with multiple offices and, in larger agencies, with several Copy Writers on staff, employment possibilities abound. Moreover, while there are many aspiring writers in the world, few have the necessary training and understanding of television production, the talent to create concise copy, and the marketing insights. A good writer who has training or experience in these areas should find employment prospects very bright.

Copy writing is often considered an entry-level position, but many Copy Writers have found their niches, and stay in the position for years. Those who show a particular flair for the creative side of advertising are often promoted to more responsible agency positions, and still others move to larger agencies, at higher salaries, thus creating openings for beginners.

Secretarial or clerical work in an agency can be a springboard to a copy writing position for a talented, ambitious individual.

Advancement Prospects

The inspired, creative Copy Writer with an instinct for advertising, who has grasped the broader concepts of creating a commercial, a plan, or a campaign, should have no trouble advancing within the industry.

Advertising professionals generally pursue highly mobile careers, either moving upward within the same company or joining another agency at a higher salary or in a more prestigious position. At larger agencies, Copy Writers may be promoted to copy chief and be in charge of writing for all media. Smaller agencies with fewer employees may offer opportunities for promotion to another job, such as Media Buyer.

Higher level positions are also becoming more available in the fast-growing communications area outside the advertising agency field. Although most ad copy is written by agency writers, many television stations still employ Copy Writers within the sales department. In addition, as they move into advertising sales in the future, independent production firms, cable TV systems, and multipoint distribution service (MDS) and subscription TV (STV) companies will need the talents of good Copy Writers.

Education

Today, a bachelor's degree is virtually a requirement for obtaining a job as a Copy Writer at an advertising agency. About 80 colleges and universities offer a degree in advertising. English and journalism studies are important to aspiring Copy Writers, but a broad liberal arts education is equally valuable. Some course work that includes mass communications, television writing or production, marketing, and psychology is also desirable.

Experience/Skills

Extracurricular activities and part-time jobs while in college can be helpful. Campus radio and television stations, school newspapers, yearbooks, and literary magazines often provide practical experience. Any project involving selling, persuasion, marketing, or distribution gives insights into advertising skills. Any writing experience, but particularly in short, concise stories, is extremely valuable. The abilities to use language well and to communicate ideas in an original, convincing way are essential. Copy Writers must also be able to work well under pressure.

Minority/Women's Opportunities

Industry sources maintain that advancement opportunities within the advertising industry are as available to women and minority-group members as they are to white males.

Unions/Associations

There is no union that specifically represents or negotiates labor contracts on behalf of Advertising Copy Writers. Some people maintain membership in the Writers Guild of America (WGA) as a way to improve their skills or gain representation in negotiations in other areas of writing.

PERFORMER

CAREER PROFILE

Duties: Acting, singing, dancing, or otherwise performing in a television production

Alternate Title(s): Talent

Salary Range: Under $10,000 to $1,000,000+

Employment Prospects: Poor

Advancement Prospects: Poor

Prerequisites:

Education—Post-high school training in performing arts; undergraduate theater arts degree helpful

Experience—Extensive amateur or professional work

Special Skills—Talent; commitment; perseverance; showmanship; poise; imagination; aggressiveness; versatility

CAREER LADDER

Featured Performer; Star

↑

Performer

↑

Performer (Other Media); College/Professional School

Position Description

A television Performer appears on camera to entertain or inform the viewer and can be an actor, singer, dancer, comedian, emcee, magician, talk or game show host, pantomimist, juggler, ventriloquist, and more. Performers are most often actors who play supporting or featured roles in dramatic series, mini-series, or situation comedies, or who are singers or dancers in entertainment specials. Many obtain work as extras or walk-ons in dramatic productions or comedy sketches. Others work in variety shows, made-for-TV movies, dramas, and soap operas.

There are a number of television environments in which Performers work. They can appear in programs for commercial and public TV stations and networks, independent production companies, cable TV systems that do original programming, advertising agencies, and, on occasion, corporate, government, health, and educational TV operations. Performers work in rehearsal halls, studios, and, occasionally, on location. Although they are usually cast and employed by a Producer, they report to the Director of a TV program for rehearsals and the performance.

When appearing on a TV show, a Performer uses the skills and techniques of the theater but plays to a camera rather than to a live audience. Action, reaction, and movement must conform to the camera's mobility and the type of shots used by the Director. Although movies make similar demands of Performers, television shows are made in such a tight time frame that Performers must learn their roles, lines, and movements much more quickly to accommodate the fast pace and shorter production schedule under which TV operates.

Many Performers work in television commercials which can be an important source of income and exposure. More than one-half of the income of the 40,000 members of the Screen Actors Guild (SAG) in 1980 came from TV commercials, according to industry reports. Most Performers do not limit themselves to TV, but perform at every opportunity, in the theater, movies, industrial shows, or nightclubs.

Additionally, the Performer:

- attends scheduled readings, camera run-throughs, blocking (position placement) sessions, and rehearsals;
- rehearses independently for his or her specific role or appearance;
- continues to attend performing arts classes.

Salaries

Most Performers work on a project-by-project, and often a day-to-day, basis. Union pay scales for Performers are complicated, and related to the type and length of a program and to the importance of the part. The American Federation of Television and Radio Artists (AFTRA) defines three categories of performance: extra (non-speaking), five-lines-or-less, and principal. In a half-hour daytime soap opera (usually one day's work), non-speaking extras are guaranteed a minimum of $81.75, five-lines-or-less Performers receive $141.74, and principals are paid $300.00. Recognized Performers often command fees considerably higher than the minimum.

Payment for work in TV commercials is even more complicated. A base session fee of $104.00 (SAG minimum), is paid whether or not a commercial is aired and a complex formula defines additional payments once it is broadcast.

A few individuals get work on continuing series where a featured Performer or star can sometimes earn more than $1,000,000 annually from TV work. Most Performers, however, are fortunate to work even a few weeks each year and their annual income from television is usually under $10,000.

Employment Prospects

Opportunities are poor. It is difficult to become a Performer of any kind. Unemployment in all performing media was estimated at 85 percent in 1981, according to *Backstage.* Except for roles in long-running shows, nearly all commercial TV opportunities are limited to one or two performances.

Performers occasionally find work in the non-broadcast areas including government, health, corporate, and educational telecommunications operations. Cable TV systems currently use Performers infrequently, but opportunities will grow as production increases.

Employment is usually obtained through contacts and by personal recommendations. Acclaimed professional or amateur performances in other media can occasionally lead to a role of some kind in a television production.

Advancement Prospects

In most professions, a good worker is usually rewarded with a promotion, increased responsibility, and a higher salary. This is not generally true of the performing arts, where Performers themselves must get the next job. Although agents are helpful, the Performer must be successful at try-outs, auditions, and call backs. Talent, exposure, and impressive credits all contribute to the continuing success of a Performer but, in general, opportunities for advancement are poor.

Very few Performers achieve feature or star status. A few experienced individuals with particular talents become well-known and are cast in supporting roles, but even those with TV, movie, and stage credits, find it necessary to supplement their incomes with jobs unrelated to performing.

Education

Post-high school training in the performing arts is becoming increasingly necessary, and is available from private teachers, schools, workshops, studios, and colleges. Courses in television production techniques are useful. A broad liberal arts education is helpful and many Performers do obtain an undergraduate degree in theater arts.

Experience/Skills

Experience is more important than education, and an aspiring Performer should work at every opportunity, in school and college plays, and in variety shows, benefits, local productions, community theater, and workshops.

Talent, of course, is the pre-eminent essential skill, but a Performer must also be extremely committed, aggressive, poised, and imaginative. A person who expects to work in TV must develop a knowledge of, and an appreciation for, television production techniques. Most Performers are versatile and adapt well to a variety of casting needs. Above all, they are persistent in seeking employment.

Minority/Women's Opportunities

Men and women are in equal demand as Performers in most TV productions. Members of minority groups have found increasing opportunities in television performance during the past ten years.

Unions/Associations

Joining an appropriate union is often difficult. Individuals sometimes cannot obtain a union card without work but cannot obtain work without a union card.

AFTRA, with 50,000 members, is an open union and can be joined by paying an initiation fee. Membership, however, does not guarantee work, and many people wait until they land their first job to join. SAG represents Performers who work in shows that are filmed (rather than videotaped) for television release. Actors must join SAG after their first film job in a principal role.

Three other guilds represent television Performers: Actors Equity Association (Equity), the American Guild of Musical Artists (AGMA), and the American Guild of Variety Artists (AGVA).

WRITER

CAREER PROFILE

Duties: Writing of scripts for television programs and productions

Alternate Title(s): Scriptwriter

Salary Range: $8,300 to $50,000+

Employment Prospects: Poor

Advancement Prospects: Poor

Prerequisites:

Education—Undergraduate degree in English, theater, radio-TV

Experience—Writing for any medium; some television or film production helpful

Special Skills—Writing talent; creativity; television production knowledge

CAREER LADDER

Producer; Director

↑

Writer

↑

Writer (Other Media); TV Production Position

Position Description

A TV Writer creates a visually-oriented script for a commercial or public television drama, documentary, variety, talk show, situation comedy, or other production. He or she may also develop and write scripts for programs produced by health, education, government, and corporate TV or video organizations. A Writer is either on staff or hired for a particular program or series, usually by a Producer or Director.

It is essential that the Writer understands the television medium. He or she must think visually and sense the way the camera will "see" each element.

The television script is a blueprint for production. The complete script is typed in two columns, "audio" on the right, and "video" on the left. The audio column, or content portion, contains dialogue or narration, as well as instructions for music and sound effects. The video side contains major visual instructions such as camera shots, and key production notes such as timing, set descriptions, and directions to Performers. The specific details are filled in by the Director. Most good TV scripts make little sense when only the audio column is read.

When developing scripts, a Writer first determines the subject matter, the purpose, and the intended audience of the program. He or she then researches the subject, organizes the material, and develops it in a format that the audience will understand and enjoy.

Writing television scripts for government, private industry, health care, and educational organizations is particularly demanding. It requires creating the appeal of an entertainment script while presenting information that trains, educates, or informs. Considerable research and particular attention to accuracy and detail are often needed.

A general TV Writer usually does not write for a news program. This activity demands a particular combination of skills and background and is performed by a News Writer assigned to the news department.

Additionally, the television Writer:

- develops a sequence-by-sequence narrative describing events in their proper continuity (called a treatment);
- attends story conferences and planning sessions to change, delete, or add to the treatment;
- attends production meetings and rehearsals, and follows up with script rewrites, additions, and deletions.

Salaries

In the entertainment field, the Writers Guild of America (WGA) minimum scale for a 30-minute commercial network script in 1981 was $7,600 and $10,600 for a 60-minute show. By negotiating rates above union scale, a recognized Writer can earn $50,000 a year or more. The young beginning freelance Writer in commercial TV, however, makes very little money, and most work at other jobs for a source of regular income.

A Corporation for Public Broadcasting (CPB) study indicated a salary range of $8,300 to $32,500, with $16,200 as the average, during 1980 for the few full-time Writers at public television stations. Staff Writers in educational and training environments earned from $12,000 to $22,000 per year in 1980, according to industry sources.

Employment Prospects

The chances of obtaining a regular full-time position as a Writer are poor. Staff Writers or script supervisors are employed at some middle and large market commercial and public stations, but there is little turnover in these positions.

A TV Writer can get a start in many ways. Those who write creative scripts may come from another, related medium such as movies or theater. Some, with production jobs, are asked to fulfill a writing assignment in the course of their regular duties. In business, government, or education, some people become Writers because they have already shown such talent (as copywriters, publicists, or journalists) or because, as subject specialists, they are best able to write the appropriate scripts to educate, train, and inform.

A very few Writers are steadily employed by personalities (Bob Hope), on network shows (*60 Minutes*), or on shows created by independent production firms (*Hill Street Blues*). Most Writers, however, freelance while pursuing a full-time career in another field, related or not to television.

Advancement Prospects

Good credits, a reputation for reliability and solid scripts, and personal connections are the ways to ensure continuing assignments. A few established individuals become head writers for commercial TV series. Some Writers become Producers or Directors, but most dedicated individuals continue to do some writing, for their own or for other shows. In the non-commercial areas, some Writers become Producers or Directors of informational and instructional programs. Overall, advancement opportunities are poor.

Education

An undergraduate degree, usually in English, is virtually essential, even though it is not always required for employment. Many individuals have an undergraduate or graduate degree in theater or radio-TV. Courses in writing, play writing, and composition, in addition to film and television production are helpful. Most writers have a broad liberal arts education.

Experience/Skills

Many Writers begin writing at an early age and continue to do so all of their lives. Part- or full-time experience at a TV station is helpful to master the special techniques, opportunities, and limitations inherent in writing for TV.

Creativity and imagination are important, as are a mastery of the language and knowledge of basic construction. A good television Writer is well-organized and disciplined, has a perpetual curiosity, and is usually in love with the English language.

Minority/Women's Opportunities

The majority of Writers in the entertainment field are white males, according to industry observers.

In the few positions existing at public stations, 43 percent of the Writers and script supervisors were women according to a 1980 CPB survey. The same study, however, showed minority-group members occupying only 5 percent of the writing positions, a decline of 12 percent from the preceding year.

Unions/Associations

Writers in educational or training environments are seldom represented by a union for bargaining purposes. Professional freelance Writers in the entertainment field may be members of WGA. Union membership is required of Writers who submit a script for network television production. The Screenwriters Guild and the Authors Guild also represent Writers. They may also hold membership in the National Writers Club, as well as in specific content-area professional associations, for purposes of professional growth and support.

MUSIC DIRECTOR

CAREER PROFILE

Duties: Evaluation, selection, and supervision of music for a television production

Alternate Title(s): Music Supervisor

Salary Range: $25,000 to $75,000+

Employment Prospects: Poor

Advancement Prospects: Poor

Prerequisites:

Education—Extensive training in music; undergraduate degree preferable

Experience—Minimum of five years in television or film music

Special Skills—Musical talent; creativity; taste

CAREER LADDER

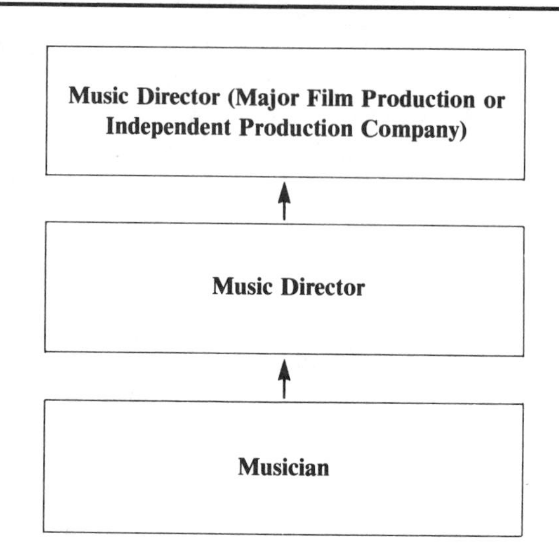

Position Description

A Music Director is responsible for evaluating, selecting, and supervising all music for a television program. Music Directors are most often employed in major television film productions or on situation comedy, variety, or dramatic series. They also work on documentary programs and, occasionally, on large-budget education, government, or corporate television productions. Sometimes, the title Music Director is used to describe the individual who is in charge of all music during an entertainment or variety special, including conducting the orchestra.

Various types and styles of music are used in television productions to enhance and underscore the visual images. The Music Director determines when and if music should be used for specific scenes to create a mood, capture the attention of the audience, or complement the action. He or she analyzes the needs of each segment or scene, and auditions, previews, and selects musical bridges and segments that appropriately accompany the video portion of the production. After the show has been recorded, he or she adds music to the visual, voice, and non-musical sound effects material on the existing tape or film, choosing from a variety of music styles and sources.

On major television film programs, the Music Director might commission original compositions from composers who usually work specifically in the film and television industries. In such cases, the Music Director supervises the arrangements of the music, chooses the orchestra size, and oversees the recording, often conducting the orchestra during the recording session. He or she sometimes composes, arranges, and orchestrates the music as well.

Sometimes, a Music Director uses existing musical bridges, cues, themes, and other short sections supplied by established music production companies. He or she selects specific portions from the tapes and records maintained by such music libraries and supervises their insertion into the sound track of the television program.

At times, a Music Director works in recording studios, but he or she is most often found in the control room and in videotape and film editing rooms of major production firms or television studios. The Music Director supervises sound editors, audio engineers, and recording engineers. He or she reports to the Producer or Director of a television production.

Additionally, the Music Director:

- oversees the audio editing of the music sound track for a television production;
- maintains a file of existing music and music

sources for possible future use;

- establishes contacts with musicians, composers, music researchers, and other potential resources.

Salaries

Music Directors are usually employed on a project-by-project basis. Their salaries vary considerably and depend on the scope and extent of music used in a production and on the size of the enterprise. Some top Music Directors with impressive credentials command more than $75,000 for the complete scoring of a major two-hour television drama. More often, however, a Music Director works on simple productions, using prerecorded library music, and will be paid from $5,000 to $10,000. He or she may work on three or four such projects each year.

Most industry observers indicate that the total yearly income for a Music Director generally ranges from $25,000 for individuals with a modest reputation to well over $50,000 for the established professional with impressive credits.

Employment Prospects

Opportunities for employment in television are poor. Most talk shows, newscasts, and other formats make little use of music. Although many television programs do include music, the duties of a Music Director in lower-budget productions are often performed by the Producer, Director, or the sound or music editor. Major market commercial and public television stations use Music Directors only occasionally. The majority of the positions are in Hollywood or New York City, usually on a program-by-program or series-by-series basis. Employment often depends on contacts within the industry. With the proper education, training, and experience, some Musicians may become Music Directors.

Advancement Prospects

Chances for advancement depend on the talent and continuing reputation of the individual. With peer recognition, critical acclaim, and film or theatrical credits, come new opportunities for more projects at larger fees.

Some Music Directors increase their earnings by working on major motion pictures. A few, in Hollywood, are employed on long-lasting television series or become staff Music Directors for independent production companies. A very few Music Directors become Producers of television films or musical theater productions, after gaining the necessary experience in TV, film, and theatrical production.

In general, opportunities for advancement are poor and the competition is heavy.

Education

Almost all Music Directors are highly trained, extremely talented, musicians. They have studied music intensely from childhood, often specializing in piano but with extensive private training on other instruments, and have an educational background that includes music theory, harmony, arranging, and music history. Many have an undergraduate degree in music and a considerable number have graduate degrees.

Experience/Skills

At least five years of experience in the use of music in television or film is frequently required. A knowledge of general production techniques is also a necessity. A Music Director should be familiar with audio recording and have extensive experience in working with short, precisely-timed music segments. Music Directors are talented, creative individuals who have a broad knowledge of music styles and a finely honed musical taste.

Minority/Women's Opportunities

Although some female and minority-group employment opportunities have occurred during the past ten years in smaller budget productions, more than 95 percent of the Music Directors in television were white males in 1981, according to industry observers.

Unions/Associations

While there are no specific unions that represent or bargain for Music Directors, all belong to the American Federation of Musicians (AFM) including those who occasionally serve as arrangers or orchestrators. Music Directors who are composers and lyricists are often members of the Dramatists Guild; the American Guild of Authors and Composers (AGAC); the American Society of Composers, Authors, and Publishers (ASCAP); Broadcast Music, Inc. (BMI); or the Composers and Lyricists Guild of America.

MUSICIAN

CAREER PROFILE

Duties: Playing of an instrument as part of an ensemble for a television production

Alternate Title(s): None

Salary Range: $10,000 to $20,000

Employment Prospects: Poor

Advancement Prospects: Poor

Prerequisites:

Education—Extensive private music instruction; undergraduate degree in music preferable

Experience—Minimum of five years of recording studio, nightclub, theater, or radio-TV work

Special Skills—Musical talent; sight reading abilities; transposition and improvisation skills

CAREER LADDER

Music Director

↑

Musician

↑

Musician (Other Entertainment Field)

Position Description

A television Musician works as part of a music ensemble that provides backup accompaniment for Performers in musical variety or entertainment productions, or music for background use in television films and dramatic or comedy programs. He or she works in rehearsal halls and in television or recording studios. For many live or taped musical entertainment programs, the Musicians are in one studio and the Performers in another, and the music is brought electronically to the Performers' studio.

Musicians usually specialize in popular music for television production and accompany vocal or instrumental Performers as part of an ensemble. They are sometimes members of jazz combos, rock groups, dance bands or classical orchestras that appear on television as featured acts or in full-scale concerts.

A television Musician usually freelances on a project-by-project basis. He or she may temporarily become part of a small group of five or six pieces, a band of 17 to 23 members, or a full orchestra of more than 50 Musicians. A few individuals work under contract and play for a regularly-scheduled program such as *The Tonight Show.* On occasion, a Musician performs as a soloist during a musical number or as a featured Performer.

This is not a supervisory job, although most large bands or orchestras distinguish between a first chair (lead Musician in a section) and second or third chairs. A Musician reports to the conductor or Music Director during the production of a television program or in a recording session.

Additionally, the Musician:

- rehearses for recording sessions, run-throughs, and performances;
- keeps abreast of musical trends, new songs, and techniques;
- performs on more than one type of instrument during a session, when necessary.

Salaries

Minimum union scale for non-solo work on a half-hour television variety program was $125, and $212 for a one-hour production in 1981, according to union sources. Rehearsals and run-throughs add to the minimum scale, depending on the extent of the production. Pay is usually based on specific time segments (hour, half-day, etc.)

Salaries above the scale are negotiable and many experienced Musicians can obtain higher rates. In general, however, the pay for Musicians is lower than for most other positions in television, and ranges between $10,000 and $20,000 a year for those who work fairly regularly. Because most television work is sporadic, seasonal, and on a project-by-project basis, most Musicians perform in the theater and nightclubs and teach in order to earn a reasonable yearly salary.

Employment Prospects

The possibility for employment are poor. As in most cases in show business, professional contacts are very important in obtaining a position.

According to government and union estimates, there were about 320,000 professional Musicians in the United States in 1981. Most were located in the major urban centers. Few public or commercial stations have a need or budget for Musicians, and the majority of those working in television are in Hollywood and New York City. Of the thousands of Musicians in those cities, less than 5 percent earn any appreciable income from television work.

Advancement Prospects

There is limited potential for advancement. Very few Musicians obtain star or featured Performer status. Advancement is based on great talent, peer recognition, and superior musicianship on a particular instrument. The skills and talents necessary are seldom transferable to other television positions. A very few Musicians who are qualified obtain full-time employment at major motion picture or television studios. Some become Music Directors. Others, with well-developed interpersonal and management skills, become contractors who put together ensembles and groups for particular television productions. A very few become conductors or directors of bands or orchestras.

Education

While a degree is not a prerequisite to obtaining employment, such disciplined training is helpful in acquiring the necessary skills. Courses in theory, harmony, composition, performance, notation, and music history are useful.

The key to any success as a Musician, however, is practice, and lots of it. Most professional Musicians begin studying an instrument at a very early age and continue for the rest of their lives. To acquire great technical skill, musical knowledge, and interpretative ability, intensive and continuous training is required. As a rule, such expertise is developed through years of study with an accomplished teacher, usually in a combination of private classes and academic study at any of the 540 colleges and conservatories of music that offer degrees in performance, theory, and harmony.

Experience/Skills

A minimum of five years of previous experience in recording studios, nightclubs, theater, or radio-TV work is usually required of a Musician.

Most television Musicians are professionals who "know their horns"—a description that implies complete technical mastery of an instrument. Musicians must be able to sight read and transpose from one key to another with ease. An ability to improvise is required, as well as a knowledge of instrumental literature. A Musician in television must possess skill in ensemble playing and be alert and dependable.

Minority/Women's Opportunities

The majority of Musicians employed in television production and programming in 1981 were white males, according to industry observers, although white female representation in the string and woodwind sections of small orchestras was significant. Employment of minority-group males in small bands on television was approximately 25 percent, according to industry sources.

Unions/Associations

All professional Musicians are members of the American Federation of Musicians (AFM), which represents them in bargaining and salary negotiations. There are 604 locals of that union throughout the United States, representing Musicians for all types of employment, including television. Some Musicians who perform as soloists belong to the American Guild of Musical Artists (AGMA).

Musicians who are also composers and lyricists are often members of one or more of the organizations listed in the entry for Music Director.

COSTUME DESIGNER

CAREER PROFILE

Duties: Designing of costumes for major television productions

Alternate Title(s): Costume Director

Salary Range: $20,000 to $40,000

Employment Prospects: Poor

Advancement Prospects: Poor

Prerequisites:

Education—Undergraduate degree in fashion or costume design helpful

Experience—Several years of costuming for theater or television productions

Special Skills—Design talent; sewing ability; creative flair; TV production knowledge

CAREER LADDER

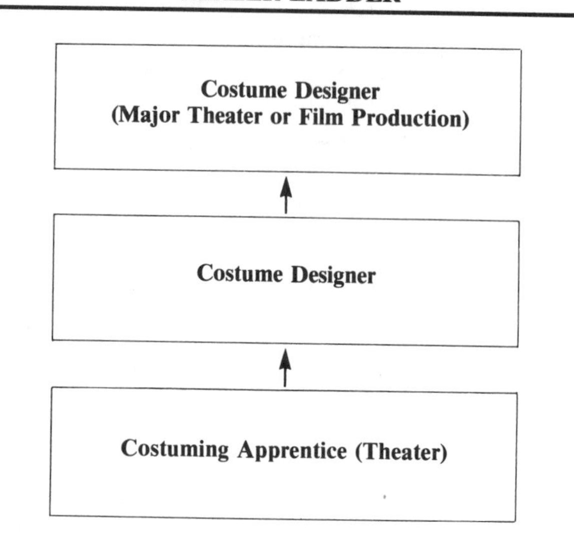

Position Description

The Costume Designer is responsible for creating the designs for all costumes worn in a television production. Since many TV productions call for Performers to wear their own street clothes, the Costume Designer is sometimes called on to approve their on-camera wardrobes. Costume Designers often work at drawing boards in their own studios, but they also spend a considerable amount of time in dressing or fitting rooms and in costume and tailor shops. During television productions, they are usually on hand in the studio and control room.

In most instances, a Costume Designer does not actually sew and construct costumes, but these activities are always under his or her supervision. Depending on the size of the project, a Costume Designer may supervise from three to eight workers, including costumers, tailors, and fitters. He or she usually reports to the Director of the production.

The Costume Designer must be adept at drawing and painting, composition and perspective, and light and shadow renderings. He or she makes sketches, models, and silhouettes, chooses fabrics, and matches clothing swatches so that the costumes will help provide the visual effect necessary for the television drama or entertainment special. The Costume Designer considers lighting, makeup, hair styles, and set design in determining the colors and styles of costumes. He or she must solve many problems, ranging from too much or too little set light to widely differing body shapes.

A Costume Designer must be capable of designing costumes for a variety of assignments. For dramatic programs, he or she may design costumes that are appropriate for a primitive cave man, or a 20th century career woman, or a futuristic alien. For variety programs or entertainment specials, he or she must be capable of creating exciting costumes for the chorus, supporting players, and featured Performers that complement and add to the overall production. He or she solves the problems of dancers' costumes (which require movement) and singers' costumes (which require throat and diaphragm freedom). When the performance calls for constant movement of a large cast, he or she adjusts costumes so that the visual effect remains pleasing and clashes are avoided.

Additionally, the Costume Designer:

- studies the script to gain an understanding of the costume requirements of a particular program or series;
- researches the styles of a specific geographic area or historical period, when necessary;
- locates costumes from stock, redesigns existing costumes, or rents appropriate costumes from a theatrical costume supply house;

• works within an allocated budget when determining the creation or acquisition of costumes.

Salaries

When they are working, top Costume Designers are relatively highly paid, as befits the talent, ability, and training essential for the position. The standard per diem union minimum for television costume designing is $106.75 a day, and for positions that are negotiated for an extended period of time, $1,100 a week. A costume design job with a network television series that runs all season will therefore generate an annual income of between $20,000 to $40,000. Most Costume Designers, however, do not work steadily, and their yearly income reflects long periods of unemployment. As an individual's reputation or style becomes established, higher rates can be negotiated. Most Costume Designers also work for theatrical productions, films, industrial shows, or nightclub acts where the pay is usually the minimum union scale, but is always negotiable.

Employment Prospects

Opportunities for employment are poor. Costume Designers usually freelance on a project-by-project basis. Major market stations do not have the budget or the need for a full-time staff position in this area.

Costume Designers are employed only by commercial networks and independent television and film production organizations, usually for specific productions. In most cases, only one person is hired for a program or series of programs. When costuming demands are extensive and complicated, the production may employ one or more costumers, who function as assistants to the Costume Designer. Most Costume Designers have been costumers in the early stages of their careers.

Opportunities often become available through connections with other theatrical and television workers. Most of the available positions are in Hollywood or New York City.

Advancement Prospects

A successful television Costume Designer who has worked for relatively small productions on meager budgets may move into costume design for motion pictures in Hollywood or make a transition to the Broadway stage. Many Costume Designers work on both coasts, and in a variety of settings, to advance their careers. Peer recognition and reputation, and a distinctive style, lead to more important assignments and higher salaries. Opportunities are so limited, however, that prospects for advancing to more prestigious positions are very poor.

Education

Style and design abilities are more important than a formal education in obtaining a position as a Costume Designer. An undergraduate degree in fashion or costume design from an art, design, or fashion institute can, however, be useful. Instruction in life drawing, color theory, and composition are necessary, as well as studies in costume history, television production, and theater arts. Course work in staging and lighting techniques for television and theater are extremely helpful.

Experience/Skills

Most Costume Designers have spent a considerable amount of time in an informal apprenticeship in the theater as costumers, dressers, wardrobe assistants, tailors, and a wide variety of other backstage jobs. They have sometimes served as Carpenters, scene painters, and occasionally performed in school or college in plays and summer stock.

A candidate for the job should have a balanced portfolio that includes design sketches for musicals, operas, Shakespearean dramas, and entertainment or nightclub shows. The individual must be expert at sewing and possess excellent drawing, painting, and graphic art skills. An understanding of the techniques, limitations, and capabilities of television production is required. He or she must also have a creative flair and sense of style and be excellent at interpersonal relationships, inasmuch as costume design is a collaborative effort.

Minority/Women's Opportunities

In the past, costuming and design have frequently been considered the domain of women. The percentage of women in the profession in 1981 was extremely high, according to industry observers. There are, however, some male Costume Designers and their number is growing. Minority-group members have been less well represented, and that representation is not increasing at a rapid pace, according to industry sources.

Unions/Associations

For bargaining purposes, Costume Designers on the east coast are represented by United Scenic Artists and, on the west coast, by the Costume Designers Guild.

MAKEUP ARTIST

CAREER PROFILE

Duties: Designing and applying makeup needed by Performers in a television production

Alternate Title(s): None

Salary Range: $11,000 to $30,000

Employment Prospects: Poor

Advancement Prospects: Poor

Prerequisites:

Education—Post-high school TV and theater arts courses in makeup required; undergraduate degree preferable

Experience—College or amateur theater; apprenticeship in legitimate theater

Special Skills—Creativity; visual imagination; familiarity with television techniques

CAREER LADDER

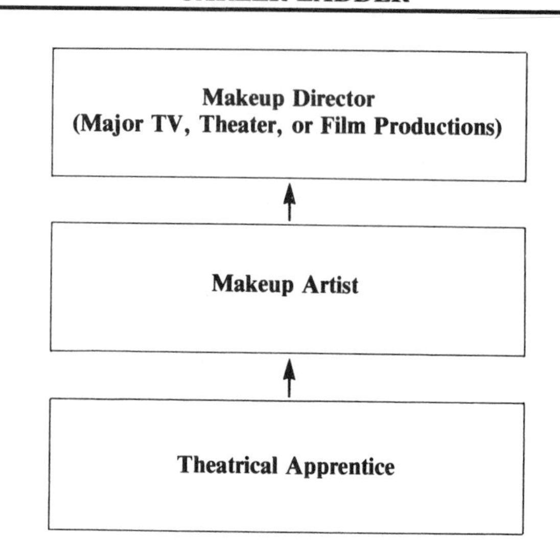

Position Description

A Makeup Artist is responsible for ensuring that the physical appearance of on-camera talent is in keeping with the style and needs of a television program. He or she consults with the Director to determine the proper effects that makeup should achieve for a particular production.

The application of makeup to Performers in a television production usually serves one of two purposes. The first is to enhance the natural features of the Performer, compensate for the washed-out effects that lighting can produce, and generally improve the Performer's appearance in much the same way as street makeup is intended to do. The TV camera can exaggerate blemishes and skin tones, and most Makeup Artists working in TV devote the greater portion of their time to applying straight (non-character) makeup that will minimize flaws and make the Performer appear natural to the viewer. Makeup Artists work on entertainment programs, network newscasts, series and soap operas, and television films, putting makeup on supporting players, stars, dancers, singers, and other on-camera talent.

The second function of TV makeup is to distinctly alter a Performer's appearance so that it conforms to the character being portrayed. Applying character, or corrective makeup is a more ambitious and challenging undertaking than is straight makeup work, sometimes requiring the Makeup Artist to devise and apply artificial features. Such challenges, however, are only required for the most complex dramatic or entertainment productions on television. Individuals may be made to look older or younger. They may be changed to assume particular racial characteristics, or their appearance may be altered to resemble a historical figure. Through the skillful use of makeup and other props, a Makeup Artist can create an illusion of swollen limbs, hunchbacks, added weight, or other body and facial changes appropriate to the character being portrayed.

Occasionally, a Makeup Artist must create the special effects that simulate the results of excessive violence, such as blood, broken noses, or knife wounds. Less frequently, he or she is called upon to transform a Performer into a character appropriate to a science fiction or horror production. Many of these specialized makeup effects have been accomplished only after elaborate testing and experimentation. The skillful use of makeup gives a TV Performer an additional advantage in portraying a character.

When no Hair Stylist has been employed for a TV production, the Makeup Artist may also assume those responsibilities. The job is not normally a supervisory one, and the Makeup Artist usually reports to the Director of a TV production.

Additionally, the Makeup Artist:

- maintains a file of facial pictures of all historical periods and nationalities, and of unusual and interesting characters;
- develops and maintains a collection of cosmetic supplies, application devices, and other equipment.

Salaries

Most Makeup Artists are non-salaried and work on a freelance basis. According to *Time* magazine, there are about ten top Makeup Artists who earn $500 to $1,000 a day. Union scale, however, is $89.75 a day for per diem employment, and $412.00 a week for positions that are more or less permanent. It is not unusual for Makeup Artists to work a total of about six months in any one year. This generally produces a minimum annual income of approximately $11,000. People who work more steadily usually earn in excess of $21,000.

As the Makeup Artist acquires experience and a reputation, and his or her skills become more highly developed, salaries considerably higher than scale can be negotiated with an employer. It is not unusual for an experienced Makeup Artist, who has developed a good reputation and reasonable contacts in the television industry, to work with enough frequency and at high-enough fees to earn an annual income of $30,000.

Employment Prospects

Opportunities are generally considered to be poor, and the competition is heavy. Most Makeup Artists are freelance, and the completion of one job, therefore, is the beginning of a search for another.

Very few TV stations employ Makeup Artists. The networks do hire them, but employment is irregular and sporadic. Independent television production companies employ them, too, usually in Hollywood or New York City. Most Makeup Artists also work for stage, theatrical plays, films, and industrial shows on a project basis.

Advancement Prospects

Advancement and continued employment are dependent upon building a good reputation, making and developing contacts, and luck. Some Makeup Artists do advance to the position of makeup director in major television dramas or theatrical or motion picture productions. Competition, however, is heavy, and advancement prospects are poor.

Education

Courses in television and stage makeup are vital, and can be acquired at a college or university with a strong TV or theater arts curriculum or at a specialized institute, studio, or workshop. Courses in staging and lighting, television production, and costume design are helpful. While a degree is not necessary, studies leading to an undergraduate or graduate degree can provide good basic training.

Experience/Skills

Some experience with television and theatrical makeup is mandatory, and may be acquired in college or any of the many community, regional, summer stock, or dinner theaters. An apprenticeship in the theater is the best way to gain the broad makeup experience necessary.

Minority/Women's Opportunities

Women comprise about one-half of the union membership for Makeup Artists, and are well represented in television, according to industry observers. Industry sources also report a substantial degree of minority-group representation among Makeup Artists.

Unions/Associations

For bargaining purposes, Makeup Artists are represented by a local unit of the International Alliance of Theatrical Stage Employees (IATSE) in New York City and in Hollywood. They may also belong to the National Hairdressers and Cosmetologists Association, in order to meet their peers and share mutual professional concerns.

HAIR STYLIST

CAREER PROFILE

Duties: Responsibility for styling and dressing the hair of Performers in a television production or series

Alternate Title(s): None

Salary Range: $15,000 to $30,000

Employment Prospects: Poor

Advancement Prospects: Poor

Prerequisites:

Education—Minimum of high school diploma and special training; certificate from beauty school preferable

Experience—Minimum of two or three years of hair styling for men and women, commercially or in the performing arts, preferably television

Special Skills—Creativity; manual dexterity; familiarity with TV lighting and camera techniques

CAREER LADDER

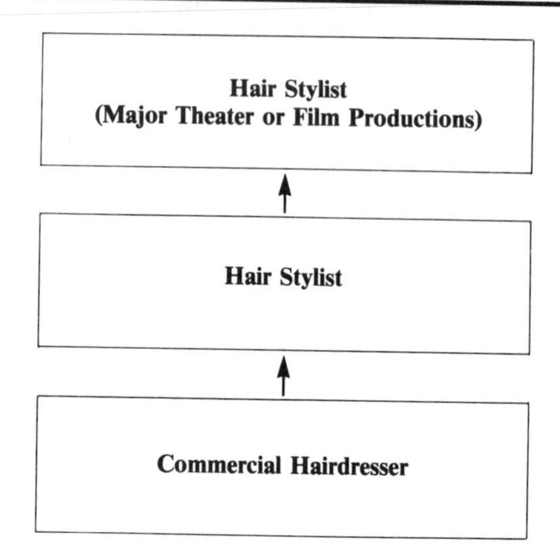

Position Description

A Hair Stylist is responsible for creating and maintaining the hair styles of Performers in a television production. The job is not a supervisory one, and he or she usually reports to the Director of the production.

While certain dramatic productions and musical-variety programs may require styles that are elaborate and extreme, most Hair Stylists mainly work with those that are considered natural, fashionable, and attractive by today's standards.

Hair styling for a television production differs from theatrical hair styling, because the television camera reveals more detail than is visible to the naked eye in a theater. As with makeup, hair styles for television are usually subtle and natural. A Hair Stylist must also solve problems caused by television lighting, which generates a considerable amount of heat often making it difficult to maintain hairdos throughout the course of a production.

Hair Stylists are employed by large independent production companies and networks for a particular television show or series. In small and middle size production organizations, hair styling is often done by the individual who serves as the Makeup Artist.

Lavish, large-scale TV productions that involve a great many feature and supporting Performers may require more than one Hair Stylist. Many super stars negotiate a contractual arrangement that provides them with a personal Hair Stylist. Other stars employ their own person who travels with them on a regular basis. But in most instances, and for relatively simple productions, a Hair Stylist works alone on a project-by-project basis.

A Hair Stylist works in the dressing room before a television show and is available in the studio for touch-ups and adjustments during the actual production.

Additionally, the Hair Stylist:

- shampoos, cuts, shapes, and trims hair, when necessary, in the course of styling;
- works with the Costume Designer and Makeup Artist when designing particular hair styles;
- designs hair styles to complement and enhance Performers' specific facial shapes;
- creates hair styles that suggest personality traits of specific characters;
- uses hair coloring agents, accepted styling techniques, standard and specialized equipment, wigs, hairpieces, falls, and other accessories to achieve desired effects.

Salaries

Hair Stylists usually freelance on specific projects, although a few are steadily employed as part of the production team for a network series or soap opera.

Minimum union scale is $89.75 per day on a per diem basis and $412.00 a week if employment is continuous. Most Hair Stylists work in television irregularly during the year and they generally earn a minimum of about $15,000 annually. People who work steadily may have an income of more than $21,000 per year. After Hair Stylists become established in the industry, fees may be negotiated that are higher than union scale. If they work regularly, they may earn as much as $30,000 a year.

Employment Prospects

The prospects for employment at the network level or at an independent production firm are generally poor. Relatively few positions exist in the television industry, turnover is extremely low, and employment is traditionally periodic and often seasonal. Industry observers are generally not optimistic about increased opportunities in the near future, although there will always be a need for some Hair Stylists in major television production operations.

Personal contacts and connections account for much of the employment of successful Hair Stylists. These contacts are usually acquired as a result of previous assignments in television or through the recommendation of an influential friend or relative. Most Hair Stylists have additional opportunities for employment by working in other show business environments, including theaters and nightclubs.

Advancement Prospects

Advancement is achieved by developing a reputation, thus ensuring continued employment and higher pay. Television Hair Stylists occasionally move to major motion picture or Broadway theater productions. A very few individuals become personal Hair Stylists for stars.

It is generally recognized that getting started in the profession is the most difficult step, and that as one's credits and reputation grow, additional and more important assignments follow more easily. Overall, however, advancement prospects are poor.

Education

A high school diploma with some additional specialized training in hair styling is usually the minimum educational requirement. A certificate or diploma from an accredited beauty culture school that specializes in hair styling is also a common prerequisite. Most vocational schools also offer courses leading to a hair styling specialty.

Experience/Skills

Virtually all Hair Stylists begin their careers by working on a daily basis in a commercial beauty salon. The opportunities for employment in television production increase in proportion to the experience an individual acquires. A minimum of two or three years of previous professional work in the theater or in television production is usually necessary, and experience in hair styling for commercial photography studios or in modeling is also useful. Because of the particular technical requirements of television, some experience in staging, lighting, and production is helpful. Participation in school or community theater productions can also serve as good training.

A television Hair Stylist needs all of the skills of a creative commercial hair stylist, including style, artistic flair, and manual dexterity. Good interpersonal skills are also necessary. In some instances, graphic arts skills and the ability to research and copy period hair styles are helpful.

Minority/Women's Opportunities

Industry sources indicate that hair styling today is a field that is equally accessible to men and women, whether members of minority groups or not. About an equal number of men and women held the position in 1981, according to industry observers.

Unions/Associations

For bargaining purposes, Hair Stylists are represented by a local unit of the International Alliance of Theatrical Stage Employees (IATSE) in New York City and Hollywood. Membership in the union, however, is not required for employment in the television industry.

Some Hair Stylists belong to the National Hairdressers and Cosmetologists Association for purposes of professional growth and support.

SCENIC DESIGNER

CAREER PROFILE

Duties: Designing of the scenery and sets for a television production

Alternate Title(s): Scenic Director; Set Designer

Salary Range: $20,000 to $40,000+

Employment Prospects: Poor

Advancement Prospects: Poor

Prerequisites:

Education—Undergraduate degree; master's degree in theater helpful

Experience—Minimum of five years in theatrical and television set design

Special Skills—Creativity; graphic arts talent; artistic flair; knowledge of TV production; interpersonal abilities

CAREER LADDER

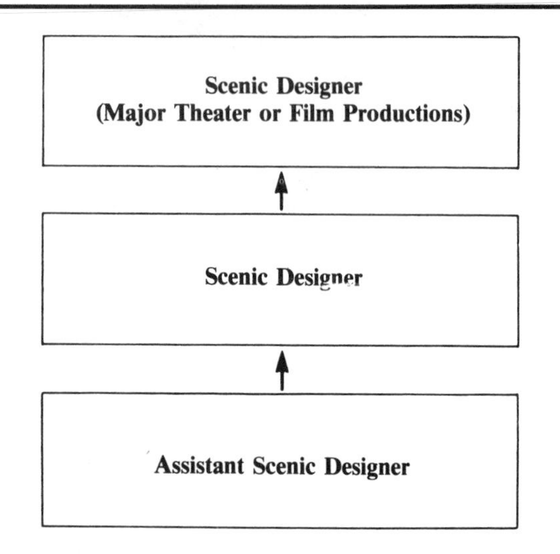

Position Description

A Scenic Designer for a TV production designs the sets and scenery, oversees their construction, and determines how they will be dressed (the addition of finishing touches). He or she conceives, designs, and supervises the creation or acquisition of backdrops, exteriors, interiors, furniture, and the decorative and functional details of all sets and scenery. The Scenic Designer decides the background color and brightness, as well as the material to be used, such as paint or wallpaper.

After reading the script and working closely with the Producer and Director, the Scenic Designer ascertains the purpose of the production, as well as the place, time, and mood of the action, and translates their concepts into physical settings. He or she makes preliminary sketches and, in the case of major productions, sometimes 3-dimensional models of sets.

The Scenic Designer must visualize the set as it will appear in various camera shots, including close-ups and long shots from specific angles. Building on the Director's concept, he or she takes into account the activity and physical movement that the sets and scenery will have to accommodate. During pre-production stages, he or she creates a floor plan, which indicates where each set will be situated, where major equipment (such as cameras) are positioned, where permanent lighting fixtures and outlets are located, and where scenery and furniture will be placed.

Scenic Designers are usually employed only by networks and independent production companies for entertainment or dramatic programs, such as musicals, specials, game shows, commercials, ongoing situation comedies, dramatic series, or soap operas.

A Scenic Designer works at a drafting board in his or her studio during the design period, in carpentry and scene shops while overseeing construction, and in a television studio during set-up and production. He or she occasionally supervises an assistant, set dressers, a Property Master, and others on a project-by-project basis. A Scenic Designer usually reports to the Director or Producer of the TV program.

Additionally, the Scenic Designer:

- works closely with the Costume Designer and Lighting Director;
- researches particular historical periods and geographic locations to ensure accuracy of set detail;
- maintains inventory of sets, scenery, backdrops, and furniture;
- supervises the dismantling of sets.

Salaries

Scenic Designers are usually paid on a per diem

basis. Minimum union scale is $100 a day, with time-and-a-half paid for all work in excess of seven hours. Recognized professionals, with a reputation for skill and talent, command considerably higher fees. While employment is seasonal and irregular, an experienced Scenic Designer often earns between $20,000 and $40,000 or more a year.

Employment Prospects

Most employment is for one-time-only jobs, but a Scenic Designer may occasionally get continuing employment on a weekly series or a daily soap opera, which often use several different sets in each show.

Even after being an assistant scenic designer for years, it is very hard to land that critical first job as a full Scenic Designer, whether on a specific project or as a staff member. With good personal contacts and a high degree of talent, however, it becomes somewhat easier to get work. Opportunities are usually confined to Hollywood and New York City and union sources estimate that one-third of their membership is unemployed at any given time. Although prospects may improve through work in cable TV and related areas, the possibilities are generally poor at present, and the competition is severe.

Advancement Prospects

This is a highly competitive field and advancement opportunities are poor. A few individuals move into more responsible, higher-paying scenic design jobs on a regular staff basis. Others try to advance by contracting for larger and more frequent projects. The very best Scenic Designers advance their careers by broadening their base to include work for major theatrical and motion picture productions along with television programs.

Education

Style and design capabilities are more important than a formal education for getting a job. An undergraduate degree in art and design, radio-TV production, theater arts, or architectural design is usually necessary, however, and a master's degree in theater arts can be of additional help. Courses in stagecraft, TV production, drafting, architecture, and the principles of color and composition are essential as well as study in drawing, painting, sculpture, and graphic arts.

Experience/Skills

A minimum of five years in television and theatrical scene design on a professional level is usually required for the job. A number of people work as assistant scenic designers before they move to the top position. Most have spent a considerable amount of time as apprentices in the theater, working in a variety of backstage roles such as scene painters and carpenters

A Scenic Designer must have creativity, fine graphic arts skills, an artistic flair and style and, since the work is a collaborative effort, excellent interpersonal abilities. A knowledge of television production techniques is also required. Most important, perhaps, he or she must possess the talent to understand the relationship among time, space, and movement.

Minority/Women's Opportunities

The number of women working as Scenic Designers is quite high, and increasing each year, according to industry sources. Minority-group members are less well represented, but the number has increased within the profession in recent years and observers believe this trend will continue.

Unions/Associations

For bargaining purposes Scenic Designers are represented by United Scenic Artists. Many also join graphic and visual arts associations in order to share ideas and advance their careers.

PROPERTY MASTER

CAREER PROFILE

Duties: Providing, maintaining, and positioning of all props used in a major television production

Alternate Title(s): None

Salary Range: $12,000 to $25,000+

Employment Prospects: Poor

Advancement Prospects: Poor

Prerequisites:

Education—College or other training in TV or theater production

Experience—Minimum of one year as a set dresser or Stagehand/Grip

Special Skills—Artistic flair; knowledge of TV/theater carpentry and production; detail orientation; precision; "scavenging" ability

CAREER LADDER

Property Master (Theater or Film Production)

↑

Property Master

↑

Set Dresser; Stagehand/Grip

Position Description

A Property Master is in charge of the acquisition, maintenance, use, storage, and disposition of properties (props) used in a major TV production or series, or a television film. He or she arranges to get the small items that add to and visually enhance a set, such as lamps, ash trays, pictures, and other decorative or functional pieces. He or she also supplies hand articles essential to Performers, hosts, guests, or the plot of a drama, such as magazines, canes, guns, drinks, and so on.

Props are an integral part of any performance, and the TV camera reveals, with closeup clarity, any flaw or error in any part of the set or its props. The Property Master determines exactly what props are required by studying the script and consulting with the Scenic Designer, the Director, the Performers, and, occasionally, the Producer. He or she purchases or rents properties, or obtains them from storage. He or she maintains a list of props, and sees that they are in place, in good condition, and available for each scene. When props cannot otherwise be obtained because of special requirements or budgetary restrictions, he or she supervises their construction.

When selecting props for a period or historical program, the Property Master first researches the era or geographic location to ensure authenticity. For contemporary entertainment programs or dramas, he or she adds the finishing touches to a set or scene by providing props that complement the presentation and stylistic setting.

The title Property Master, a term borrowed from the legitimate theater, is normally used only in major dramatic TV programs. Property Masters are employed at network television studios, independent production companies, or film sound stages in New York or Hollywood and occasionally they work on location. At most TV stations, the Property Master's duties are handled by the Floor Manager and a Production Assistant.

Property Masters sometimes supervise assistants called set dressers, who physically place and arrange props in the appropriate locations on a set. For some major TV productions, a Property Master may supervise a property buyer, who purchases necessary items, or a property maker, who works with the scenic design crew, and Carpenters in constructing special items. Property Masters usually report to a Scenic Designer or the Director.

Additionally, the Property Master:

- maintains and cleans props that are scheduled for repeated use;
- monitors expenditures and stays within budget;
- packs and unpacks props when accompanying a

show on location;

• arranges for the shipping and transportation of props;

• strikes (removes) props from the set after production, returns rented or borrowed items, and arranges for storage or appropriate disposition of others.

Salaries

Union scale is about $120 a day for Property Masters who are employed for a specific project on a per diem basis. For continuing employment, the minimum is $390 a week, a rate that is often negotiated upward by experienced Property Masters with a solid reputation for efficiency and capability. At the higher daily rate, however, opportunities are not as common, and long periods between jobs are normal. Annual incomes usually range from a low of $12,000 to a high of $25,000 or more.

Employment Prospects

Opportunities for employment are poor, and depend heavily on contacts and relationships in the entertainment field. Most Property Masters freelance on a project-by-project basis.

Set dressers, who function as assistants to the Property Master in major television productions, and Stagehand/Grips are the most likely candidates for promotion to the job of Property Master.

The seasonal nature of television production results in lengthy periods of unemployment. Most individuals work in theatrical and other entertainment fields to augment their income from television.

Advancement Prospects

Property Master is usually the top rung of a career ladder that may have begun at one of the entry-level jobs in television, film, or theatrical production. Because many individuals consider working in the Broadway theater or motion pictures to be prestigious, they often seek to become the staff Property Master with a theater or film studio to advance their careers. Prospects for getting one of these jobs, however, are poor, since they usually go to people with years of experience and many contacts.

Education

Some post-high school training in television or theatrical production at a college, university, or special theatrical workshop is strongly recommended. Studies in art, theater history, scenic design, and construction are useful. An undergraduate degree in theater arts is also helpful. Apprenticeships of two to three years are sometimes available through one of the appropriate local branches of the International Alliance of Theatrical Stage Employees (IATSE).

Experience/Skills

At least one year of experience as a set dresser or Stagehand/Grip is usually necessary before an individual can assume the position of Property Master.

Artistic flair is a must, as are a basic knowledge of theatrical and television carpentry and construction and TV production techniques. Successful Property Masters are also detail-oriented and precise.

A Property Master is a master "scavenger." He or she has the contacts, abilities, and skills to find and obtain the perfect item from stores, theatrical supply houses, thrift shops, museums, art galleries, and the homes of friends and relatives.

Minority/Women's Opportunities

Both women and members of minority groups are well represented in the field. Industry sources indicate that women occupied about half of the positions in 1981.

Unions/Associations

Property Masters are represented by local units of IATSE, in New York City and Los Angeles, for negotiating and bargaining purposes.

STAGEHAND/GRIP

CAREER PROFILE

Duties: Setting up, moving, and dismantling of sets for a major television production

Alternate Title(s): Staging/Lighting Assistant

Salary Range: $10,000 to $18,400

Employment Prospects: Fair

Advancement Prospects: Fair

Prerequisites:

Education—High school diploma; undergraduate degree in radio-TV or theater preferable

Experience—College/amateur scenic work in TV or theater helpful

Special Skills—Knowledge of TV or film production techniques; physical strength; esthetic sense

CAREER LADDER

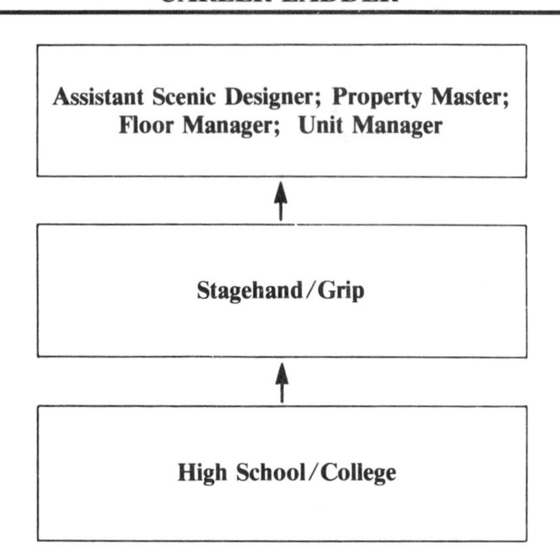

Position Description

After sets have been constructed or requisitioned and delivered to a television studio, the Stagehand/Grip assembles them according to the designated plan created by the Scenic Designer. He or she puts together and sets up backdrops, scenery, and set pieces for rehearsals and production. He or she positions furniture and places props and technical equipment in their proper studio locations. The Stagehand/Grip also sees that the furniture and sets are moved or changed as required by the script. As necessary, he or she assists in maintenance, adjustments, and alterations of scenery and set pieces. When a production is completed, the Stagehand/Grip strikes (dismantles) and stores them in the scene shop or prepares them for transport to outside storage.

The terms "Stagehand" and "Grip" are borrowed from the theatrical and film worlds, respectively, and the responsibilities of the job in TV are similar to those in the theater or motion pictures. There are differences, however, that are directly related to television production techniques. For example, unlike the theater, complete removal of one set and set-up of another during production is rarely required in TV.

A Stagehand/Grip performs strenuous, physical tasks such as moving heavy furniture as well as less arduous chores such as rearranging small props. In some major market TV stations, he or she may also hold time or cue cards, place microphones, or handle camera cables during production.

As a rule, a Stagehand/Grip works at independent television and film production companies. Work is often confined to Hollywood or New York City, but some major market stations also employ people in this position. Occasionally, a Stagehand/Grip will work on location. He or she usually reports to a Unit Manager prior to production, or to a Floor Manager during production. In large productions, he or she may be assigned to a Property Master. Some major film and TV programs designate a senior individual as a key or chief grip, with some supervisory duties over other staging employees.

Additionally, the Stagehand/Grip:

- helps the Lighting Director in preparing or positioning lighting equipment;
- assists in moving camera cranes, booms, and dollies during production;
- keeps the studio, staging, and storage areas neat and orderly.

Salaries

Salaries for the union position of Stagehand/Grip are relatively good. Freelance people working on a per diem basis are paid a minimum of $116.88 a day when covered by an International Alliance of Theatrical Stage Employees (IATSE) contract, with double wages for overtime. Per diem work, however, is not regular or continuous, and long periods of unemployment are common. When represented by the National Association of Broadcast Employees and Technicians AFL-CIO (NABET) in a full-time regular position, they receive from $12,300 to $18,400 annually, depending on length of service. Non-union positions are paid less, ranging from $10,000 to $15,000 at most stations.

Employment Prospects

Opportunities for this entry-level job are only fair. Employment is normally confined to major productions in New York City or Hollywood or at one of the top 20 major market commercial or public TV stations. Freelance work is sporadic, and unemployment has been prevalent recently. As a lower-paid entry-level job at some major market stations, however, the turnover is relatively high and openings do occur from time to time, as those in the position are promoted or leave for other jobs.

Advancement Prospects

Opportunities for bright, ambitious individuals are fair, and depend largely upon personal initiative and talents. The job can sometimes serve as an informal apprenticeship for scenic designing, and with the proper education, experience, talent, and aggressiveness, a bright Stagehand/Grip can advance to a position of assistant scenic designer. Some Stagehand/Grips become Property Masters for major film and television productions. A Stagehand/Grip who becomes extremely proficient at his or her job might move to a larger production company at a high salary, where he or she is involved with more ambitious and complex programs. In non-union shops and at some major market TV stations, the Stagehand/Grip may have the opportunity to move to the position of Floor Manager or Unit Manager.

Education

Education credentials beyond high school are not required, but an ambitious candidate who views the job as an entry into television production should have a college degree in radio-TV or theater. Courses in theatrical staging, lighting, and television and film production are extremely helpful.

Experience/Skills

Professional experience in television production is not required. Any experience in set construction or related scenic responsibilities in college television or theater or in any amateur or community theater production is a decided advantage.

A Stagehand/Grip should have some knowledge of television or film production techniques, physical strength, and a sense of esthetics.

Minority/Women's Opportunities

Industry sources indicate that both women and members of minorities are well represented in this position. Women comprised approximately 15 percent of union membership in 1981, while employment of minority-group members was somewhat higher.

Unions/Associations

Stagehand/Grips are sometimes represented by a New York or Hollywood unit of IATSE. Some individuals at major market television stations belong to NABET for representation in labor negotiations.

CARPENTER

CAREER PROFILE

Duties: Construction of sets, backdrops, props, and other scenic pieces for use in a television production

Alternate Title(s): None

Salary Range: $12,000 to $25,000+

Employment Prospects: Poor

Advancement Prospects: Poor

Prerequisites:

Education—Trade or vocational school

Experience—Apprenticeship in carpentry; some background in TV or theatrical production or construction

Special Skills—Good carpentry skills; manual dexterity; logical mind; ability to work rapidly and under pressure

CAREER LADDER

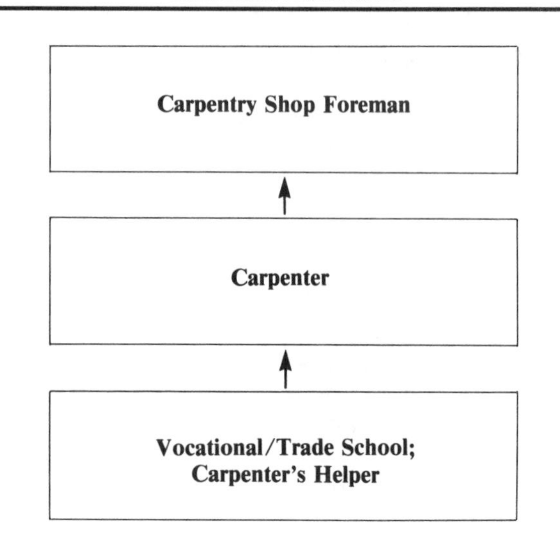

Position Description

A Carpenter is responsible for constructing the sets used in television programs. He or she is normally employed in the scene shop of a television network or independent production company, or in a commercial scenery studio that specializes in building sets for television, theatrical, and motion picture production. Although most such studios are located in Hollywood and New York City, a few are found in other major urban centers that have significant theatrical and television activities.

Carpenters work from blueprints, sketches, floor plans, and instructions from the Scenic Designer and a foreman. A Carpenter works with scene painters, assistant scenic designers, and other carpenters and skilled workers in constructing sets. He or she uses a variety of hand and power tools and works with different materials, such as wood, plastic, fiberglass, and muslin. In the construction industry, carpentry is commonly divided into two categories—rough and finish. Carpenters who build television and theatrical scenery deal only with finish carpentry.

Carpenters work on various types of set construction, including three-sided rooms of all kinds, backdrops, panels, furnishings for game shows and news broadcasts, and outdoor set (exterior) pieces, as well as special furniture and props. Most television scenery is realistic, but on occasion may be fanciful or abstract. Sets are usually not constructed for permanence or structural integrity. They are often built in units or sections to facilitate storage and re-assembly.

Building television scenery is somewhat different from theatrical set construction. Because TV frequently uses closeup shots, the scenery usually has a finished and natural appearance, and the set generally seems to be realistic and detailed. Since in a theater the audience is further removed from the scenery, specific detail may be omitted while certain significant features may be exaggerated.

Carpentry work in television is generally completed on a tight deadline. When constructing scenery for a TV production, a Carpenter usually has little time to spare. For *Saturday Night Live*, for example, preliminary sketches are ready on Tuesday, final instructions on Wednesday, and six or eight sets are completed by early Saturday.

Additionally, the Carpenter:

- checks the accuracy of construction with levels, rulers, and framing squares;
- paints colors and designs on backdrops and panels;
- devises ways to solve special construction problems, in consultation with the Scenic Designer and foreman.

Salaries

Television Carpenters are reasonably well paid. The union scale for a freelance worker begins at $125.51 a day, with double time paid for any overtime work. At that rate, a Carpenter employed for less than half a year (not an uncommon occurence), would earn an annual minimum of about $12,000. An individual who works steadily can easily earn $25,000 in a year, and with seniority, even more. The union minimum for a full-time Carpenter is $374.52 a week, or nearly $20,000 a year to start. Non-union shops generally pay considerably less on an hourly and project-by-project basis.

Employment Prospects

The opportunities for employment are poor. The demand for Carpenters is periodic and seasonal, and depends on production schedules. Most scenery shops employ some Carpenters on a full-time basis, and augment the crew with temporary help to accommodate a heavy schedule. Because there are often slack periods for television Carpenters, they usually supplement their television income with general carpentry and construction work. A very few major market television stations employ a Carpenter on a regular staff basis. In most instances, the relatively little scenic construction required is handled by the Floor Manager and Production Assistant(s).

Advancement Prospects

The opportunities for promotion are poor. Some union Carpenters advance their careers through promotion to shop foreman. This position is scarce, however, and there is very little turnover. The freelance (temporary) Carpenter generally considers frequent or continuous employment to be the equivalent of advancement.

Education

Basic carpentry skills may be learned at a vocational or trade school, or through apprenticeship. The latter is usually a four-year program of on-the-job training, accompanied by classroom instruction. Industry sources, however, recommend that an individual obtain special training in motion picture or television production to be adequately prepared for the special requirements of TV scenery construction.

Experience/Skills

Competence in basic carpentry techniques, which may be acquired as an apprentice or Carpenter's helper, is required. Additional experience in television or theatrical production or construction is generally considered a prerequisite. An apprenticeship in theatrical carpentry, which is sometimes available through a local union branch, is excellent training.

Carpenters must have good basic carpentry skills, a considerable degree of manual dexterity, and a logical approach to construction problems. They should be able to work rapidly and under some pressure.

Minority/Women's Opportunities

Industry sources report that many women and members of minority groups are becoming active as Carpenters. Women reportedly comprised approximately 10 to 20 percent of union membership in 1980. Estimates regarding minority involvement vary, but industry sources indicate that opportunities for members of minorities have more than doubled over the past ten years.

Unions/Associations

Carpenters employed on a freelance basis in television set construction are represented by a local New York or Hollywood unit of the International Alliance of Theatrical Stage Employees (IATSE) for negotiating and bargaining purposes. Some individuals are also members of the United Brotherhood of Carpenters and Joiners of America, which is a bargaining agent for Carpenters in all areas of construction.

PART II
VIDEO AND TELECOMMUNICATIONS
Introduction

A remarkable change is occurring in communications. Rapid technological advances are transforming a television industry based on the physics of scarcity* into one of electronic diversity. New ways of transmitting and receiving messages abound, mostly based on non-broadcast technology. Some observers feel that we are undergoing a communications revolution. To others, it's an evolution in which newer technologies have developed from the older forms to create unprecedented new possibilities in the ways we transmit and receive information.

The number of technological systems, devices, and methods of reception that can be—and are—used with a modern television set is staggering: broadcast TV; basic cable television; pay cable; home computers; video games; videodiscs; videocassettes; low-power television (LPT); multipoint distribution service (MDS); subscription television (STV); direct broadcast satellites (DBS); teletext; viewdata; cable audio. Only broadcast television and basic cable TV were in existence ten years ago!

The television set is beginning to play a dramatically different role in our lives. From a limited number of programs broadcast to a mass audience, we are entering an era of "narrowcasting." Specific messages or programs that reach small targeted audiences with specialized interests are now available to an increasing number of professionals, students, and consumers. We can play back instructional programs on videodiscs or cassettes; watch first-run movies transmitted by STV, MDS, or pay cable TV operations; listen to a special music-only cable channel; play a video game alone or with others; receive important information in written form (teletext); shop by selecting products displayed on the screen (through viewdata); and receive entertainment or information programs via direct broadcast satellite (DBS) to our homes or schools. We can now use the TV screen with computers, and in the near future, we will be able to receive programs from low-power television (LPT) stations which will offer specialized programs to viewers in small geographic areas.

*This phenomenon effectively limits the number of over-the-air transmissions that can be made without mutual interference.

The standard 21-inch television screen may be replaced by a larger device in which the image can be a full 4-by-5-feet in size. New electronic line formations that will produce exceptionally clear pictures are in the offing, along with stereo TV sound.

These new devices and technologies are not limited to home viewers. Schools, colleges, private industry, government, and health agencies are using them with increasing frequency. This astonishing proliferation has created new businesses along with a need for people to develop, manage, market, service, and operate their systems and products. While employment statistics for most of these newer industries are not available, all observers predict an increasing number of career opportunities for all types of workers.

Definitions

Video is often used to describe the use of television production in non-broadcast settings. When used in the context of the home video industry, it includes the manufacture and distribution of consumer electronics products and programs such as videocassette recorders (VCRs), videodisc machines, and programs on cassettes and discs which are used with the family TV set.

Most observers use the general term telecommunications to describe all means of audio and visual transmission of signals from one point to another. Broadcast television, radio, telephone, and telegraph are part of the telecommunications industries. The term has commonly been used to describe an interconnection method, often by cable TV, MDS, or STV operations, linking two or more electronic points in which an exchange of information takes place. Partly to distinguish their use of technology from the show business image of broadcast television, educators and other non-broadcast specialists use the term to imply a more serious use of the electronic media for information and instruction. Many professionals in non-commercial operations such as closed circuit television (CCTV), instructional television fixed service (ITFS), instructional television (ITV), and media centers in health, government,

Chart II-1 — Yearly Sales of Videocassette and Videodisc Machines

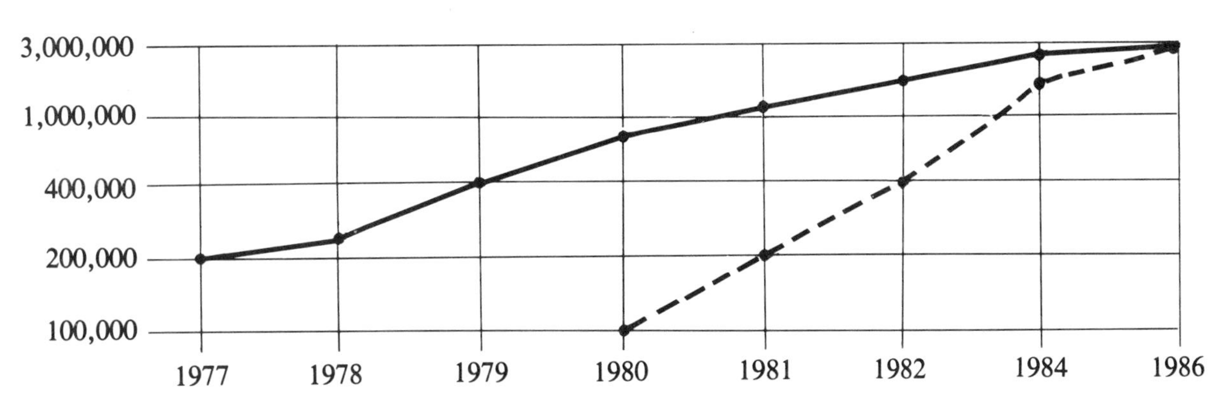

Figures represent sales through 1981 and projections thereafter. Estimates derived from the International Tape/Disc Association, Electronic Industries Association, and Argus Research.

and private industry, use the term telecommunications to describe their services.

What follows is a discussion of the six major areas of video and telecommunications highlighting their environments, employment opportunities, and trends for the 1980s. Important positions in these areas are depicted in the 29 job descriptions that follow this introduction.

Section 1: Consumer Electronics and Home Video

The consumer electronics industry has created a host of entertainment and educational products that have had an incredible impact on our society. The items for sale include radios, television sets, video games, videocassette recorders/players, videodisc machines, programs on videocassettes and discs, personal computers, stereo systems, large-screen TV projectors, and many other technical marvels. Retail sales of these items amounted to $15 billion in 1980, according to the Electronic Industries Association (EIA).

An estimated 35,000 retail outlets in the U.S. handled consumer electronics products in 1981. About 500 national wholesalers and original equipment manufacturers (OEMs) design, manufacture, duplicate, and distribute to retailers all of this equipment and programming as well as a multitude of accessories. More than 1.9 million people were employed in creating or distributing these products in 1979, according to the EIA. Of them, more than 200,000 were service technicians who did maintenance and repair on technical equipment.

Although videotape recording has been used to record and play back TV programs for broadcast since the late 1950s, it was not generally available for home or school use until the 1970s. Small format videocassette machines became available in 1975, and have now settled into two basic systems known as Beta and VHS. Both use half-inch videotape on cassettes inserted into the machine, which is attached to a TV set. Both can record programs on blank cassettes from a TV, record sound and pictures from a home video camera and microphone, and play back pre-recorded programs on the television set. The formats, however, are incompatible. A Beta videocassette will not play back on a VHS machine, and vice versa.

Unlike VCRs, which can record, the newer videodisc machines, first introduced in 1978, only play back pre-recorded programs on a TV screen. But with thousands of programs becoming available, that is an important function. There are two systems currently on the market. The Laser Optical Videodisc (LV), sold under the names MagnaVision and LaserDisc, and a newer arrival, the Capacitance Electronic Disc (CED), which was introduced by RCA under the name SelectaVision in 1981. A third system, Video High Density (VHD) is similar to CED and is scheduled for introduction in 1982. It will be sold by Panasonic and General Electric, among others. All three videodisc systems are incompatible. None will play back the pre-recorded discs of the others.

Videodisc players usually produce sharper resolution than VCRs and they are generally less expensive. The American consumer will make a choice but some observers speculate that many people will buy both types.

Since their introduction, sales of both videocassette and videodisc machines have been growing at a phenomenal rate, and are expected to increase at an even greater pace in the future (see Chart II-1). By 1986, cumulative VCR sales are predicted to total 15 million sets, or penetrate 18 percent of the U.S. homes that have televisions. It is expected that more than 14 percent of U.S. homes will probably have a videodisc machine by that time.

According to the National Video Clearinghouse (NVC), there are currently more than 35,000 pre-recorded programs available on videocassette and videodisc, of which more than 4,000 are the same popular movies and entertainment specials often offered by cable TV systems and STV and MDS stations. These programs are sold or rented through the mail or through retail outlets.

By 1986, most experts predict an annual retail market of nearly $5 billion in videocassette machines and blank and pre-recorded cassette sales, and a $3.2 billion market in videodisc machines and pre-recorded discs.

This burgeoning new industry, in which almost every prediction has proven to be conservative, offers many opportunities for employment at all levels of production, distribution, maintenance, and retail sales in the 1980s.

Section 2: Cable Television, Subscription Television, and Multipoint Distribution Service

Several types of alternative television transmission systems have been introduced during the past two decades which promise increased opportunities for growth and employment. These include cable television (CATV), subscription television (STV), and multipoint distribution service (MDS) operations. All of these services are designed to reach specific targeted audiences and, unlike broadcast television, the viewer pays a fee to receive the programs. These new industries have experienced an astounding growth in a few short years.

Cable TV

The oldest form of alternative transmission service is cable TV, begun as a means of providing better reception of regular broadcast television channels. It is generally believed that cable TV started in the mountains of Oregon and Pennsylvania more than 30 years ago, concurrent with the expansion of broadcast television. A master antenna was raised on a mountain peak and the houses in the valleys were linked to it by coaxial cable, enabling home viewers to receive a clear picture. The system was called Community Antenna Television (CATV), but the acronym now generally stands for cable TV.

By 1950, there were 70 CATV systems serving 14,000 subscribers who paid a fee to the owner of the antennas and coaxial cables and received broadcasts of television channels. By 1981, there were more than 4,400 operating systems serving over 19.1 million homes, according to the National Cable Television Association (NCTA). At that point, more than 25 percent of the nation's homes were wired to receive cable TV, and the number is expected to grow to well over 50 percent by 1990. Between 1976 and 1979, the pre-tax profits in cable television grew from $47.7 million to $199.3 million.

The technical emphasis in cable TV is on the electronic and physical connection of a subscriber's TV set to the "head end" (control center) by coaxial cable. The system transmits signals received in any of four ways: from satellites, from off-air signals of TV stations, via microwave, and from local origination (whether live, videotape, or computer generation). The signals are sent from processing equipment and amplifiers at the head end to a series of major trunk cable lines, then to feeder trunk lines that run to clusters of homes, and finally to drop lines that connect the individual home by coaxial cable directly to the system.

Most cable systems offer 12 channels to subscribers, but any system constructed after March 1972 must have at least a 20-channel capacity, according to Federal Communications Commission (FCC) regulations. Systems under construction today often offer 50 or even 120 channels in various program or channel groupings called tiers, and the subscriber can select from among a number of packaging arrangements called availabilities.

All subscribers pay a fee to receive basic cable service that covers the reception of local stations, two or three stations from outside their geographic area, and special religious, arts, or music channels. Nearly 50 percent of the subscribers pay additional fees to receive pay cable service which usually includes first-run Hollywood movies and other special entertainment programs.

More than 40,000 people were employed in cable TV in 1980. Most worked at local cable companies, but more than 3,000 were employed at the national headquarters of multiple system operators (MSOs) that own more than one cable system. Of all employees in cable TV, 17.2 percent were officials, managers, and professionals; 5.1 percent were in sales; 26.7 percent held office or clerical positions; 23.9 percent were technicians; and 26.9 percent were operators, crafts people, and laborers. Women occupied 32.3 percent, and members of minority groups 12.8 percent, of all positions.

There has been rapid growth in employment and the number of jobs will continue to increase as more communities award franchises and are wired for cable. Opportunities are generally good in sales, installation, engineering, and administrative positions. As cable systems increase their original programming, opportunities may develop in production.

STV and MDS Operations

Cable TV operations are referred to as systems or plants. STV and MDS operations are often referred to as stations, because they are licensed to broadcast through the air. The video signal either arrives at the station from a satellite or originates locally at the station. It is then broadcast in a modified form on a specially-assigned frequency or by an existing UHF TV station to subscribers' homes, where a special antenna relays the signal to a device ("box") at the TV set. The signal is either unscrambled (STV) or down-converted to a lower frequency (MDS) by the box, and is sent to the subscriber's TV set for normal viewing. Because these stations broadcast over the air waves, they are licensed by the FCC as common carriers or broadcast stations.

STV operations began in a very limited and experimental way during the mid-1960s, and have had considerable growth in the past few years. These stations transmit a scrambled signal to the TV sets of subscribers, who pay to receive the service. The programs received are movies and special entertainment shows, usually on only one channel rather than the multiple channels of a cable system. Operating mainly in areas where cable TV is not entrenched or not available, STV has grown from six scattered stations in 1979 to more than 24 in 1981, serving almost 1.2 million homes. Most analysts predict 45 or more stations by 1984, serving 3.2 million subscribers. Gross revenues should exceed $768 million by then.

MDS operations have only recently appeared as an alternate system. Originally a business communications service, MDS organizations have grown more slowly than STV companies. This technique of transmission broadcasts a signal by microwave over a 20- to 30-mile radius to private homes, multi-family dwellings, and office buildings. MDS stations are less expensive to build than CATV systems or STV operations. Although the FCC currently limits MDS operations to two channels, it is technically feasible for such an operation to transmit 20 or 30 signals simultaneously on adjacent channels. If the FCC authorizes such transmissions in the future, MDS operations could offer a tiered program service, as does cable TV, and become a "cable of the air."

In 1981, about 480,000 subscribers received MDS signals in their homes from 61 operating stations, according to industry analysts at Paul Kagan Associates. Another 62 stations were licensed but not in operation. There are more MDS than STV stations, but MDS stations usually have fewer subscribers.

In both of these telecommunications services, new job openings are already available. As new stations are licensed by the FCC and begin transmission during this decade, employment opportunities will continue to increase, particularly for technical and sales people.

Section 3: Education

Simple audio-visual devices have been used in the United States for more than a century in formal and informal educational settings by teachers who have sought new ways to enrich and supplement traditional classroom and individual learning. Since 1900 the phonograph, radio, magnetic wire and tape recording, stereo sound, programmed learning machines, and, more recently, television, video, and computers have been used for educational purposes. The use of media in schools and colleges grew quite rapidly until the end of World War II, when a veritable explosion occurred. Sparked by the threat of Russian dominance in space, the U.S. Congress passed legislation in 1958 that made the use of educational media to improve instruction a national policy.

Since that time, many traditional AV centers have expanded their functions to embrace all media, including the newer electronic forms, and act as service units for their parent institutions by providing equipment, assisting in curriculum development, producing or acquiring appropriate materials, and maintaining media libraries. Although the emphasis has diminished somewhat in the past ten years, almost every school system, college, and university in the U.S. makes some use of audio-visual technology. Television and video, as well as traditional AV tools, serve the needs of teachers and students at every level of education.

About 2,500 elementary- and secondary-level school district systems produced and used their own instructional television (ITV) programs by the mid-1970s. ITV was available to more than 72 percent of all elementary- and secondary-school teachers in the U.S., and 59 percent used the medium in their classrooms at that time. The main source of TV programming in most elementary and secondary schools today is public television. Nearly all public TV stations transmit a full schedule of instructional programs during the day which are used by teachers as part of their classroom teaching. Videotape machines are also commonly available to many educators. Most school systems and colleges operate circulating libraries of films and videocassettes for teachers to use.

More than 2,100 institutions of higher education used TV for instructional purposes at the close of the 1970s, according to the Corporation for Public Broadcasting. The CPB study indicated that 42 percent of the colleges received their TV programming through a closed circuit television (CCTV) system, linking buildings and classrooms by a closed-loop coaxial cable, somewhat like a commercial cable television system. CCTV systems transmit instructional programs in specific subject areas during the day to classrooms on campus. Some programs are produced in campus television studios by staff production personnel for later transmission over the system. Instructional television fixed service (ITFS) stations are operated by about 500 schools and colleges to transmit and receive educational programs. These stations broadcast to a limited geographic area in which the schools or classrooms are located.

Not surprisingly, Communications and Media Instructors at colleges and universities use video and television in their classroom teaching. Communications Instructors teach television production, introduction to broadcasting, mass communications, and other broadcast-related courses. Media Instructors teach media production, instructional technology, educational television, and related topics. For a list of degree and non-degree programs, see Appendix I.

The opportunities for employment in educational telecommunications have decreased somewhat in the past ten years because of lack of support for education. Still, challenging jobs are available, including ITV Coordinators and Media Librarians who help teachers to maximize the use of instructional television and media in the classroom. There are increased opportunities for Instructional Designers to help faculty improve the teaching-learning process through the use of the new technology. Some of the larger school systems and colleges that operate video production facilities have openings for production, engineering, and technical employees.

Section 4: Private Industry

Industrial and corporate instructors have used traditional audio-visual tools since the early part of this century as aids for employee training, for improving inter-office communications, and for public relations purposes. Corporate use of new technologies grew rapidly during the late 1970s. Many companies, in virtually all industries, have established a media center to house, purchase, and produce media programs, usually with a modest staff. By 1980 there were as many as 13,500 significant users of video programming in business and industry (*Video in the 80s,* Dranov, Moore, and Hickey, Knowledge Industry Publications, White Plains, NY, 1980). In their respected report, Judith and Douglas Brush identified more than 1,000 companies that originate television and video programs on a regular basis (*Private Television Communications: Into the Eighties,* International Television Association, Berkeley Heights, NJ, 1981).

The largest corporate use of TV and video is in the production of training programs. National companies use videocassettes to transmit product information, new sales techniques, and motivational lectures to employees in various locations. Many companies use video and telecommunications to train production-line workers in new methods or to instruct middle-managers in supervisory skills. Still others use the new technology for such internal corporate communications as periodic company video newsletters, on-the-job safety programs, or new employee orientations. Many large companies use television and video for annual stockholders' reports or for public and community relations.

The expenditure for TV and video equipment and programs by private industry is more than the combined total of such expenditures by educational, government, and health organizations. In 1980, the business community spent $1.5 billion for television and video equipment, and that amount is expected to double by 1982. The increase in the use of telecommunications and video by private industry is reflected in the increase in membership of the International Television Association (ITVA)—which consists of individuals involved in corporate television—from 160 in 1971 to more than 2,300 in 1980. With further growth predicted in corporate use of telecommunications and video, there are increasing opportunities for employment in all areas, including production, engineering, and administration.

Section 5: Government

Almost every government agency of any size at the local, state, or federal level operates and maintains audio-visual equipment, often only an overhead or 16mm film projector used for internal training purposes or public information activities. Most establish a media center with small staffs to acquire, store, circulate, and, occasionally, produce media programs. Some government agencies maintain extensive telecommunications and video operations.

There were approximately 3,000 active users of video programs in federal, state, county, and local government in 1980 (*Video in the 80s,* previously cited). The federal government was the largest user, spending nearly $100 million for audio-visual equipment, programs, production, and distribution. At least 28 federal agencies produced 18,815 videotape or videodisc titles during fiscal year 1980, and, of the $61.3 million spent just for media production at the federal level, 54 percent was for videotape and videodisc programs. For the first time, video production was more than twice the volume of motion picture production. According to the National AudioVisual Center (NAC), the executive branch of the federal government owned or leased 1,612 audio-visual facilities in 1980, many of which were used for video production.

Government agencies use TV and video for internal training and public information. The Department of Defense (the largest producer and purchaser of AV materials) produces programs to reach its military and civilian employees around the world with new policy information, medical and health data, and plans for training in specific tasks. At the state level, public health, agriculture, youth services, law enforcement, and transportation agencies use TV and video to reach and instruct employees spread throughout the state, for internal training purposes and as public information tools.

Some state agencies employ as few as five people in all areas of media, with perhaps only one individual serving as the local Audio-Visual Program Officer. Many of the larger federal agencies operate their own production units and employ production, engineering, and other employees in jobs similar to those in commercial TV. With the general cutback in public services, however, the opportunities for employment in government telecommunications are relatively poor.

Section 6: Health

Health care specialists have been using audio-visual materials and, particularly, video and television for a number of years and in a variety of ways. Medical schools make extensive use of videocassettes to demonstrate surgical and medical procedures and to present patient case histories. Hospitals use video and telecommunications along with other media for patient education and entertainment, and for public and community relations. Individual physicians often subscribe to telecommunications networks for continuing education, conferences, demonstrations, and seminar programs that help them to complete state, medical association, or board requirements for Continuing Medical Education (CME) credits.

Many health care facilities and hospitals operate media centers which serve as central acquisitions, storage, and distribution libraries for media support materials. The media center often produces materials. According to some estimates, the field of health care accounts for the fifth highest use of audio-visual and media products, with more than $250 million spent at the close of the 1970s. Almost every institution or hospital unit with more than 250 beds makes some use of media. Nearly one half of the hospitals in the U.S. use TV and video in some form. More than 500 operated closed circuit telecommunications systems for patient use, according to the American Hospital Association (AHA). Other estimates indicate that 80 percent of the 7,000 hospitals in the nation are involved in telecommunications and video programming in some way, and 75 percent produce some of their own programs (Steve Cartwright, "Health Care: a Growing Field," *Educational and Industrial Television,* August 1979). Production is often minimal, however, and limited to a single camera.

Larger medical schools and health care facilities may run a full-scale TV production unit and employ production and engineering as well as administrative staff. Employment opportunities in the health area of telecommunications are only fair.

The Job Descriptions

Part II contains 29 descriptions of jobs in the areas of Video and Telecommunications. The positions are divided among six sections: Consumer Electronics and Home Video; Cable Television, Subscription Television, and Multipoint Distribution Service; Education; Private Industry; Government; and Health.

VIDEO MARKETING MANAGER

CAREER PROFILE

Duties: Responsibility for all national marketing and distribution for a major consumer electronics product or program company

Alternate Title(s): VP, Marketing; National Sales Manager; Director of Marketing

Salary Range: $50,000 to $100,000+

Employment Prospects: Poor

Advancement Prospects: Poor

Prerequisites:

Education—Undergraduate degree in business administration or marketing

Experience—Minimum of five years in consumer electronics marketing

Special Skills—Leadership qualities; organizational ability; good judgment; persuasiveness; business acumen

CAREER LADDER

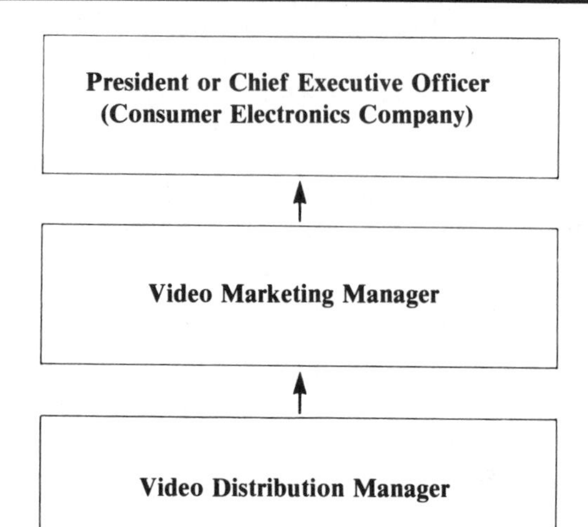

Position Description

A Video Marketing Manager is the top sales executive for a large video-related company. As such, he or she is in charge of all national and, sometimes, international marketing, sales, advertising, and promotion activities associated with the company's consumer electronics products. It is his or her responsibility to aggressively and successfully market the product line(s) or video program label(s) to the home consumer.

There are more than 500 such companies in the United States, producing videocassette and videodisc machines, audio/stereo outfits, video games, electronic accessories, video programs and films, and traditional audio-visual equipment. The majority of these companies distribute their equipment or programs to retailers through a field staff or branch office, or through independent distribution companies of various types. The retailers ultimately sell or rent the products to consumers. Some firms, however, distribute, sell, or rent their products directly to the home user.

The success or failure of the company's total sales efforts depends on the ability of the Video Marketing Manager. It is his or her job to strengthen marketing strategies, improve the visibility and consumer awareness of the equipment or video programs, and to maximize the volume of sales, while maintaining costs. He or she provides marketing research to assist in product development, and creates the marketing strategies to introduce products and support sales. The Video Marketing Manager establishes company-wide policies on pricing, discounts, special sales, rentals, returns and exchanges, and other terms of sale.

The Video Marketing Manager usually reports directly to the company's president or chief executive officer, and is normally located at its national headquarters.

Additionally, the Video Marketing Manager:

- supervises retail sales and distribution channels through company branch offices, independent firms, and authorized dealers;
- oversees all print and radio-TV advertising, promotional literature, billboards, and point-of-purchase displays and other in-store merchandising;
- manages all marketing and sales employees, and often, the advertising and promotion staffs as well.

Salaries

The financial rewards for Video Marketing Managers are considerable, although they vary with the size of the company, its sales volume, and the type of products or programs manufactured or distributed. Salaries range from $50,000 at relatively small companies with limited product lines to well over

$100,000 for senior and experienced individuals at major national and international firms. The total yearly earnings usually include a base salary, bonus arrangement, liberal expense account, stock options, profit-sharing, and extensive life insurance, health, and retirement plans. Some also receive a small commission on overall yearly sales. Membership in social or country clubs, a company car, and other perquisites are often offered as part of a compensation package for a Video Marketing Manager.

Employment Prospects

The opportunities are poor and limited to the 500 or so companies that employ Video Marketing Managers. Not surprisingly, the competition is severe. Only the most able and successful marketing professionals reach the position, and then only after many years of demonstrated ability. Most have had extensive experience as a Video Distribution Manager within the company, or in a related consumer electronics firm.

Advancement Prospects

The opportunities for advancement are poor. For many, the assumption of a Video Marketing Manager position at a major company is the peak of a successful marketing career. Some may move upward to assume overall control of a company as its president or chief executive officer. The majority of such positions in the United States in 1980 were held by individuals with strong marketing backgrounds. Some Video Marketing Managers advance their careers by forming their own consumer electronics firms.

Education

Most Video Marketing Managers have an undergraduate degree in business administration or marketing and many have a master's degree in the field.

Experience/Skills

Most Video Marketing Managers at major companies have obtained the job after five to ten years of product or program marketing experience. In some smaller companies, less experience is required, but an extensive background in advertising and promotion is still necessary.

Video Marketing Managers are poised, confident, business leaders with good organizational skills. Many have a flair for promotion and public speaking, are good conceptualizers, and are extremely persuasive. The successful individual must be capable of making sound business judgments.

Minority/Women's Opportunities

The percentage of women holding top marketing management positions in consumer electronics has increased during the past ten years. Minority-member employment has also risen somewhat in smaller companies, according to industry sources. The position, however, is usually occupied by a white male.

Unions/Associations

There are no unions that represent Video Marketing Managers. Many companies belong to and participate in industry trade associations, including the International Tape/Disc Association (ITA), the Electronic Industries Association (EIA), the National Audio-Visual Association (NAVA), and the Motion Picture Association of America (MPAA). Video Marketing Managers often represent their companies at trade meetings of these groups. In addition, many belong to the American Management Association (AMA) and similar associations, to share mutual concerns and develop policies and procedures.

VIDEO DISTRIBUTION MANAGER

CAREER PROFILE

Duties: Management of an office or firm that distributes video software, hardware, and accessories to retail stores

Alternate Title(s): Branch Office Manager; Independent Representative

Salary Range: $30,000 to $50,000+

Employment Prospects: Poor

Advancement Prospects: Poor

Prerequisites:

Education—Minimum of high school diploma; undergraduate degree in marketing, advertising, or business administration preferable

Experience—Minimum of five years in product or program sales or marketing

Special Skills—Business acumen; leadership ability; organizational aptitude; aggressiveness; persuasiveness; persistence

CAREER LADDER

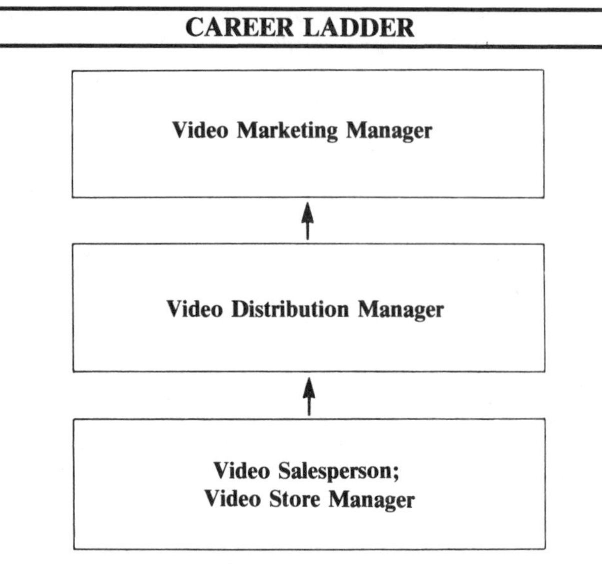

Position Description

The title, Video Distribution Manager, is a general term used to describe a sales executive in the middle of the chain of distribution in the consumer electronics industry. He or she acquires products from manufacturers or wholesalers and sells or rents them to retail stores. He or she is an executive in charge of a sales organization that serves as an intermediary between either a national original equipment manufacturer (OEM), or a national video program wholesaler, and a retail outlet. His or her main responsibility is to establish and meet the sales goals in the territory covered and supervise the activities of Video Salespersons.

The majority of national manufacturers and wholesalers distribute their equipment or programs to retail stores in one of two ways: through company-owned district branches; or through non-affiliated, regional, independent distribution companies of various types. The stores then sell or rent the products to consumers.

A Video Distribution Manager who is employed full time and exclusively by the national wholesaler or manufacturer (but is stationed away from the home office) is usually called a Branch Office Manager. He or she supervises a staff that takes orders, makes sales calls, provides promotional materials, and sometimes stocks the products in retail stores.

Independent companies often act as distribution firms and are under contract by one or more national wholesalers or manufacturers to represent them and to distribute their products or labels. These independent companies may or may not stock any or all of the lines they represent. The Video Distribution Manager in charge of such a company is known as an Independent Representative (Rep), and his or her company usually has exclusive distribution in a given territory or region. He or she supervises a staff that sells and distributes to the retail store.

Most OEMs or national video program wholesalers employ from eight to ten district Branch Office Managers. These executives supervise sales to retail stores in a specific geographic territory (Southwest, Tri-state, etc.). Independent distribution companies also usually operate in a prescribed area.

Additionally, the Video Distribution Manager:

- provides overnight or special catalog order fulfillment to retailers in his or her territory;
- maintains and balances inventory in warehouses or distribution points for re-distribution to retail stores;
- supervises the servicing of all retail accounts and actively seeks and acquires new ones;
- recommends stocking of products, accessories, hardware, and programs to Video Store Managers;

• suggests new products, product lines, or product improvements to a manufacturer or wholesaler.

Salaries

The salary for a Video Distribution Manager who serves as a Branch Office Manager varies according to the extent of his or her territory, the size of the company, its policies, and the individual's experience. Earnings range from $30,000 to $50,000, according to industry sources. Many Branch Office Managers are paid a base salary and a relatively small commission for sales within their territories. Some work on a sales quota and receive a bonus for exceeding it. Others work largely on a commission basis. Programming (software) positions usually pay more than equipment (hardware) positions. Fringe benefits are quite liberal and usually include a reasonable expense account, a company car, and the retirement and health benefits of the parent company.

Most Independent Reps own all or a portion of their distribution company, and receive all or a portion of the profits. They earn considerably more than Branch Office Managers.

Employment Prospects

This is not an entry-level position. Most job openings for Branch Office Managers occur when people move to another industry job. While there is considerable turnover in these sales management positions, the competition is severe, and the prospects for employment are poor. National manufacturers or wholesalers often select bright and aggressive Video Salespersons for the position. Some successful Video Store Managers establish their own distribution operations to service other retailers in their geographic area, thereby becoming Independent Reps.

Advancement Prospects

The opportunities for moving up are generally poor. Some Branch Managers are promoted to a more responsible marketing or sales position such as Video Marketing Manager in their companies' national headquarters. Others may take similar sales jobs at increased salaries at other industry-related firms. Some start their own consumer electronics operation. Independent Reps usually advance their careers by expanding the territories they cover and the product lines they represent.

Education

All national companies that employ Branch Office Managers require a minimum of a high school diploma. Most prefer additional training in sales, sales management, or marketing and sales distribution. Most large national employers require an undergraduate or graduate degree in marketing, advertising, or business administration. The ability to motivate and lead a sales staff, however, is usually more important than a formal education for both Branch Office Managers and Independent Reps.

Experience/Skills

Employers at national companies usually require a minimum of five years of sales or marketing experience in appointing Branch Office Managers. Middle management experience in sales is preferred.

Branch Office Managers and Independent Reps must be good business people with sales and marketing backgrounds. They must display leadership qualities, be aggressive, competitive, and persistent. Most are persuasive and well organized and have respect for profits.

Minority/Women's Opportunities

The percentage of women and minority-group member Video Distribution Managers employed as Branch Office Managers has increased in the past ten years, according to industry sources. Most Branch Office Managers as well as Independent Reps in 1980, however, were white males.

Unions/Associations

There are no unions that serve as bargaining agents for Video Distribution Managers. Some belong to the National Electronic Distributors Association (NEDA) to share trade concerns and advance their careers. Some Independent Reps belong to the Electronic Representatives Association (ERA) for the same purposes.

VIDEO SALESPERSON

CAREER PROFILE

Duties: Selling video programs, consumer electronics equipment, and accessories to retail stores

Alternate Title(s): Sales Representative; Sales Agent; Marketing Representative

Salary Range: $10,000 to $30,000+

Employment Prospects: Good

Advancement Prospects: Poor

Prerequisites:

Education—High school diploma; some college or vocational school courses in sales or marketing preferable

Experience—Minimum of two years in consumer electronics or other sales

Special Skills—Dependability; persistence; persuasiveness; aggressiveness; initiative

CAREER LADDER

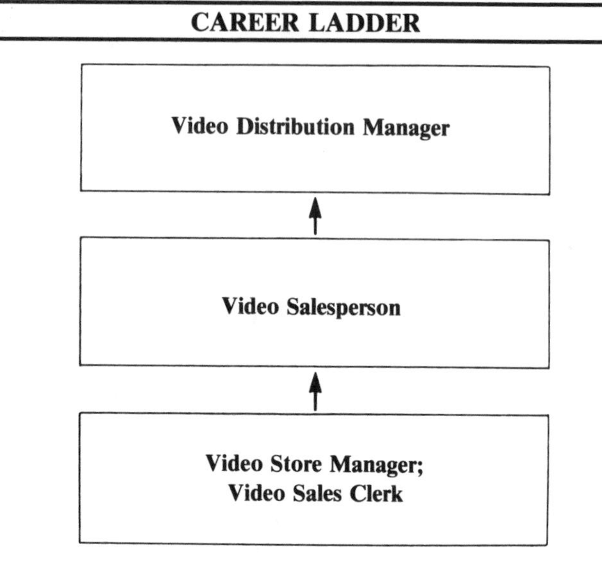

Position Description

A Video Salesperson sells and services video programming, video equipment, and consumer electronics accessories to retail outlets for resale to consumers. Most Salespersons are employed at independent video distribution companies or at a branch office of a national video-related wholesaler or manufacturer. Video Salespersons sell programming, videocassette or videodisc machines, video cameras, projection television units, cables, connectors, and traditional audio-visual equipment, including 8mm and 16mm cameras, projectors, and photographic supplies.

At a branch office, the Video Salesperson concentrates on selling the products of an individual national original equipment manufacturer (OEM) or of a video program wholesaler. A Salesperson for a hardware (equipment) company may handle cameras or videocassette machines and their accessories for a particular equipment company (such as Sony or RCA), whereas a Salesperson for a video program wholesaler sells the videocassette or videodisc releases of a particular label or line (such as Magnetic Video, Paramount, or Disney).

A Video Salesperson working for an independent distribution company usually sells programs or equipment for a variety of national manufacturers or program wholesalers. Many such firms handle products from more than 15 national companies.

A Video Salesperson usually reports to a Video Distribution Manager (either a branch office manager or an independent rep), and either works in a prescribed geographic territory, or is assigned specific retail accounts. In some cases, Video Salespersons report directly to the home office of the national wholesaler or manufacturer.

Most Video Salespersons travel quite a bit, visiting and servicing from 20 to more than 40 retail accounts in their given territory. Some are assigned to phone duty at the branch office or headquarters of the Video Distribution Manager.

Additionally, the Video Salesperson:

- acquires new accounts in his or her territory;
- delivers merchandise to and collects payment from some stores;
- trains retail Video Sales Clerks in the operation of new equipment and accessories;
- distributes sales literature, point-of-purchase displays, brochures, posters, and other merchandising materials to stores.

Salaries

Video Salespersons are paid either a straight salary, a small salary plus a commission, or a draw against commission. There is no industry-wide pattern, but relatively few sales employees work on a straight salary basis. Some companies offer commissions on gross sales, others on gross profits from an individual's sales, and still others on a sliding scale tied to the gross profit received by the company for a particular item.

Compensation ranges from $10,000 for beginners with a small company or distribution firm to $30,000 and more for experienced individuals who deal in expensive products. Since payment is usually tied to sales results in one way or another, yearly earnings are a direct result of the sales success of the individual. As a rule, sales commissions are slightly less than 10 percent of the gross income received by the company for the sale of the product. When commissions are calculated on the gross profit to the company, the return to the Salesperson ranges from 20 percent to 40 percent of that profit on an item.

Most companies provide sales samples, a small entertainment account, and telephone and auto expenses. Some provide a company car, and good hospitalization and pension plans. A few offer the individual an opportunity to participate in a company profit-sharing plan.

Employment Prospects

Many branch offices of the larger national manufacturers employ seven or more Video Salespersons. Independent distribution companies within a region usually employ three to five. In addition, there is considerable turnover of sales employees in consumer electronics. As a result, the job opportunities for the bright, capable sales-oriented individual are good.

Advancement Prospects

Despite the relatively high turnover rate of sales employees in the consumer electronics industry, the opportunities for advancement are relatively poor. The competition for more responsible positions is heavy. Many Video Salespersons use their skills and experience in one product line or label to obtain a better-paying sales job at a competitive company. Some advance within their own company and become Video Distribution Managers (at branch offices) or move to middle-management marketing positions at the home office. Others seek sales positions in other industries.

Education

A high school education and some evidence of course work in sales or marketing at a community college or vocational school is usually required. A few employers prefer to hire individuals with an undergraduate degree. While a major in marketing is desirable, an undergraduate degree in any field is helpful. Some training in motivation, sales techniques, and marketing is also valuable. The ability to sell, however, is more important than a formal education.

Experience/Skills

Although the position of Video Salesperson is occasionally an entry-level job, individuals who have at least two years of direct sales experience are usually preferred. Some Video Salespersons gain this experience in fields other than consumer electronics, while others have worked as Video Store Managers or Video Sales Clerks.

A Video Salesperson must be dependable and display a considerable amount of initiative. Persistence and persuasiveness in sales efforts, an outgoing, gregarious personality, and a service orientation, are required. Above all, a Video Salesperson must like to sell and to make money.

Minority/Women's Opportunities

The percentage of women occupying video sales positions has increased dramatically in the past few years, according to industry observers. While the number has increased more in the programming (or software) areas, a number of women are now also Video Salespersons for equipment manufacturers. The percentage of minority-group employees, however, has not increased as dramatically, except in urban areas.

Unions/Associations

There are no unions that serve as bargaining agents for Video Salespersons. Many individuals attend company-sponsored sales training sessions, or those sponsored by the Consumer Electronics Group (CEG) of the Electronic Industries Association (EIA) to learn new sales techniques and methods.

VIDEO STORE MANAGER

CAREER PROFILE

Duties: Management and operation of a video or consumer electronics retail store

Alternate Title(s): Video Retailer; Video Dealer

Salary Range: $15,000 to $30,000 +

Employment Prospects: Fair

Advancement Prospects: Poor

Prerequisites:

Education—Minimum of high school diploma

Experience—Minimum of two years of experience in retail sales

Special Skills—Video product knowledge; sales talent; business acumen; merchandising instinct

CAREER LADDER

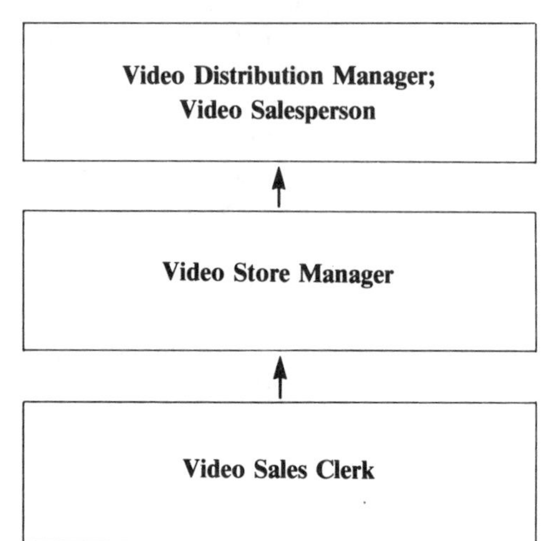

Position Description

A Video Store Manager is responsible for the day-to-day operation and management of a retail store that deals in consumer electronics and related accessories and supplies. He or she is in charge of all income and expenses, all short- and long-range planning, and the general profitability of the store. A Video Store Manager may be the owner or part-owner of an independent store. He or she may manage the daily operation of an outlet that is part of a chain, or may be the head of a video-TV-record department within a department store.

More than 5,000 stores specializing in video have begun operations within the last four years. Many emphasize prerecorded programs and films on videocassette and videodisc (software), and carry equipment as a sideline. Others concentrate on equipment (hardware), and carry programming and accessories as adjuncts. In addition, there are several types of consumer stores that handle video equipment, including photographic stores serving both school and home markets, traditional music record outlets, audio component or appliance stores, and major department stores.

In any situation, however, a Video Store Manager directs the operation and supervises all Video Sales Clerks and Video Service Technicians. In some stores, the Manager acts as the video buyer, selecting what products are carried. He or she is ultimately responsible for the success or failure of the enterprise.

Additionally, the Video Store Manager:

- develops advertising and promotional campaigns, and in-store displays;
- establishes special sales and discount policies;
- determines the level of investment in inventory of hardware, accessories, and programming, and maintains inventory controls;
- stays informed of industry trends, new products, and new rental, sales, and returns policies.

Salaries

Earnings for Video Store Managers vary considerably, according to the geographical location and size and profitability of the store. The Manager who is also the owner usually collects a weekly salary and shares in or retains any profits the store returns during the year. Such individuals earn $30,000 or more annually, depending on the store's success in a given year. Those who do not own a share of the store are usually compensated by a combination of salary and commission or yearly bonus. The relatively inexperienced Video Store Manager receives approximately $15,000 yearly as compensation.

Employment Prospects

The opportunities for employment as a Video

Store Manager are fair. It is not an entry-level position and most Managers are owners or co-owners of a store and have financial and sales experience in the retail trade. Although some young entrepreneurs have operated video stores successfully, many such outlets have folded in the past year because of undercapitalization, competition, or their owners' lack of experience. Some successful owners of consumer electronics stores, however, expand by opening other stores and promoting capable Video Sales Clerks to the new positions. Department stores sometimes promote bright and successful Video Sales Clerks with leadership qualities to handle the video, audio-visual, or record departments.

Advancement Prospects

The prospects for advancement from the position are generally poor; there is considerable competition. For many Video Store Managers, the position is the successful realization of a desire to be one's own boss by owning and operating a store. They may seek advancement and additional success by owning and operating more stores. Some salaried managers use the contacts they have acquired to advance their careers by becoming Video Salespersons of particular consumer electronic products or labels. A few become Video Distribution Managers for national wholesalers and manufacturers. Managers who work for large department stores sometimes move to more responsible middle-management positions in their store or elsewhere in the chain.

Some owner/managers in the audio-visual industry participate in the rigorous certification program of the National Audio-Visual Association (NAVA). Becoming a NAVA Certified Media Specialist can help in career advancement.

Education

Most employers require a minimum of a high school diploma in assigning or promoting people to this position. In some department store chains, an undergraduate degree in management or retailing is required.

Experience/Skills

The majority of Video Store Managers in the consumer electronics industry have had at least two years of experience in retail sales prior to assuming the position. Day-to-day experience as a Video Sales Clerk is a prerequisite for the job.

Video Store Managers must be well-organized and display leadership qualities. In addition, an entrepreneurial spirit, initiative, and sound business judgment are required. Video Store Managers should have a thorough knowledge of the products they sell. They are experts in audio-visual and video equipment and programming, understand local tastes and desires, and are good merchandisers. Above all, they are excellent sales people who enjoy selling to and serving retail customers.

Minority/Women's Opportunities

Women have become increasingly involved in video retail management in the past three years. While the majority of positions are held by white males, women have assumed more leadership roles, particularly as co-owners, according to industry observers. The percentage of minority-group members in video store management, however, remains rather low, except in major cities.

Unions/Associations

There are no unions that represent Video Store Managers. Many, however, belong to such associations as NAVA, the National Association of Recording Merchandisers (NARM), or the National Association of Retail Dealers of America (NARDA) to keep abreast of the industry and share common concerns. Many Managers and store owners attend the annual trade exhibitions and seminars of the Consumer Electronics Group (CEG) of the Electronic Industries Association (EIA). The semi-annual Consumer Electronic Shows (CES) sponsored by the EIA offer the largest display of new electronics products and programs available for resale.

VIDEO SALES CLERK

CAREER PROFILE

Duties: Selling video hardware and software at the retail level

Alternate Title(s): None

Salary Range: $8,000 to $14,000

Employment Prospects: Good

Advancement Prospects: Fair

Prerequisites:
Education—High school diploma
Experience—Some retail sales background preferable
Special Skills—Good interpersonal skills; sales aptitude; initiative

CAREER LADDER

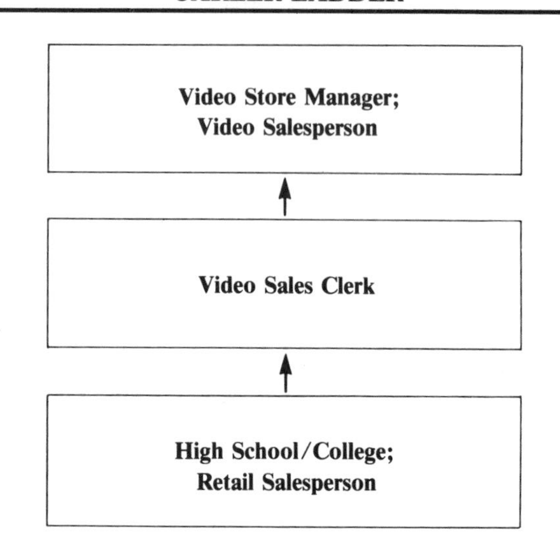

Position Description

A Video Sales Clerk is responsible for selling video hardware (equipment) and software (programming) to customers at a retail store. In many instances, the Sales Clerk also sells video and audio-visual accessories, including connecting boxes, cables, blank videocassettes, bulbs, reels, and other supplies. In stores that specialize in video programming, the Sales Clerk is involved in the rental of videocassettes and videodiscs for customer use.

A Video Sales Clerk usually reports directly to the Video Store Manager or owner. In small stores, the Clerk is responsible for selling or renting all displayed merchandise. In larger operations, he or she may specialize in either software or hardware or in a particular line of equipment.

It is the Video Sales Clerk's responsibility to service customer requests and to inform the customer about the merchandise offered by the store. He or she outlines the various options available, demonstrates equipment and programs, and persuades the customer to rent or buy.

Video Sales Clerks operate in a dynamic environment that can be challenging and rewarding. New products, new equipment, or new label lines or programs are constantly being introduced, as are new merchandising techniques, discounts, and special sales. In addition, different people move through the store daily. It is the Video Sales Clerk's responsibility to balance all of these factors and devise new sales methods that meet customers' needs and that increase sales for the store.

Video stores are located in both suburban shopping malls and in major urban centers. A Video Sales Clerk usually works in one location, but in multiple-store companies, he or she may be transferred periodically to another branch.

Additionally, the Video Sales Clerk:

- studies the operation or user characteristics of each new product offered for sale or rental by the store;
- assists in setting up in-store and window merchandising displays;
- maintains inventory records and sales files.

Salaries

Salaries for Video Sales Clerks are relatively low. Most employers start beginners with a small salary and a relatively large commission on each sale. As the individual progresses, the salary is raised and commissions are lowered. Some stores operate on a draw-against-salary plan, while others offer a straight salary. Most stores have various incentive sales plans and rewards in order to motivate employees. Most Video Sales Clerks are paid on an hourly basis.

In 1981, the compensation for Video Sales Clerks ranged from $8,000 for beginners and inexperienced full-time individuals to $14,000 for experienced employees with good track records, according to industry sources. Benefits are usually minimal, with only the larger stores and chains offering medical and retirement plans.

Employment Prospects

The Video Sales Clerk's position is usually an entry-level job at most retail stores. The majority of stores employ from two to four workers in sales for various shifts and assignments. There is a considerable amount of turnover among Video Sales Clerks and as a result, employment opportunities are good. Many large department stores provide in-service training in sales merchandising for apprentices. Others seek bright, aggressive high school or college students to work part-time or on weekends. After graduation, they are often employed full-time.

Advancement Prospects

Most Video Sales Clerks seek advancement in video or audio-visual retail sales or in other consumer electronics fields. The opportunities for advancement are fair. Some diligent individuals are appointed video or appliance department managers at larger chain operations, and others move into branch management for a video store chain that specializes in particular hardware or software products. Some aggressive Video Sales Clerks use their retail sales experience to move to more responsible positions as sales manager at a store that sells products unrelated to video or audio-visual equipment. A few advance their careers by becoming Video Salepersons for a particular consumer electronics product or manufacturer's line. Some sales personnel in the audio-visual field participate in the Certified Media Specialist Program of the National Audio-Visual Association (NAVA). Such certification is valuable in obtaining more responsible positions within the industry.

Education

A high school diploma is usually required. Some junior college or college course work is preferred by a few larger department store chains. The ability to sell, however, is more important than a formal education.

Experience/Skills

Some previous experience in retail sales in any field is often required. Most Video Sales Clerks have an outgoing personality, an ability to relate to others, initiative, and an aptitude for selling. A neat appearance and a gregarious manner are helpful. The most successful individuals are friendly and service-oriented and somewhat competitive and aggressive. Above all, a good Video Sales Clerk must like to sell and be interested in making money.

Minority/Women's Opportunities

The opportunities for women and minority-group members as Video Sales Clerks are good. There has been a marked increase in employment opportunities for both in the past ten years, according to industry observers. While minority males are more often found in retail stores in urban centers that emphasize video and audio-visual equipment, more than 50 percent of all employees in stores that deal mainly in video software, 8mm/16mm films, or photographic supplies are white or minority females, according to industry sources.

Unions/Associations

There are no unions that represent Video Sales Clerks. Some individuals, however, belong to NAVA to advance their careers. In addition to the certification program mentioned above, NAVA (in association with Indiana University) holds yearly training seminars in sales and new products. Other Video Sales Clerks attend periodic sales conferences and seminars held by the National Association of Retail Dealers of America (NARDA) or the Electronic Representatives Association (ERA) to improve their skills.

VIDEO SERVICE TECHNICIAN

CAREER PROFILE

Duties: Repairing and servicing of equipment at a consumer electronics retail store

Alternate Title(s): Maintenance Technician; Bench Technician

Salary Range: $11,000 to $20,000+

Employment Prospects: Excellent

Advancement Prospects: Good

Prerequisites:

Education—High school diploma; some vocational or technical school training

Experience—Minimum of six months of maintenance experience preferable

Special Skills—Mechanical ability; electronic aptitude; manual dexterity; inquisitiveness

CAREER LADDER

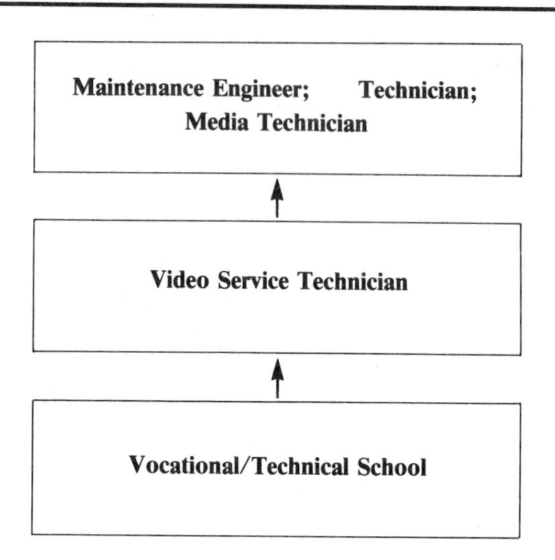

Position Description

A Video Service Technician is responsible for the maintenance and repair of equipment in a consumer electronics store and usually reports directly to the store's owner. He or she works on customers' videocassette recorders and videodisc machines as well as other audio-visual equipment such as 8mm and 16mm film cameras, slide projectors, and stereo systems. This individual also maintains the technical equipment owned by the store. Today's Video Service Technician must be familiar with the digital electronics, microprocessors, and integrated circuitry used in highly complex electronics gear.

He or she is also responsible for setting up and maintaining a comprehensive repair service that can become a profit center for the store. The presence of a good repair department can bring customers back for new purchases and help to build the store's reputation. From one to five Service Technicians are employed in an average retail outlet. In some instances, a central group of Video Service Technicians operates out of one main store and repairs equipment sent to them from branch locations.

Additionally, the Video Service Technician:

- interprets diagrams and schematic charts of video equipment;
- attends training programs to keep abreast of new equipment and service techniques;
- works with manufacturers' field service representatives or design engineers to isolate equipment problems;
- maintains an inventory of supplies, accessories, and test gear used in servicing equipment.

Salaries

Video Service Technicians are usually paid on an hourly basis. The range is from a low of under $5.00 an hour to a high of $12.00, according to the Electronics Industry Association (EIA). The national average in 1981 was $7.65 an hour. The annual range is from $11,000 for beginners to more than $20,000 for experienced employees in large cities.

Salaries often depend on the geographical location of the store and the competition from other outlets in the area. Video Service Technicians are also usually rewarded financially for completing factory repair certification training or other in-service training programs. A few large chains offer excellent benefit programs. There are often many opportunities for overtime work.

Employment Prospects

The opportunities for employment are excellent. According to the EIA there are more than 200,000 repair people who work with consumer electronics equipment. There are far more positions open than there are candidates and, with the burgeoning home video market, there will be even more opportunities in the future. Many Video Service Technicians move to other firms at a higher salary, leaving positions open for promising newcomers. With more than 35,000 retail stores, chains, department stores, and other outlets selling video, audio-visual, stereo, and photographic equipment, the opportunities for employment are bright.

Advancement Prospects

Some Video Service Technicians with a talent for sales work, considerable experience, and an entrepreneurial bent open their own retail stores. Many improve their skills and, with further training, advance their careers by moving into broadcast-related electronics as a Maintenance Engineer or become Technicians at cable TV systems or multipoint distribution service (MDS) or subscription television (STV) operations. A few seek advancement and security as Media Technicians in educational institutions.

Education

A high school diploma and some evidence of vocational or technical school training is usually required. Course work in the basic sciences, algebra, and physics is helpful. Drafting courses that help an individual understand spatial relationships and physical characteristics are also useful. There are more than 475 public, private, and correspondence schools offering consumer electronics servicing courses, according to the EIA.

Experience/Skills

Although six months to one year in electronic or audio-visual repair work is preferred, lengthy experience is not necessary in obtaining an entry-level position. Many employers hire recent vocational or technical school graduates and train them on the job.

Most Video Service Technicians serve an apprenticeship. After two years of technical training or four years of work experience, an individual can qualify for a voluntary examination leading to certification as one of the 10,000 Journeyman Certified Electronic Technicians working under the auspices of the International Society of Certified Electronic Technicians (ISCET). Another 5,000 students and beginning employees who do not have the experience for a Journeyman certificate have passed an Associate Level test and are so certified by the ISCET.

Some experience in the operation of video, photographic, audio, and audio-visual equipment is necessary. A Video Service Technician needs an inquisitive mind and should enjoy solving technical problems. He or she should also have good manual dexterity and an aptitude for electronics.

Minority/Women's Opportunities

Although historically the job has been filled by white males, the percentage of minority-group members holding the position has increased dramatically in the past few years, according to industry observers. The Consumer Electronics Group of the EIA has sponsored minority training centers for beginning technicians as part of its Service Technician Development Program. The opportunities for women are excellent, but relatively few train or apply for positions as Video Service Technicians.

Unions/Associations

There are no unions that serve as bargaining agents for Video Service Technicians. Some belong to the National Audio-Visual Association (NAVA) and attend seminars and certification programs held at Indiana University to advance their careers. Other people are members of the Institute of Electrical and Electronics Engineers (IEEE), the National Association of Television and Electronic Servicers of America (NATESA), or the National Electronic Service Dealers Association (NESDA). Some also belong to the Audio-Visual Technicians' Association (AVTA) to share concerns and advance their careers.

SYSTEM MANAGER

CAREER PROFILE

Duties: Responsibility for the daily management and operation of a cable TV system or an MDS or STV station

Alternate Title(s): General Manager; Station Manager

Salary Range: $18,000 to $50,000

Employment Prospects: Good

Advancement Prospects: Good

Prerequisites:

Education—Post-high school training; undergraduate degree preferable

Experience—Minimum of three years in business management, preferably in telecommunications

Special Skills—Organizational skills; leadership qualities; good judgment; technical knowledge; sense of responsibility

CAREER LADDER

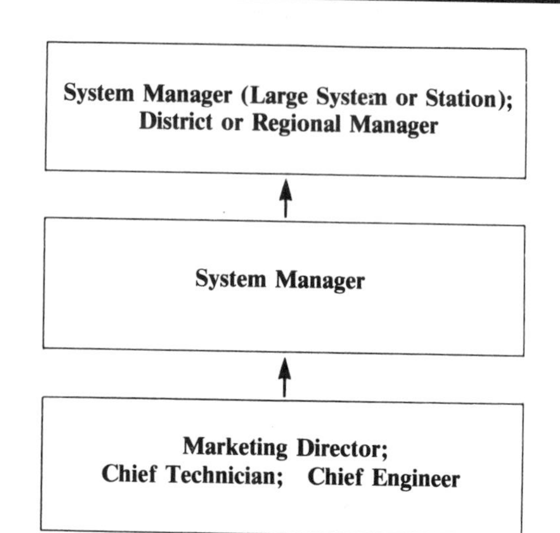

Position Description

A System Manager of a cable television system or a multipoint distribution service (MDS) or subscription television (STV) station is in charge of the overall management of the operation, including income and expenses, programming, sales, and engineering. He or she is responsible for maintaining good relationships with local governmental agencies, the community, and all outside contacts, and directly supervises all departments of the organization. As the chief operating officer of a small telecommunications firm, a System Manager is responsible for managing a program service that may reach from 2,500 to more than 50,000 subscribers. In cable TV, the person in this position is referred to as the System Manager and in STV or MDS operations as the General Manager or Station Manager, but the functions are similar.

The System Manager determines the types of basic program and pay TV services to be offered to customers, the number of channels to be used, and the charges for services. Another important function is the planning for new facilities such as satellite receiving stations, increased channel capacity, and production equipment. He or she must plan and budget so that the operation can keep up with technology, increase its subscriber base, and earn a healthy profit.

A System Manager oversees the operation of a cable system or an MDS or STV broadcast station in accordance with all Federal Communications Commission (FCC) regulations and with all other federal, state, and local laws. He or she keeps informed about construction laws, tariffs, and access charges, and works closely with local utility companies and governmental agencies to provide a profitable public service to the community.

A System Manager is sometimes an owner or part-owner of the company. In other cases, he or she is a salaried employee who reports to a board of directors. In many organizations, he or she is part of a management team that represents a group owner or, in cable TV, a multiple system operator (MSO). A System Manager may also serve as the local manager for one or two of a company's systems in a particular geographic area.

Additionally, the System Manager:

- forecasts monthly and annual income and expenses;
- evaluates and approves all marketing and sales efforts;
- hires all employees and monitors their performance.

Salaries

Salaries for System Managers depend on the size

of the operation and its geographic location. Salaries in cable TV systems ranged from $18,000 in the smaller systems to $50,000 in major markets with more than 40,000 subscribers in 1980, according to the National Cable Television Association (NCTA). Station Managers in STV and MDS companies were paid somewhat less in 1980, according to industry observers. Many companies offer excellent benefit plans, including a company car, health and pension plans, and profit sharing. Most offer bonus plans and reasonable expense accounts.

Employment Prospects

Opportunities for bright, diligent individuals are good. An estimated 7,000 U.S. communities do not have a cable system or an STV or MDS station, and most observers predict that such operations will be installed in many of them in the next few years. Moreover, the turnover and attrition rate in the fast-growing industry is high. The position is not an entry-level job, however, and while many new operations are opening, it is usually only the person with some experience who is selected as System Manager.

A talented Marketing Director, Chief Technician, or (from broadcasting) Chief Engineer can often become a System Manager.

Advancement Prospects

The opportunities for advancement are good, but the competition is heavy. Many System Managers move to larger, competing systems or stations in other markets with more responsibilities and at higher salaries. Some successful Managers are transferred within their own organization to larger existing systems or to other communities to establish new franchises or licenses. Other people are promoted to district or regional manager positions within the company, where they are in overall charge of all systems or stations in a geographic area. Still others move to middle management positions at their parent companies' headquarters.

Education

A high school diploma is required and some vocational, community college, or college training is recommended. As in most management positions in telecommunications, an undergraduate degree is extremely useful. Most System Managers of national MSOs have undergraduate degrees in business administration or marketing. A few have degrees in communications, computer science, or electrical engineering.

Experience/Skills

A minimum of three years in general business management, but preferably in telecommunications, is required. Many System Managers have been Chief Technicians in cable, MDS, or STV or Chief Engineers at broadcast stations where they gained considerable experience in engineering practice. Increasingly, individuals with marketing backgrounds are being sought as System Managers.

A candidate should be a self-starter with initiative and leadership qualities. Sound business judgment, acceptance of responsibility, a respect for profits, and organizational skills are required, as are poise and social skills for dealing with a variety of community leaders and agencies.

Minority/Women's Opportunities

According to FCC statistics, 23.9 percent of all officials and managers including System Managers in cable television were white females in 1980, and 4.3 percent were minority-group males. The percentage of women in management positions is much higher in cable television than in other segments of the telecommunications industry. Because cable is still largely a rural operation, however, there is a relatively low percentage of minority employment. Since STV and MDS companies are located in urban areas, the percentage of minority representation has risen in such firms, according to industry observers.

Unions/Associations

There are no unions that serve as bargaining agents for System Managers. Most cable companies belong to the NCTA, and many System Managers represent their companies in that association. Many cable System Managers also belong to a state or regional cable association, and some belong to the Cable Television Administration and Marketing Society (CTAM) and the Community Antenna Television Association (CATA) to share mutual concerns and advance their careers.

Managers of STV or MDS companies often belong to the Subscription Television Association (STA) or to the National Association of MDS Service Companies (NAMSCO) to share common industry concerns.

OFFICE MANAGER

CAREER PROFILE

Duties: Responsibility for the office and business functions at a cable TV system or an STV or MDS station

Alternate Title(s): Accounts Receivable Clerk

Salary Range: $13,000 to $17,000

Employment Prospects: Good

Advancement Prospects: Good

Prerequisites:

Education—High school diploma and some business/vocational school

Experience—Minimum of one year in office management and bookkeeping

Special Skills—Accuracy; reliability; organizational ability; good interpersonal skills

CAREER LADDER

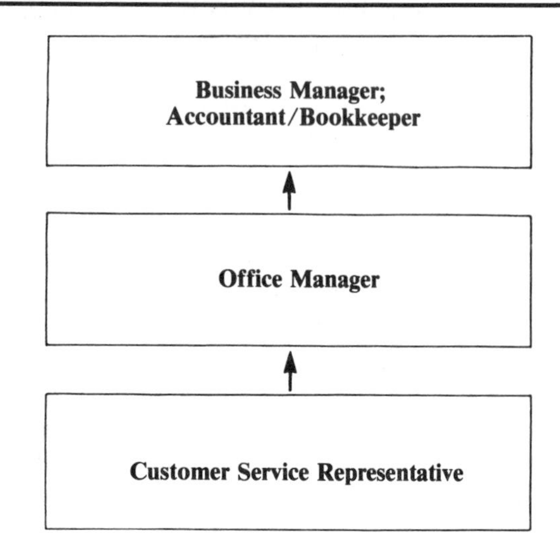

Position Description

The Office Manager of a cable TV system or a subscription television (STV) or multipoint distribution service (MDS) company is responsible for the smooth operation of daily business activity. The major function of the position is to relieve the System Manager from as many routine day-to-day office obligations as possible, and to coordinate all financial and office functions so that they support the marketing and engineering staffs. The Office Manager also works as a troubleshooter in solving specific office problems.

The Office Manager is second in command of the business aspects of the operation and is in charge of many of the financial transactions of the company. One of his or her primary responsibilities is to monitor the accounts receivable status to ensure that money that is owed to the company is paid as promptly as possible. He or she prepares regular accounts receivable reports in association with the marketing and sales department. The Office Manager also supervises monthly billing of subscribers and authorizes service disconnections for non-payment. The accurate maintenance of all of the company's financial records and journals is an additional function of this position.

In most circumstances, the Office Manager reports directly to the System Manager. In some larger organizations, he or she reports to a business manager or the senior accountant/bookkeeper.

The Office Manager usually directly supervises from one to more than ten people, depending on the size of the operation. The majority of local cable systems and STV and MDS organizations are small business enterprises, employing as few as five full-time workers. At such companies, the Office Manager, along with a Customer Service Representative often performs the duties of an accounts receivable clerk, a service dispatcher, a billing clerk, and an accountant/bookkeeper. In larger operations or station organizations, as many as six individuals may perform these functions.

In whatever situation, the Office Manager is vital to the day-to-day operation of the organization, and is the principal middle-management coordinator, ensuring that the company's overall business activities mesh smoothly with the efforts of the Marketing Director and sales staff.

Additionally, the Office Manager:

- supervises the activities of the customer service department, assuring prompt handling of all accounts and complaints;
- selects, hires, trains, and supervises office staff;
- maintains all employee payroll records;
- schedules the work of Installers and Technicians, in association with the Chief Technician.

Salaries

Despite the responsibility of the position, salaries for this job in cable TV are relatively low, ranging from $13,000 to $17,000 in 1980, according to the National Cable Television Association (NCTA). Salaries at MDS and STV companies in 1980 were comparable, according to industry sources.

Employment Prospects

While this is not an entry-level job, there are openings for capable individuals. Positions open up frequently as Office Managers are promoted or move to more responsible jobs in broadcast, media-related or non-media related companies. Some experienced, talented Customer Service Representatives are promoted to the position. The opportunities in general for agressive, bright individuals are good, and the skills required are readily transferable to other businesses.

Advancement Prospects

The possibilities for advancement for bright individuals are good. Some Office Managers at large firms are promoted to the positions of business manager or accountant/bookkeeper at their own company or at other stations or systems owned by the parent organization. Some take more responsible financial or office supervisory jobs at new systems or STV or MDS stations. Still others use their office management skills to obtain better-paying positions in non-media-related industries.

Education

A high school diploma and a year or two of training at a business or vocational school are required. Courses in bookkeeping and accounting are essential. Some training in electronic data processing or management information systems is helpful, inasmuch as the sales, billing, and accounting records in most of the companies are computerized.

Experience/Skills

One or two years of actual office management and bookkeeping is required. Experience in television, cable, or the newer telecommunications industries is desirable, but background in a responsible position in a non-video-related industry is useful.

Office Managers are usually bright, young, well organized, and aggressive individuals who can handle complicated tasks easily and accurately. The ability to balance conflicting demands and requests routinely is a must, as is reliability. The Office Manager must also have good interpersonal skills and be capable of supervising office workers. A familiarity with office calculating machines is mandatory.

Minority/Women's Opportunities

Opportunities for women are excellent. According to Federal Communications Commission (FCC), 81.1 percent of the office and clerical employees (including Office Managers) in cable TV in 1980 were white females. While this figure is slightly lower than for comparable positions in broadcasting, the vast majority of office workers in cable are women. The percentage of minority males in these positions was 1.1 percent in 1980.

Unions/Associations

There are no unions or professional organizations that represent Office Managers in cable television or STV or MDS operations.

CUSTOMER SERVICE REPRESENTATIVE

CAREER PROFILE

Duties: Coordinating installation and service requests and complaints at a cable TV system or STV or MDS company; converting inquiries into orders

Alternate Title(s): None

Salary Range: $11,000 to $17,000

Employment Prospects: Good

Advancement Prospects: Good

Prerequisites:

Education—Minimum of high school diploma

Experience—Minimum of one year of office work

Special Skills—Detail ability; numerical aptitude; organizational capability; service orientation; sales talent

CAREER LADDER

Office Manager

↑

Customer Service Representative

↑

High School/Business School; Clerk; Secretary;

Position Description

A Customer Service Representative coordinates all customer requests for service with the sales, marketing, and engineering staffs of a cable television system or a subscription television (STV) or multipoint distribution service (MDS) station. The Representative is essential in helping to maintain successful and profitable rapport with potential and current subscribers. He or she usually reports directly to the Office Manager in small organizations, and, increasingly, to the Marketing Director in larger ones. In either case, this is usually not a supervisory position.

The Representative receives all requests and complaints concerning information, prices, and services from subscribers and makes certain they are dealt with promptly. He or she must also be able to convert a common inquiry into a firm sale and to upgrade any installation request into a higher-priced service, if possible. The Representative also discourages disconnect requests through the application and reinforcement of sales strategies.

At a small system or station, a Customer Service Representative may deal with as many as 2,500 subscriber orders during a year. In many small firms, he or she also performs the duties of a billing clerk (invoicing subscribers), a work order control clerk (expediting installations and service calls), or service dispatcher (scheduling Installers and Technicians). In larger companies, the Representative concentrates on providing over-the-counter and telephone responses to potential customers or current subscribers. Larger companies often have more than one Customer Service Representative.

In small systems with few or no sales personnel, Customer Service Representatives are crucial to subscriber growth. In addition to performing public relations, sales, and service roles, they ensure the timely coordination between subscribers' orders and requests and the company's proper fulfillment and response. The Representative often provides a vital and primary link between the marketing/sales and business/financial departments of a large telecommunications company.

Additionally, the Customer Service Representative:

- maintains accurate records of all calls, and dispatcher and work orders;
- processes all phone-in or over-the-counter orders from new customers and assigns them to appropriate Salespersons;
- prepares preliminary invoices for monthly billings to subscribers and calls delinquent accounts for payment.

Salaries

Salaries for Customer Service Representatives in

cable TV are relatively high. Experienced individuals earned from $11,000 to $16,000 per year in 1980, according to the National Cable Television Association (NCTA). For those who also performed billing clerk functions, the 1980 salaries were $14,000 to $17,000, also according to the NCTA. Salaries in STV and MDS stations were comparable, according to industry sources.

Employment Prospects

The position is often an entry-level job. Many diligent clerks and secretaries are promoted to Customer Service Representative. Because of the growing number of telecommunications companies, and the relatively high turnover rate in this job, there are many opportunities for employment. The recent high school or business school graduate should also find it relatively easy to become a Customer Service Representative.

Advancement Prospects

The opportunities for advancement are reasonably good for those with some initiative and experience. A person can learn many of the fundamental aspects of customer service, telecommunications operations, and general office functions. As the company grows, many Customer Service Representatives are eventually promoted to more responsible jobs in various intermediate supervisory roles. After some years of experience, an individual may move into the position of Office Manager within the company or at a competing firm. Still others obtain more responsible jobs at higher salaries in non-media, service-related industries.

Education

A high school diploma is a minimum requirement for the job. Individuals with some business or vocational school training are preferred. Courses in computer operations and training in management information systems can be extremely helpful.

Experience/Skills

Although the position is often considered an entry-level job, candidates are usually expected to have had at least one year of office experience. A working knowledge of standard office equipment is required and some experience with computers or word processors is helpful.

Most Customer Service Representatives are bright young people who have a penchant for detail and an ability to work with numbers. They are organized and capable of dealing with a variety of tasks and people on a day-to-day basis.

Interpersonal skills in managing and coordinating internal logistics between departments are necessary and a service-oriented and pleasing personality is helpful. A Customer Service Representative should possess good telephone manners, tact, and courtesy in dealing with subscribers and potential customers. He or she also must be able to reinforce sales strategies during over-the-counter or telephone contacts. An ability to sell a service is mandatory in obtaining the position.

Minority/Women's Opportunities

In the majority of cable TV systems and STV or MDS firms, the position of Customer Service Representative is held by white females. Of all workers in the office clerical category (which includes Customer Service Representatives) in cable TV in 1980, 81.7 percent were white females, according to the Federal Communications Commission (FCC). Minority-group members of both sexes occupy fewer positions, but their representation is strongest in urban areas.

Unions/Associations

There are no unions or professional organizations that represent Customer Service Representatives at cable TV systems or at MDS or STV companies.

MARKETING DIRECTOR

CAREER PROFILE

Duties: Developing and coordinating of all marketing activities for a cable TV system or an MDS or STV station

Alternate Title(s): Marketing Manager

Salary Range: $20,000 to $35,000+

Employment Prospects: Good

Advancement Prospects: Good

Prerequisites:

Education—Undergraduate degree in marketing, advertising, or communications

Experience—Minimum of three years of experience in telecommunications sales or marketing

Special Skills—Aggressiveness; leadership qualities; organizational skills; outgoing personality; promotional talent

CAREER LADDER

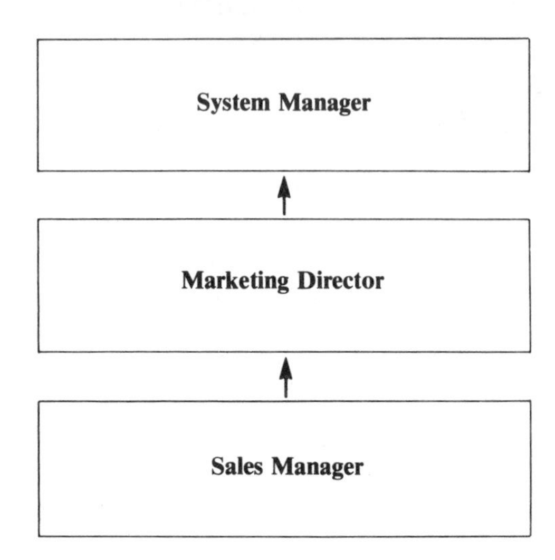

Position Description

A Marketing Director of a cable TV system or of a subscription TV (STV) or multipoint distribution service (MDS) operation is responsible for providing the income that allows the company to operate profitably. He or she is responsible for all marketing, promotion, publicity, and advertising activities and for all direct sales to subscribers.

The major objective of a Marketing Director is to obtain—and retain—as many customers as possible for the organization. He or she develops sales campaigns that will reach a reasonable percentage of the potential subscribers in the system's or station's area. The byword of these consumer-oriented organizations is customer satisfaction. The viewer "votes" every month by the payment of his or her bill or by a request for a disconnect. It is an on-going and complicated marketing challenge to attract, connect, satisfy, and retain subscribers, and thus to maximize the volume of sales, without increasing the costs to the company.

One of the most important duties of a Marketing Director is to select and train the sales staff. He or she is in day-to-day charge of from two to ten Salespersons, and is responsible for coordinating and monitoring all sales activities. The Marketing Director also conceives and develops sales tools, including print brochures and advertising displays, to assist all sales efforts.

The Marketing Director often performs important programming duties. In cooperation with the System Manager, he or she researches and selects the types of basic and premium pay television services the company will offer subscribers. The Marketing Director may also pick programs from existing national satellite services, negotiate with pay-TV program companies, or deal directly with suppliers to obtain particular programs.

Marketing Directors report directly to the System Managers. Not all companies, however, employ Marketing Directors. They are usually found only at systems that have 10,000 or more subscribers. At smaller operations, the duties are either performed by the System Manager alone or are divided between the System Manager and the Sales Manager.

Additionally, the Marketing Director:

- establishes rates, discounts, and special offers for program services, in consultation with the System Manager;
- recommends and researches new subscriber services, such as pay-per-view programming, home security, and interactive customer response systems;
- designs and monitors research to determine subscriber satisfaction.

Salaries

Salaries for Marketing Directors in 1980 ranged from $20,000 at small and medium size cable TV operations to more than $35,000 for experienced individuals with seniority at systems with a large number of subscribers, according to the National Cable Television Association (NCTA). A company car, a bonus or profit-sharing plan, a reasonable expense account and a health and pension plan are usually provided. In some instances, the Marketing Director is also paid a small commission on overall yearly sales above quotas. Salaries at STV and MDS operations were somewhat lower during 1980, according to industry observers.

Employment Prospects

Opportunities at a cable TV system or at an STV or MDS station are good. While the position is not an entry-level job and requires some experience in telecommunications marketing, there is considerable turnover. In addition, the dramatic growth of these industries creates opportunities at new systems and stations each year for alert and qualified individuals. Some talented, experienced Sales Managers are promoted to Marketing Directors.

Advancement Prospects

Opportunities are good, but the competition is strong. Most Marketing Directors are ambitious and seek new challenges and opportunities in the telecommunications arena. After achieving a successful sales record and high profits for one company, some move to a similar but higher-paid job at a new operation. Still others become System Managers in similar or smaller markets. Successful Marketing Directors also take more responsible marketing or programming positions at the headquarters of cable TV multiple system operators (MSOs), or join pay television service companies such as Home Box Office and Showtime.

Education

An undergraduate degree in marketing, advertising, or communications is essential at all companies. A master of business administration (MBA) in marketing is virtually a requirement at large urban organizations. Courses in sales management can be useful. In addition, some training in computer science is helpful, inasmuch as most of the customer service and billing operations make extensive use of computers, management information systems, or word processing systems to track and service subscribers.

Experience/Skills

A minimum of three years of previous experience in telecommunications sales or marketing is usually required. This can often be gained in the Sales Manager position.

A Marketing Director should be aggressive and capable of providing motivation and leadership to a diverse group of Salespersons. He or she should be well-organized and have some background in actual sales work. An outgoing and positive personality; a good feel for logistics, statistics, and research; and a flair for promotion and publicity are required.

Minority/Women's Opportunities

The number of women and minority-group members occupying the position has been increasing during the past ten years, according to industry observers. Many are joining the marketing and program staffs of the special pay-cable national program suppliers. Those that offer special opportunities for members of minorities include the Black Entertainment Network (BET), the long-established Spanish International Network (SIN), and the new Spanish-language Galavision.

Unions/Associations

There are no unions that represent Marketing Directors. Many individuals belong to the organizations listed in the System Manager entry, particularly the Cable Television Administration and Marketing Society (CTAM). At least 52 percent of cable systems' marketing personnel subscribe to and use the research capabilities of the new Cable Advertising Bureau (CAB).

Some female Marketing Directors (and women in other positions) belong to Women in Cable (WIC). A very few Marketing Directors (and other cable management employees) belong to Men in Cable Everywhere (MICE), a strictly social organization that also admits women, or to the newest cable social group, Former Local Origination Programmers (FLOPS).

SALES MANAGER

CAREER PROFILE

Duties: Direct responsibility for all sales at a cable TV system or STV or MDS company

Alternate Title(s): Director of Sales

Salary Range: $15,000 to $20,000

Employment Prospects: Good

Advancement Prospects: Fair

Prerequisites:

Education—High school diploma and some college

Experience—Minimum of two years in sales

Special Skills—Aggressiveness; leadership qualities; persistence; organizational skills; dependability; sales ability

CAREER LADDER

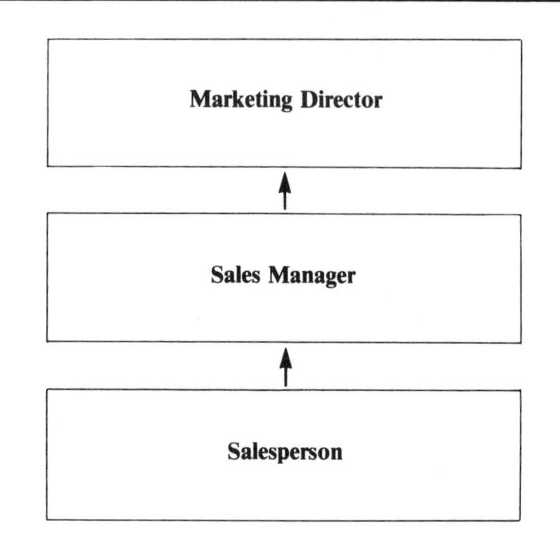

Position Description

A Sales Manager is in charge of all of the sales activities of a cable television system or a subscription TV (STV) or multipoint distribution service (MDS) operation. It is his or her responsibility to direct a sales operation that tries to convince anywhere from 2,500 to more than 50,000 individual customers to purchase the company's program services.

The Sales Manager usually reports directly to the Marketing Director and recruits and trains all Salespersons employed by the company. Depending on the size of the operation, he or she may have a staff of from three to more than ten Salespersons who sell by telephone or in the field. He or she assigns these individuals to particular areas or accounts and directly supervises their activities on a daily basis.

A successful sales force is essential for a profitable cable, STV, or MDS operation. The Sales Manager's role in training Salespersons is, therefore, a vital one. He or she works with new individuals to make sure they clearly understand the company's services and how to sell them effectively to potential subscribers. The Sales Manager usually works with each Salesperson to develop a selling technique most appropriate for that individual. The Sales Manager's success is directly tied to that of the sales staff. It is in his or her best interest, therefore, to train a highly-motivated and successful selling team.

The job is more than simply a supervisory one. A Sales Manager is usually involved in consumer sales on an individual, even door-to-door, basis. He or she actively sells to and services current customers as well as new subscribers, and initiates bulk sales to apartment houses or other multi-unit buildings. A Sales Manager is expected to spend at least three-quarters of his or her time aggressively selling program services to potential subscribers. He or she should set the performance standards for all Salespersons. A Sales Manager must successfully meet his or her own goals as well as other established quotas for contacts and actual sales.

In conjunction with the Marketing Director, the Sales Manager maps out sales campaigns and establishes overall goals and quotas for both specific geographic areas and time periods. He or she devises the best strategies for reaching potential customers, and convincing them to subscribe. The Marketing Manager also develops ways to persuade current customers to purchase added program services. Through constant contact with the public, the Sales Manager can judge the effectiveness of the company's sales pieces and displays. He or she makes recommendations to the Marketing Director as to what kinds of promotional materials and sales messages are likely to be successful.

Additionally, the Sales Manager:

• keeps all Salespersons informed of new changes in service prices, discounts, special offers, and organizational policies;
• maintains records of Salespersons' progress and activities;
• recommends price adjustments and new program services to the Marketing Director.

Salaries

Sales Managers at STV, MDS, and cable TV companies are usually paid in any or all of three major ways: a relatively small salary, a commission on each sale, and an "override" payment for each commissionable sale made by all Salespersons. The compensation for Sales Managers in cable TV ranged from $15,000 to $20,000 in 1980 from all sources of income, according to the National Cable Television Association (NCTA). Salaries at STV and MDS companies were comparable, according to industry observers. Sales Managers usually receive health, vacation and insurance benefits, and a company car or automobile allowance.

Employment Prospects

Opportunities for employment are good. Although the position is not an entry-level job, there is a constant need for qualified Sales Managers with good leadership abilities in cable TV, STV, and MDS companies. In addition to the substantial number of new telecommunications operations opening each year, there is a considerable turnover in existing sales positions. Many experienced and talented Salespersons are promoted to this more responsible position. Aggressive individuals with some "outside" (door-to-door) sales experience should be able to find jobs.

Advancement Prospects

There is a relatively high turnover rate in the new telecommunications industries, (particularly in sales), and some Sales Managers obtain more responsible sales or marketing positions within the industry. The competition is severe, however, and advancement is achieved only by those with initiative and strong managerial skills who obtain further education and training.

Sales Managers are usually ambitious and upwardly mobile. They often seek higher-paid jobs at larger market cable or telecommunications organizations. Some are promoted to the position of Marketing Director within their parent organization and are assigned to new systems or stations. Still others may apply their sales experience to companies in non-media fields. Overall, opportunities for advancement are only fair.

Education

A high school diploma and some college education are required. Courses in business, marketing, interpersonal relations, and psychology are helpful. A formal education is less important, however, than the ability to motivate, train, and lead a sales staff.

Experience/Skills

The position is not an entry-level job. A minimum of two years of sales experience with a consistent and reliable performance record is required. An additional year of sales management experience or a thorough knowledge of cable, STV, or MDS services and sales operations is also a common requirement.

A candidate should be capable of being an aggressive supervisor who leads by example. He or she must be persuasive, persistent, enthusiastic, and dependable. The individual must also be well-organized and consistent, and have a genuine love of, and talent for, sales work.

Minority/Women's Opportunities

According to statistics compiled by the Federal Communications Commission (FCC), 22.8 percent of the sales workers (including Sales Managers) in cable television in 1980 were white females, and 10.7 percent were minority males. The employment level of women is lower in this field than in comparable positions in commercial broadcasting, but the percentage of minority-group members is higher.

Unions/Associations

There are no unions or professional organizations that represent Sales Managers in cable TV, STV, or MDS operations.

SALESPERSON

CAREER PROFILE

Duties: Selling of cable television, STV, and MDS program services directly to homes

Alternate Title(s): Sales Representative

Salary Range: $12,000 to $20,000

Employment Prospects: Good

Advancement Prospects: Excellent

Prerequisites:

Education—Minimum of high school diploma; college courses in business preferable

Experience—Some door-to-door, retail, or phone sales

Special Skills—Sales ability; persistence; dependability; persuasiveness; initiative; competitive drive

CAREER LADDER

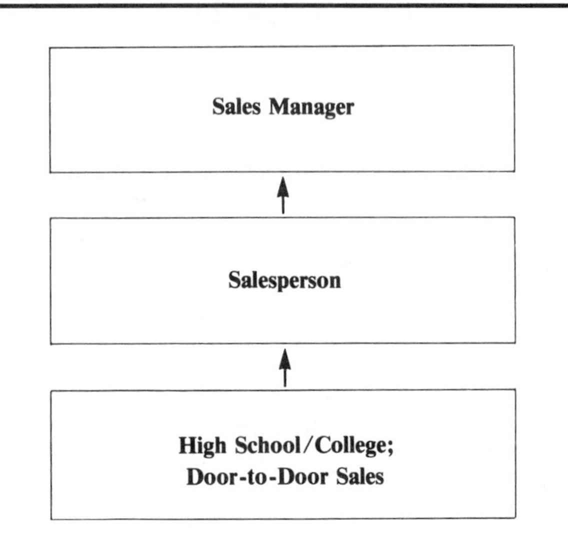

Position Description

A Salesperson for a cable TV system or a subscription TV (STV) or multipoint distribution service (MDS) station is responsible for selling the company's program service to individuals in homes within the franchise or coverage area. He or she sells in person or on the telephone.

A Salesperson is the backbone of the cable TV, STV, and MDS industries, and is immediately responsible for the financial success or failure of a company. He or she must be able to describe the various program services offered in order to help the customers understand what they are purchasing. A good Salesperson represents the company in day-to-day contacts with potential subscribers in an accurate, honest, and personable manner.

Most telecommunications operations employ from two to ten Salespersons, depending on the time of year, work load, and market situation. At some organizations, Salespersons work part-time, or a small full-time sales staff is augmented by additional help during a particular sales drive. At other organizations, the majority of the sales work is contracted to an outside firm that specializes in door-to-door selling of telecommunications services.

Sales quotas and goals are set for each Salesperson by the Sales Manager and Marketing Director. The Salesperson is expected to achieve specific levels of performance, such as contacting 90 percent of the individuals in the assigned territory. Some industry professionals feel that at least 30 percent of the potential customers contacted will purchase the basic program service and another 20 percent will purchase one premium service. The Salesperson's job is to reach and sell to those customers thus increasing both the company's subscriber base and income.

Although most sales efforts are directed toward individual homeowners, some Salespersons are assigned to multiple unit dwellings (apartment houses, hotels, and condominiums) and make sales calls on the managers of those buildings, who arrange for cable TV, STV, or MDS program services for all tenants. While the majority of a Salesperson's efforts are person-to-person, some sales are made by telephone from the local office, either by answering inquiries or through "cold" calls to prospective subscribers. Other sales are made through follow-up calls designed to interest current customers in subscribing to additional program services.

In most instances, a direct sales effort follows an advertising campaign that is launched on radio or television, by direct mail or outdoor billboards, or by doorknob or mailbox drops. Salespersons are then provided with brochures, program guides, pamphlets, teaser material, and other sales tools to use in their pitch.

Additionally, the Salesperson:

- attends daily sales meetings to receive assignments, new policies, procedures, goals, quotas, and sales literature;
- maintains accurate sales logs, field sales order forms, and other records.

Salaries

Salespersons are usually paid in either, or both, of two ways: a small base monthly salary, and a graduated retroactive commission structure based on the sales made each week. Some work strictly on a commission basis. Commissions are often paid at the close of the sale, although some companies pay after the actual installation or hookup of equipment at the subscriber's home. Most companies offer incentive bonus plans in addition to commissions and salaries.

In 1980, total compensation—including salary, bonus, and commissions—ranged from $12,000 to $20,000 a year, depending on the skill and ability of the individual, according to the National Cable Television Association (NCTA). The salary range at MDS and STV companies was the same during the year, according to industry observers. Most Salespersons who work on a contract basis do not receive health, insurance, or pension benefits, however.

Employment Prospects

The position is usually considered an entry-level job and overall opportunities for employment are good. Qualified Salespersons are in demand at all STV, MDS, and cable TV companies. With the dramatic growth expected in these industries in the 1980s, and with the relatively high turnover of sales employees, the prospects for Salespersons should remain bright.

Advancement Prospects

Opportunities are excellent for successful Salespersons to move up. There is a constant need for qualified sales management personnel in all areas of the new telecommunications industries. Many Sales Managers leave companies for opportunities elsewhere, creating vacancies that must be filled. Some aggressive and successful Salespersons move to higher sales positions at competitive companies in larger markets. Others find sales jobs in fields unrelated to telecommunications.

Education

A high school diploma is a minimum requirement for employment. Some college training with courses in business is often preferable. A candidate's aptitude for and ability at selling, however, are more important than a formal education.

Experience/Skills

Although the Salesperson is often an entry-level job, most employers prefer to hire people with some sales background. Experience in door-to-door selling is most helpful, but experience in retail store or telephone sales is also useful.

A gregarious nature, competitive drive, and evidence of initiative and perseverance are important qualities. An ability to relate to others, to listen, and to be quietly persuasive are very important. A neat appearance is also essential as is the ability to express ideas well, and the possession of imagination. Above all, the candidate must be dependable, must like to sell, and must be interested in making money.

Minority/Women's Opportunities

Employment of women and minority-group members in cable TV sales jobs is relatively high. According to the Federal Communications Commission (FCC), 22.8 percent of the people designated as sales workers (which includes Salespersons) were women and 10.7 percent were members of minorities in 1980. The opportunities in STV and MDS companies will increase as urban markets develop more fully, according to industry observers.

Unions/Associations

There are no unions or professional organizations that represent Salespersons in cable TV, STV, or MDS companies.

CHIEF TECHNICIAN

CAREER PROFILE

Duties: Responsibility for all on-site technical installation, transmission, and operation at a cable TV system or STV or MDS company

Alternate Title(s): Chief Engineer

Salary Range: $18,000 to $50,000

Employment Prospects: Good

Advancement Prospects: Good

Prerequisites:

Education—High school diploma and some technical training required; college work in electrical engineering, physics, or a related field preferable; FCC First Class License

Experience—Five years in broadcast or cable TV engineering

Special Skills—Technical aptitude; administrative ability; leadership qualities; design skills

CAREER LADDER

Chief Technician or Chief Engineer (Large System or Station); Director of Engineering

↑

Chief Technician

↑

Technician

Position Description

A Chief Technician is responsible for the day-to-day leadership and supervision of the technical staff of a cable TV system or a multipoint distribution service (MDS) or subscription television (STV) operation. The foundation of any such company rests on that technical staff. A Chief Technician may supervise from three to more than 30 employees.

In some STV and MDS companies, a Chief Technician is sometimes called a Chief Engineer. In cable television, however, the title of Chief Engineer usually implies broader responsibilities, including more administrative duties and planning and developing new services. Many small cable systems, and STV and MDS companies, contract with outside professional engineering firms for the services of a Chief Engineer.

In the absence of an on-site Chief Engineer, the Chief Technician is the senior and most highly skilled member of the engineering staff. As such, he or she is responsible for all engineering aspects of the operation, including the construction and installation of antenna towers, satellite receiving stations (dishes), amplifiers, signal processing equipment, and the head end (electronic control center).

The Chief Technician also aligns the various electronic elements of incoming satellite or microwave signals and processes them for re-transmission to subscribers' homes. At a cable TV system, he or she is also responsible for obtaining the necessary permits from local electrical and telephone utility companies for the use of their poles or underground facilitites. The Chief Technician determines the number of connections for each pole and diagrams the layout of the cable system. He or she also supervises the connection, installation, and servicing of the system's equipment in subscribers' homes.

In addition to the engineering duties, a Chief Technician often serves as the system or station manager. He or she may be an owner or part owner of the operation, or may be assigned to a particular locality by the parent company or multiple system operator (MSO).

Additionally, the Chief Technician:

- evaluates and purchases all technical and transmission equipment;
- ensures the system's compliance with all federal, state, and local utility and safety regulations;
- designs and installs systems to deliver new program services, including interactive (two-way) communications, teletext, and data transmission;
- prepares all necessary technical applications, including construction permits, proof of performance tests, and license renewals and modifications for the Federal Communications Commission (FCC);
- supervises the preventive maintenance program

for all technical and electronic equipment.

Salaries

While the salaries for Chief Technicians in cable TV and STV or MDS operations are somewhat less than for their peers in the broadcast industry, they are still relatively high. Chief Technicians were paid from $18,000 to $28,000 in 1980, according to the National Cable Television Association (NCTA). Chief Engineers' salaries at the larger cable systems ranged from $35,000 to $50,000 in 1980, also according to the NCTA. Salaries for Chief Engineers at STV and MDS stations, however, were slightly lower.

Employment Prospects

The chances of obtaining this position are good. In 1981, there were more Chief Technician positions available than there were qualified candidates to fill them. The rapid growth of the new telecommunications industries during the past five years has created a general shortage of engineering and technical personnel. Experienced and capable engineers should have little difficulty in obtaining positions, particularly if they have some background in managing a small business. Some Technicians who have pursued further study and training, and who possess the requisite leadership qualities, are promoted to the position.

Advancement Prospects

The lack of qualified engineering personnel in cable TV systems and in STV and MDS stations has created good opportunities for advancement, even though there is heavy competition for top jobs. Some Chief Technicians in cable television are promoted to new and larger systems, or to positions of Chief Engineer of regional or system-wide operations within their own MSO companies. In STV and MDS operations, some obtain director of engineering positions in the parent company, in charge of the planning and construction of other stations. Others advance their careers by joining competitive operations at higher salaries. Still others join manufacturing companies as design engineers, field service managers, or construction/engineering managers.

Education

A high school diploma and one or two years of training in electronics are minimum requirements for the position. Most large systems or stations prefer that candidates have some college education or an undergraduate degree in electrical engineering, physics, or a related science. A First Class Radiotelephone License from the FCC is required for Chief Engineers in STV and MDS station operations.

Experience/Skills

A minimum of five years of experience in broadcast electronics or a related field involving highly complex transmission equipment is a prerequisite for the position. Solid experience with all types and models of technical gear, a good understanding of the principles of electronics, and experience with FCC rules and regulations are also required.

A Chief Technician must be able to combine good technical design skills with hands-on experience. He or she must also exhibit leadership qualities, be budget conscious, and be a capable administrator.

Minority/Women's Opportunities

As of this writing, there are no female Chief Technicians or Chief Engineers at any cable system or STV or MDS station. Less than 2 percent of such positions were held by minority males in 1980, according to industry estimates.

Unions/Associations

There are no unions that serve as bargaining agents for Chief Technicians or Chief Engineers in cable, STV, or MDS systems and stations. Many Chief Technicians belong to the Society of Cable Television Engineers (SCTE) or to the Community Antenna Television Association (CATA) to share common concerns and advance their careers. Some Chief Engineers belong to the Society for Motion Picture and Television Engineers (SMPTE), the Association for Maximum Telecasting (AMT), or the Society of Broadcast Engineers (SBE) for the same purposes.

TECHNICIAN

CAREER PROFILE

Duties: Maintaining and servicing of technical equipment at a cable TV system or an STV or MDS station

Alternate Title(s): Plant Technician; Bench Technician; Maintenance Technician; Service Technician; Trunk Technician

Salary Range: $13,000 to $21,000

Employment Prospects: Excellent

Advancement Prospects: Excellent

Prerequisites:

Education—High school diploma and technical school training

Experience—Minimum of six months on-the-job training

Special Skills—Technical aptitude; reliability; conscientiousness; curiosity

CAREER LADDER

Chief Technician

↑

Technician

↑

Installer

Position Description

Technician (or Plant Technician in cable television) is the general title for the person who maintains, repairs, and installs equipment at a cable system or a multipoint distribution service (MDS) or subscription television (STV) station. He or she reports to a Chief Technician or Chief Engineer. In smaller operations, the Technician may perform a variety of technical duties. In large settings, he or she may concentrate in one of several areas.

In cable TV, the technical emphasis is on the electronic and physical connection of subscribers' homes to the head end (the central originating point of the system) by coaxial cable. Technicians are responsible for maintaining equipment at both ends as well as the cable itself and all support gear along its route. Since MDS and STV stations transmit their signals over the air, their operations are somewhat less complicated than cable systems. Technicians in MDS and STV are responsible for the transmission equipment as well as the receiving devices in subscribers' homes.

The specialized Technician positions include:

Bench Technician—(in cable) diagnoses and repairs broken or malfunctioning subscriber converter boxes, prepares equipment for installation in homes, and repairs testing devices.

Maintenance Technician—(in cable) repairs damaged cable between telephone poles and fixes equipment damages caused by adverse weather conditions; (in MDS and STV) repairs testing equipment, amplifiers, transmitting antennas, and pay TV signal scrambling units.

Service Technician—(in cable) repairs malfunctions in subscribers' equipment, fixes signal amplifiers, and electronically scans the system to spot problems before breakdowns occur; (in MDS and STV) repairs faulty receiving antennas and subscriber descrambling units.

Trunk Technician—(in cable) corrects problems in the main cable lines that transmit the signal and in amplifiers that enhance it, and wires and maintains electronic components in the field.

Additionally, the Technician:

- services television production equipment (cameras, program switchers, audio consoles, etc.);
- maintains videotape and film equipment used to send programs to subscribers.

Salaries

Salaries in 1980 for Technicians in the cable TV industry ranged from $13,000 for beginners to $21,000 for experienced individuals, according to the National Cable Television Association (NCTA).

In general, Bench, Trunk, and Service Technicians are paid at a slightly higher rate than are Maintenance Technicians. The salary levels for Technicians at STV and MDS stations are comparable, according to industry sources.

Salaries at cable TV systems and at STV and MDS companies have risen in the past few years because of the supply-demand factor, and are now only slightly lower than those for similar positions in the broadcast field. As an additional incentive to prospective employees, many multiple system operators (MSOs) offer company stock purchase plans, life insurance, and medical and dental plans.

Employment Prospects

The opportunities for employment of Technicians are excellent. Most small and medium size cable systems and MDS and STV stations employ from three to five Technicians. Larger operations may have more than 30 on staff. In addition to the new systems and stations being constructed, there is generally high turnover and qualified Technicians are constantly in demand. Many Installers complete additional study and training courses and are promoted to Technicians.

Advancement Prospects

Chances for advancement for diligent individuals with some initiative are excellent. A major problem in the industry is the raiding of trained and qualified Technicians by competing companies. Most operations promote from within the organization to fill the vacancies that result. At many major MSO cable systems bright, reliable Technicians are paid to attend classes that will qualify them for better-paying positions. The experienced and highly capable Technician, with a background in all technical areas, can be promoted to Chief Technician.

Some Technicians advance their careers by moving into slightly higher-paying jobs in broadcasting, or to computer or electronics engineering positions in non-telecommunications industries.

Education

A high school diploma and some technical or electronics trade school training are minimum requirements. Some larger MSOs operate their own training facilities.

Candidates for jobs as Service and Trunk Technicians at cable TV systems and as Service Technicians at STV and MDS operations should have a Second Class Radiotelephone License from the Federal Communications Commission (FCC). The license is particularly valuable at STV and MDS companies.

Experience/Skills

Candidates should have at least six months of actual on-the-job experience in the Technician position being sought.

Technicians must be reliable and conscientious individuals who have an inquisitive mind, mechanical and technical aptitude, and a basic understanding of electronics and physics. The ability to read and interpret schematic diagrams is also necessary. Most good Technicians are excellent problem-solvers and troubleshooters. Technicians must also be able to overcome any fear of heights since roof climbing, pole climbing, and bucket truck aerial work are routine aspects of many Technician positions.

Minority/Women's Opportunities

White females occupied 2.1 percent of the Technician positions in cable TV in 1980, according to studies by the FCC. Minority women were not well represented. Minority males held 11.8 percent of all Technician jobs in cable TV in 1980. Most Technicians are white males.

Unions/Associations

There are no national trade unions that act as bargaining agents for Technicians at cable TV systems or at STV or MDS operations. Some local and state utility and technical organizations represent Technicians in particular geographic areas.

Technicians often belong to the associations detailed in the Chief Technician entry. In addition, Society of Cable Television Engineers (SCTE) has recently inaugurated a program leading to certification as a Professional Cable Television Engineer, and it is expected that many Technicians will participate in the program in order to advance their careers.

INSTALLER

CAREER PROFILE

Duties: Installation of receiving equipment for a cable, STV, or MDS operation

Alternate Title(s): None

Salary Range: $8,500 to $22,500 +

Employment Prospects: Excellent

Advancement Prospects: Good

Prerequisites:

Education—High school diploma and technical school training

Experience—School training in electronic equipment

Special Skills—Mechanical aptitude; technical ability; courtesy; physical strength

CAREER LADDER

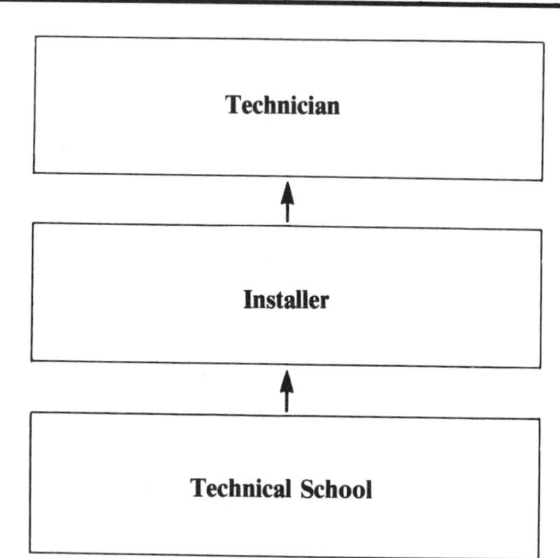

Position Description

An Installer ensures that the signal, and programming, of a telecommunications operation is received in the homes of its subscribers. It is the usual entry-level technical position at a subscription television (STV) or multipoint distribution service (MDS) station or at a cable TV system.

Most small cable, STV, and MDS companies contract with special firms that provide installation services. At some large operations, Installers are full-time employees of the cable franchise or of the MDS or STV license holder. Virtually all systems, however, use contract Installers at some time.

Although Installers usually work in private homes, they can also connect their companies' equipment in apartment houses, hotels, or office buildings.

The duties of an Installer vary, depending on whether he or she works for a cable TV system or for an STV or MDS operation. In cable, an Installer prepares the customer's home for the reception of the television signal by running wire from a telephone pole or underground terminal and attaching it to the connector box(es) and television set(s) within the home. In addition, he or she sometimes performs the initial construction work for a cable system, which includes digging trenches, relocating telephone poles, and stringing cable between poles. At least three-quarters of such work is contracted to outside companies, according to industry sources. After construction is completed, an Installer works closely with, and sometimes performs some of the duties of, a Technician.

An Installer working for STV or MDS operations attaches a special antenna to the roof of the subscriber's house, tests and adjusts it to receive the stations's signal, and connects it to a de-scrambling device and to the television set(s) in the customer's home.

Installers receive their assignments from the Customer Service Representative or the dispatcher in the main office of the company. They report, however, to the Chief Technician, Chief Engineer, or other technical supervisor.

Additionally, the Installer:

- explains and demonstrates the operation of the system to subscribers and describes available channels and programs;
- performs minor repairs and adjustments in subscribers' homes;
- (in cable TV) adds or deletes program services by adjusting customers' cable connections and disconnects their equipment, when requested.

Salaries

Installers working directly for cable systems or for

STV or MDS operations are usually paid on an hourly basis. In 1980, rates in cable TV ranged from $4.25 an hour for beginners to $7.00 an hour for experienced individuals with some seniority, according to the National Cable Television Association (NCTA). This amounted to $8,500 and $14,000 a year, respectively, for a 40-hour, 50-week work cycle. Most Installers work overtime hours, which are usually paid at a time-and-a-half rate.

Some independent contracting firms pay Installers on a piece work basis and the rate is usually $20.00 for each installation. At an average of 4.5 installations a day, over the course of 50 weeks, a person can often make $22,500 for the year. Many contract Installers earn consistently more annually.

Employment Prospects

Opportunities are excellent. The rapid growth of cable TV systems and the slower but equally promising growth of STV and MDS operations offer many prospects for employment. In most of the nation, there are more positions open than there are candidates.

Small STV or MDS companies or cable TV systems employ from three to five Installers (usually on contract), and at large cable systems, an installation staff of 20 individuals is not uncommon. The greatest number of opportunities is with independent contracting firms.

Advancement Prospects

Opportunities are good for diligent individuals who apply themselves to daily tasks in a responsible manner. Many Installers seek additional electronics training and technical improvement, and become Technicians in cable TV operations. Others move to supervisory construction positions at new systems owned by the parent multiple system operator (MSO). Still others advance their careers by joining competing companies or other independent contractors at higher wages. Installers who work for independent contracting firms are exposed to many potential employers, which creates increased advancement possibilities for them.

Education

A high school diploma and some post-high school education at a technical school is usually required. Training can also be obtained in the military service. Many major MSOs and cable TV trade associations offer training sessions and seminars. At MDS and STV operations, some study for a Second or Third Class Radiotelephone License from the Federal Communications Commission (FCC) is usually necessary.

Experience/Skills

Little previous professional experience is required. Interest and schooling in electronics and the repair, maintenance, and operation of electronic equipment are, however, essential. STV and MDS operations often require more technical education in broadcast engineering than is necessary for an Installer in a cable system.

Installers need mechanical aptitude and some technical ability with electronic equipment. Candidates should also possess physical strength and good health as well as have no fear of heights. Since there is close contact with subscribers, an Installer should be courteous and generally enjoy working with the public. Installers working for independent contractors must be willing to travel and live away from home for long periods of time because such companies move from city to city.

Minority/Women's Opportunities

The majority of Installers for STV and MDS companies and cable TV systems are white males, according to industry sources. White females were employed in 2.8 percent and minority-group males in 15.8 percent of the skilled and semiskilled positions (including Installers) in cable TV in 1980, according to the FCC. Representation by both groups should continue to grow.

Unions/Associations

There are no national unions or professional organizations that represent Installers in cable TV or in STV or MDS operations at the present time. Some local and state utility and construction unions, however, do represent them in specific locations.

DIRECTOR OF MEDIA SERVICES

CAREER PROFILE

Duties: Administering media support services, including television and video, in a school district, college, or university

Alternate Title(s): Director, Instructional Technology; Director, Learning Resources; Director, Educational Media

Salary Range: $9,400 to $30,000+

Employment Prospects: Fair

Advancement Prospects: Poor

Prerequisites:

Education—Master's or doctorate degree in education

Experience—Minimum of two to three years of educational media experience; classroom teaching

Special Skills—Verbal and writing talent; persuasiveness; leadership qualities; supervisory ability; good interpersonal skills; inquisitiveness

CAREER LADDER

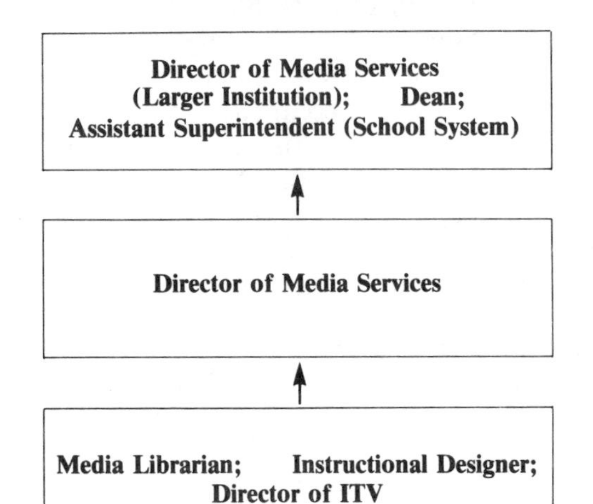

Position Description

A Director of Media Services is the senior administrative official in charge of providing instructional support materials to schools and colleges that use media as part of their curricula and training. He or she ensures that well-selected materials and equipment are easily accessible to teachers and students for classroom use. A Director of Media Services works in elementary and secondary schools, community and 4-year colleges, and universities.

In addition to the traditional audio-visual equipment and materials such as 16mm films, slides, charts, filmstrips, overhead transparencies, and audio tapes and records, there is a vast new array of technological devices. These new media include electronic learning laboratories, computers, and videotape, videocassette, and videodisc equipment and programs. It is the responsibility of the Director of Media Services to coordinate the use of this new technology within the instructional process.

The Director of Media Services is an educational and technological generalist who views television/video programming as one component of a multimedia service. As part of his or her overall responsibilities, the Director of Media Services evaluates and purchases media equipment and programs to supplement and enrich classroom instruction, and often supervises their use.

Directors of Media Services operate in a variety of administrative situations. In some cases, they have overall responsibility for an instructional television (ITV) operation, including the production of programs that are shown through a school-owned or commercial TV system. Sometimes the Director of Media Services oversees all print (library) and non-print (media) services for a school or college. He or she is usually in charge of the day-to-day distribution of video and other equipment and instructional programs.

In higher education, media services departments are complex and often employ more than 30 individuals who handle a variety of tasks. A Director of Media Services manages various units such as film production, audio production, graphic arts, technical and maintenance, and video/film booking. In smaller school systems, these functions may be handled by as few as five people.

A Director of Media Services in a school system usually reports to an assistant superintendent or, at a college or university, to a vice president of academic affairs.

Additionally, the Director of Media Services:

- provides equipment and operators for classroom use;
- supervises the maintenance of equipment;
- develops budgets, reports, research, and other

data relevant to program circulation.

Salaries

Salaries in 1978-79 ranged from $9,400 for a beginner with a bachelor's degree working in a small school system, to more than $30,000 for experienced individuals with advanced degrees and seniority at major institutions of higher education, according to professional sources.

According to an 8-year study of graduates of educational media programs conducted by Darryl Sink (reported in the *Educational Media Yearbook, 1980,* Brown and Brown, Libraries Unlimited, Littleton, CO), beginners with a bachelor's degree earned from $9,400 to $12,900. New employees with master's degrees earned from $12,000 to $18,500, with the mean salary at $15,300, and for people with a doctorate degree, beginning salaries ranged from $14,000 to $22,000, with the mean at $17,900.

Employment Prospects

With the general decline of support for education and media in recent years, the opportunities for employment are only fair.

This is not usually an entry-level job. Although some small school systems employ individuals who have just received their college degree(s), most seek more experienced people.

Most states have a unit concerned with state-wide media coordination, and many circulate classroom and in-service materials to schools. In addition, most school districts, community and 4-year colleges, and universities have media centers. There is relatively little turnover, however, and openings are usually filled from within.

In some organizations, the Media Librarian or an Instructional Designer is promoted to the position. In some cases, the Director of Instructional Television moves up to the Director of Media Services slot.

Advancement Prospects

Opportunities are poor compared to other video and telecommunications areas. For many senior-level Directors or Media Services at large school systems or at higher-education institutions, the position is the climax of a successful career in educational communications and media. A few take administrative posts at larger universities as deans of instructional resources or at larger school systems as assistant superintendents.

Most advance their careers by moving to larger school systems, colleges, or universities, which offer higher salaries and more prestige. Some obtain higher-paying positions in government, health, or business media centers.

Education

A bachelor's degree in education is a minimum requirement, and all but the very small school systems require a master's degree. A master's is a minimum requirement at colleges and universities, but most higher-education institutions prefer a doctorate.

A candidate should have his or her degree(s) in education with the major field of study in educational media, instructional technology, or learning resources. Courses in curriculum design, educational administration, and educational psychology are necessary.

Experience/Skills

While some small schools or colleges employ recent graduates, the majority require a minimum of two to three years of educational experience. A broad knowledge of all technology, including the traditional audio-visual equipment, is necessary. Equally important is classroom teaching experience that provides insight into the day-to-day instructional process and environment.

Most good Directors of Media Services are expert practitioners of various teaching and learning techniques, and are eager to discover new methods that work. They possess strong verbal and writing talents, are articulate, persuasive, and have excellent interpersonal skills. They have good leadership qualities and are able to supervise a number of individuals in various tasks.

Minority/Women's Opportunities

The opportunities for women and members of minorities are excellent, inasmuch as the vast majority of educational institutions aggressively apply and practice equal employment opportunity (EEO) and affirmative action policies. Many institutions and schools report that they have a difficult time finding women or minority-group applicants, however, and the majority of positions today are held by white males, according to professional sources.

Unions/Associations

There are no unions that specifically represent Directors of Media Services. Some individuals belong to the National Education Association (NEA) or American Association of University Professors (AAUP), which may represent them as faculty members.

Most Directors of Media Services belong to the Association for Educational Communications and Technology (AECT), the American Library Association (ALA), or the Educational Film Library Association (EFLA) in order to share mutual concerns and advance their careers.

INSTRUCTIONAL DESIGNER

CAREER PROFILE

Duties: Assisting teachers in the design of new instructional techniques using video and media

Alternate Title(s): Curriculum Specialist; Curriculum Writer

Salary Range: $12,000 to $22,000+

Employment Prospects: Good

Advancement Prospects: Good

Prerequisites:

Education—Undergraduate degree in education; master's degree preferable

Experience—Minimum of one to two years in media, training, and instruction

Special Skills—Interpersonal skills; logic; organizational ability; knowledge of media equipment; statistical aptitude

CAREER LADDER

Director of Media Services

↑

Instructional Designer

↑

Graduate Assistant; Teacher

Position Description

An Instructional Designer is a learning facilitator who assists faculty members (in college), teachers (in schools), and trainers (in business, health, or government) to improve instruction and learning. He or she takes advantage of the instructional capabilities of media to help design effective means of teaching. An Instructional Designer is alert to new instructional methods and technologies, and is familiar with the uses of television, non-broadcast video, and telecommunications systems in education.

The position of Instructional Designer is a relatively new one, incorporating the traditional functions of a curriculum specialist and curriculum writer. Instructional Designers are usually employed by large school systems, colleges, or universities. Increasingly, they work in private industry, health, and, on occasion, government training projects and programs.

In whatever setting, an Instructional Designer is responsible for assisting a subject expert to design or re-design an entire course, specific lessons, individual lectures, or professional training sessions. He or she helps to develop the content and to design the most effective method for teaching it, making use of media technology. In addition, he or she recommends and updates instructional materials, including video programs. When appropriate, he or she organizes and plans new media programs, and often writes, edits, and rewrites scripts for instructional video use.

An Instructional Designer sometimes works independently, but more often is a member of a team which may include a content or subject specialist (faculty member in a specific discipline), Media Technician, Media Librarian, Graphic Artist, and, often, a Producer or Director.

This is not generally a supervisory position and the Instructional Designer usually reports to a Director of Media Services or Director of Instructional Television.

Additionally, the Instructional Designer:

- consults with faculty and supervisors regarding instructional needs and designs;
- develops cost-benefit and accountability studies for media use;
- evaluates the effectiveness of new instructional techniques, equipment, and video programs.

Salaries

Salaries in 1980 ranged from $12,000 for beginners to more than $22,000 for experienced individuals, according to industry sources. Earnings in education were generally lower than those in health or private business and industry. Almost all employers offer excellent fringe benefits, including health and pension plans. In most instances, salaries are a reflection of the individual's level of education and of his or her particular skills and achievements.

Employment Prospects

In spite of a general decline in educational media opportunities, Instructional Designers are in some demand, and employment prospects for this specialty are good. Increasing costs have caused educational administrators to examine all possible ways to improve teaching and training and to manage it more efficiently. The sound application of new technology is one of the more viable approaches.

Available positions in education are usually at the larger school systems and universities, which are involved with extensive media programs, telecommunications, and video. Some elementary and secondary school teachers who have worked in curriculum development get further training in this new discipline and find jobs as Instructional Designers.

In addition to the opportunities in education, there are an increasing number of positions available in private business and industry and in the health areas. Most observers predict that this trend will continue and that such organizations will provide the major marketplace for graduates of instructional development programs in the future.

Advancement Prospects

Many Instructional Designers serve in joint academic appointments and teach instructional development, educational psychology, or other courses in colleges or universities. As such, their career advancement is often tied to their academic achievements—papers, research, teaching competence, etc. Some Instructional Designers advance their careers by moving to more responsible positions as Directors of Media Services at smaller colleges or school systems. Still others join private industry for better-paid positions. The general opportunities for advancement are good.

Education

An undergraduate degree in education is a minimum requirement, and a master's degree is preferable. Individuals who seek employment at a major university must have a doctorate.

Course work in educational psychology, curriculum design, and sociology are required. Courses in educational media, including television, video, and other non-broadcast technology, are mandatory.

Experience/Skills

One to two years of experience, either as a graduate assistant, or as an elementary or secondary school or college teacher is usually required.

Knowledge and experience of the various theories of learning and of the use of a wide range of media resources and technology is essential. Skill in the operation and use of video equipment, and in writing and editing video and television programs, is required.

Instructional Designers must possess good interpersonal skills, inasmuch as there is constant contact with faculty and staff members of an organization. An inquisitive mind, a penchant for creative and logical thought, and statistical aptitude are necessary. Most Instructional Designers have had some teaching experience, and are excellent motivators and cooperative leaders.

Minority/Women's Opportunities

Opportunities for women are excellent. In 1980, almost one-half of all Instructional Designers were white females, according to industry estimates. The opportunities for women and members of minorities are expected to increase in the future as the profession develops.

Unions/Associations

There are no unions that serve as bargaining agents for Instructional Designers. Some individuals who also serve as faculty members at schools and colleges belong to the American Association of University Professors (AAUP) or the National Education Association (NEA).

Many Instructional Designers belong to the Association for Educational Communications and Technology (AECT), and to its Division of Instructional Development, to exchange ideas and advance their careers. Others belong to the American Society for Training and Development (ASTD) and its Media Division, or to the National Society for Performance and Instruction (NSPI).

MEDIA LIBRARIAN

CAREER PROFILE

Duties: Acquiring, scheduling, and circulating non-print materials

Alternate Title(s): Media Specialist; Media Clerk; Media Booker; Media Scheduler

Salary Range: $9,000 to $16,000

Employment Prospects: Poor

Advancement Prospects: Poor

Prerequisites:

Education—Undergraduate degree; master's degree in education or library science preferable

Experience—Minimum of one year of cataloging, classifying, and circulating media

Special Skills—Organizational ability; detail orientation; responsible nature; interpersonal skills

CAREER LADDER

Media or Library Administrator; ITV Coordinator

↑

Media Librarian

↑

Librarian; Secretary

Position Description

A Media Librarian is responsible for acquiring, previewing, scheduling, and circulating films, videocassettes, videotapes, videodiscs, and other non-print materials for educational, public, or training use. He or she processes customer orders and maintains records detailing the location and disposition of all such material.

In the past, Media Librarians dealt predominantly with 8mm and 16mm films, audio tapes, and records. Today, they are increasingly becoming responsible for assisting in the selection of videocassettes and videodiscs that a library or media center makes available.

In some small media centers, a Media Librarian is responsible for checking-out and scheduling non-print instructional support materials. He or she serves as a "booker" or "scheduler" and performs mainly clerical support functions. Most often, a Media Librarian is a professional librarian or media specialist, charged with acquiring, evaluating, categorizing, and cataloging film and video materials and is considered a member of the professional media support staff. He or she selects video and film titles from a variety of sources, including reviews in professional literature, producer or distributor sales representatives or promotional literature, and exhibits.

Media Librarians are employed by schools and colleges, and private industry, government, and health media centers, and, increasingly, are important staff members at public libraries. They serve as collectors and distributors of non-print materials to instructors, trainers, students, and the public. As a rule, they are responsible for the circulation of video and film materials for on-campus or in-company use. In some large media centers that have extensive collections of film and video materials, or in a public library, Media Librarians are also responsible for off-campus or public circulation.

Media Librarians often supervise many part-time student workers and, in large media centers, are often responsible for the supervision of one or two full-time employees. Media Librarians employed by a school or college usually report to a Director of Media Services. Those employed by a government media operation report to the Chief, Audio-Visual Services. A Media Librarian working at a public library reports to the head librarian.

Additionally, the Media Librarian:

- monitors the frequency of circulation of videotapes and films as well as the use of the media center;
- supervises the preparation and publication of the media center's film and video rental/sale catalogs;
- responds to inquiries and requests for video and

film programs, processes complaints, and resolves circulation and user problems.

Salaries

Earnings in 1980 ranged from $9,000 for beginners to more than $16,000 for experienced individuals in large media centers or public libraries, according to professional estimates. The range often depends on a combination of seniority and education. Individuals with professional training in library science or media earn higher beginning salaries and advance more rapidly. In some cases, however, long-time secretarial employees in media centers who are promoted to or assume the position earn as much as professionally trained employees.

Employment Prospects

Opportunities in education or public library work are poor. There are usually more applicants than there are available positions.

There is relatively little turnover in the field. In addition, the current decline in support for education and public libraries has resulted in fewer new positions. Often, when openings occur, they go to other librarians within the system who have worked in circulation, cataloging, or acquisitions, and who have taken additional course work in media. Many administrators of small media centers choose a secretary experienced in media to fill the job.

While the chances for employment in industry or health media centers are better, the outlook in general is poor, particularly compared to other areas in television, telecommunications, and video.

Advancement Prospects

Possibilities for advancement are poor. In a period of declining support, the competition for jobs in the library and educational media worlds is great.

Media Librarians can sometimes advance their careers by obtaining positions at larger media centers or public libraries in other geographic areas. Some move within their own organizations into more responsible positions as administrators of media programs in education. In the library field, some move into better-paid positions with the responsibility for acquisition of print as well as non-print materials. Those who have concentrated on the distribution of instructional videocassettes and videodiscs sometimes take additional courses in educational media and curriculum design and become ITV Coordinators at slightly higher salaries.

Education

A master of education or master of library science degree is almost always required, and at libraries a master of library science degree is mandatory.

In smaller media operations where the position has less responsibility for the selection and acquisition of video and film programs, an undergraduate degree is necessary. Courses in media, cataloging, and classification of materials are required, and a broad liberal arts background is helpful.

Experience/Skills

A minimum of one year spent cataloging and classifying media and handling circulation procedures is usually mandatory. In addition, some part-time or full-time experience in the distribution of non-print materials is often necessary. Since most media centers use mini computers in their circulation operations, some familiarity with word processing equipment and computers is required.

A Media Librarian must be extremely well organized, detail oriented, and responsible. He or she should be inquisitive and able to relate to users in a friendly and helpful manner.

Minority/Women's Opportunities

More than 85 percent of Media Librarians in 1980 were white females, according to professional sources. While employment opportunities for members of minority groups have increased in the past ten years, the number of individuals in the position has not risen proportionately, except in urban settings.

Unions/Associations

There are no national unions that represent and bargain for Media Librarians. Many individuals belong to the American Library Association (ALA) and one or more of its divisions, notably the Library and Information Technology Association (LITA). Others hold membership in the Special Libraries Association (SLA). Some Media Librarians also belong to the Association for Educational Communications and Technology (AECT), the Information Film Producers of America (IFPA) and the Educational Film Library Association (EFLA) to share common concerns and advance their careers.

MEDIA TECHNICIAN

CAREER PROFILE

Duties: Repair and maintenance of media, television, and video equipment at a school or university

Alternate Title(s): Maintenance Technician

Salary Range: $9,000 to $15,000

Employment Prospects: Good

Advancement Prospects: Good

Prerequisites:

Education—High school diploma; some vocational or technical school training

Experience—Minimum of six months of audio-visual equipment repair preferable

Special Skills—Mechanical aptitude; electronics skills; inquiring mind; good problem-solving ability

CAREER LADDER

Technician; Video Service Technician

↑

Media Technician

↑

Audio-Visual Equipment Repair; Technical School

Position Description

A Media Technician maintains, repairs, and services the technical equipment at a school or college media center. He or she repairs traditional audio-visual equipment, including 16mm projectors and cameras, slide and filmstrip machines, opaque and overhead projectors, television sets, and videocassette and videodisc machines. In media centers that have television studio equipment, he or she is responsible for repairing cameras, switchers and associated equipment. A Media Technician must be familiar with a wide variety of audio-visual equipment. Many problems that arise are mechanical, but others require a knowledge of digital electronics and integrated circuit boards.

It is the Media Technician's responsibility to develop and maintain an orderly system of preventive maintenance to ensure that the video and other media equipment is in good operating order at all times. This involves developing daily and weekly work sheets and adhereing to maintenance schedules.

Small media operations at schools usually employ only one full-time Media Technician while at larger university media centers, two to three are required full-time to keep the equipment in good operating condition. In the majority of cases, high school or college students are employed part-time to augment the services of the Media Technician, who supervises and trains them.

A Media Technician usually operates out of a maintenance equipment room at the central headquarters of a media center. In some large media operations with more than one location, he or she may be assigned to a specific branch or office. On occasion, he or she is required to do on-location repair work in classrooms or other buildings where equipment is located.

A Media Technician usually reports directly to a Director of Media Services in small operations and to that person's assistant in large organizations.

Additionally, the Media Technician:

- maintains an inventory of parts, accessories, and supplies;
- interprets schematic charts and diagrams detailing television, video, and media electronic components and systems;
- works with manufacturers' field service representatives, and attends periodic seminars and training sessions to keep abreast of new equipment and techniques;
- recommends the replacement of outdated or irrepairable equipment.

Salaries

Salaries in 1980 ranged from $9,000 for beginners to more than $15,000 for experienced individuals with seniority at school and college media centers, according to most estimates. While salaries are generally lower than for similar maintenance positions at subscription TV (STV) and multipoint distribution service (MDS) companies or in the home video field, the fringe benefits and security offered are often considerably better. All schools and universities offer good employee health, life insurance, and pension plans. An employee credit union is often available and reduced ticket prices for school or university events and good vacation schedules are normal.

Employment Prospects

Opportunities are good. There is a lack of qualified people in established media centers, and the position does not usually require as much electronics knowledge and training as do the somewhat similar positions of Video Service Technician (in home video) or Technician (in Cable, STV, or MDS operations), although the basic skills are similar.

There is some turnover in the position, as individuals leave for better-paid jobs in private industry. In addition, most government, health, and corporate media centers require Media Technicians. The number of media centers in education, health, and government, however, is not likely to grow. Nevertheless, as in most technical positions in television, video, and telecommunications, there are usually more openings than there are qualified candidates.

Advancement Prospects

Opportunities for advancement are good. As in most technical positions in electronics, there is some turnover in the field. Alert and diligent individuals who combine their mechanical and electronic experience with further study, can move to usually better-paid positions as Technicians in STV, MDS or cable TV operations. Some get jobs in the consumer electronics field as Video Service Technicians, with higher pay but fewer benefits. A few move to larger markets to advance their careers.

Education

A high school diploma and some evidence of post-high school training in a trade, technical, or vocational school are usually required. Classes in the basic sciences, as well as in math and algebra, are useful. Drafting, basic electronics, and electrical engineering courses are extremely helpful. A Federal Communications Commission (FCC) license is not required.

Experience/Skills

Most employers prefer some experience in the repair of audio-visual equipment. Saturday or evening work at a television repair or consumer electronics store is helpful. Six months to one year of experience is preferred but, because of the scarcity of people for the job, technical school training often serves in lieu of experience for beginners.

A knowledge of the operation of various types of traditional audio-visual equipment and television sets is required, as well as an ability to diagnose mechanical and electronic problems. A Media Technician needs an inquisitive mind, some mechanical dexterity, and an intuitive feel for problem-solving.

Minority/Women's Opportunities

The position of Media Technician was held mostly by white males in 1980, according to industry estimates. Opportunities for women are excellent, but relatively few train or apply for the job. Prospects for minority-group members are also excellent, and the number of minority workers in the job has increased, particularly in urban areas, over the past few years. The position, however, remains largely occupied by white males.

Unions/Associations

There are no national unions that serve as bargaining agents for Media Technicians. Most of the jobs in government, school, and college settings, and in some health areas, are local or federal civil service positions, and employees are represented by the appropriate governmental unions.

Some Media Technicians belong to the National Audio Visual Association (NAVA) and attend training seminars in association with Indiana University in order to advance their careers. Others belong to the National Association of Television and Electronic Servicers of America (NATESA), the National Electronic Service Dealers Association (NESDA), or the Audio Visual Technicians Association (AVTA).

DIRECTOR OF ITV

CAREER PROFILE

Duties: Administering an instructional TV operation at an educational institution

Alternate Title(s): Director of ETV; Assistant Director of Media

Salary Range: $14,000 to $25,000+

Employment Prospects: Fair

Advancement Prospects: Poor

Prerequisites:

Education—Master's degree; doctorate preferable

Experience—Minimum of two years in classroom teaching and instructional TV

Special Skills—Sound educational judgment; good language abilities; interpersonal skills; administrative and leadership qualities

CAREER LADDER

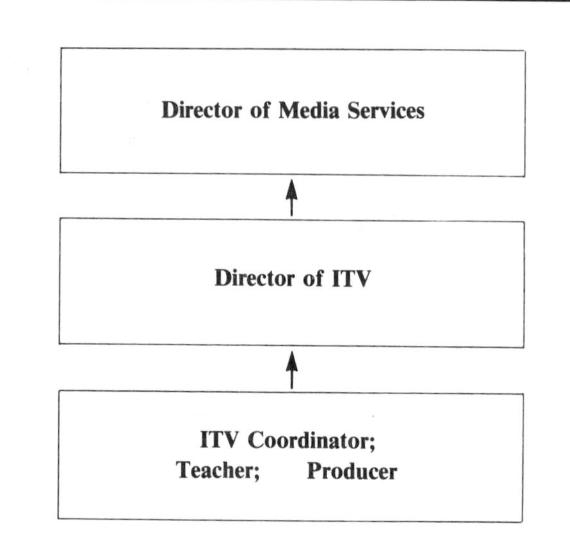

Position Description

A Director of ITV (instructional television) is responsible for the operation and administration of a station or system at a school, school district, college, or university that provides educational television programs and services to its students in a classroom environment in order to supplement and enrich regular instruction. At most school-owned operations, ITV is also used for in-service teacher training.

Most ITV operations are similar to, although smaller than, commercial or public TV broadcast stations. The Director of ITV serves in a capacity that is parallel to a General Manager of such a station, or to the System Manager at a cable TV system or a subscription TV (STV) or multipoint distribution service (MDS) company. He or she oversees the operation and its schedule in accordance with the regulations of the Federal Communications Commission (FCC) and with the policies of the institution.

A Director of ITV is as much a professional educator as he or she is a media or television manager. He or she is dedicated to the use of TV to improve classroom instruction and efficiency and helps to identify curriculum areas and courses where ITV may be used beneficially.

Depending on the size and scope of the operation, a Director of ITV may supervise from four to 20 people. The small settings usually include technical and operations employees. At such stations, the Director of ITV is responsible for selecting and renting or otherwise acquiring prerecorded instructional programs for transmission.

At the larger operations at 2- and 4-year colleges and universities and large school districts, the Director of ITV is also responsible for the original production and transmission of programs that meet defined educational objectives. He or she hires and supervises a large staff, including Producers, Directors and a number of production, operations, and engineering people, as well as the teachers who will appear on camera.

A Director of ITV often reports to a Director of Media Services. Occasionally, ITV operations are part of the duties of the General Manager of a public television station licensed to a school or college. As a rule, however, a Director of ITV heads up a separate department within an educational institution and reports to an assistant superintendent of a school system or a vice president of academic affairs at a college or university.

Additionally, the Director of ITV:

- develops and manages the budget for technical equipment and operational expenses;
- plans and develops research regarding the effec-

tiveness of instructional television;

• generates awareness of the ITV operation both within the academic system and by the public.

Salaries

Salaries in 1980 ranged from $14,000 for somewhat experienced Directors of ITV at small school operations to more than $25,000 for highly experienced, full-time individuals at institutions of higher education, according to professional sources. As is usually the case in education, payment is based on a combination of the individual's academic degree(s), seniority, and experience. Salaries are higher at the larger school systems. In general, Directors of ITV operations at 4-year and community colleges earn more than their peers at the school or district level.

Employment Prospects

Opportunities are only fair. During 1978-79, according to the Corporation for Public Broadcasting (CPB), 71 percent of all institutions of higher education (2,129 establishments) used television for instructional purposes, and most of them employed an individual who served as a Director of ITV. The situation was similar at the elementary and secondary school level.

While the number of ITV systems has grown in recent years, the number of people serving specifically as Directors of ITV is estimated at less than 3,000, according to professional sources.

In many school districts, the Director of ITV continues to have other media, teaching, or curriculum duties. Full-time positions are usually available only at colleges, universities, and larger school districts. Some Producers of ITV are promoted to the position when an opening occurs. More often, a talented ITV Coordinator or classroom teacher is promoted to Director of ITV. There is little turnover in the field, and, with expenditures for education decreasing, fewer new ITV operations are expected in the future.

Advancement Prospects

Opportunities for professional advancement are poor. There is little turnover and competition is great. Many Directors of ITV see their position as the climax to a successful career. Some, however, are promoted to Director of Media Services at their own school district, college, or university. Others move to larger systems in other areas, and a few join systems in health or government telecommunications operations.

Education

A master's degree is required for the job at the school or community college level, and a doctorate is usually necessary at 4-year colleges or universities. Most Directors of ITV have extensive schooling in educational media development, design, and utilization, as well as in television production and communications. Course work in educational psychology, curriculum design, and instruction are usually mandatory.

Experience/Skills

Some experience in classroom teaching is important, particularly in school systems, where actual experience in classroom instruction is usually a prerequisite for the position. Candidates should have at least two to three years of classroom and instructional television experience.

A Director of ITV must be a professional with sound educational judgment. In addition to good speaking and writing abilities, he or she must possess excellent interpersonal skills to deal with colleagues and subordinates, and be organized and persuasive. Strong administrative and leadership talents are also important.

Minority/Women's Opportunities

Opportunities for women are excellent. According to professional estimates, nearly 40 percent of such positions at the school district level were occupied by white females in 1980. The number of white females employed as Directors of ITV in higher education is lower and the number of minority-group members in the position was even fewer, except in urban settings.

Unions/Associations

There are no national unions that represent Directors of ITV for bargaining purposes.

As teachers or faculty members they are represented by the American Association of University Professors (AAUP), the National Education Association (NEA), or the American Federation of Teachers (AFT). Many belong to the Association for Educational Communications and Technology (AECT) and, in particular, to its Division of Telecommunication (DOT). Many Directors of ITV at community and junior colleges represent their institutions as members of the Instructional Telecommunications Consortium (ITC) of the American Association of Community and Junior Colleges (AACJC).

ITV COORDINATOR

CAREER PROFILE

Duties: Responsibility for the use of instructional television programs at a school, school district, or public broadcasting station

Alternate Title(s): ITV Utilization Specialist; Director of Instructional Services (public television)

Salary Range: $9,500 to $40,000

Employment Prospects: Fair

Advancement Prospects: Poor

Prerequisites:

Education—Minimum of an undergraduate degree in education; master's degree in media preferable

Experience—Minimum of one to two years of classroom teaching or as a Media Librarian

Special Skills—Organizational abilities; leadership qualities; interpersonal and language skills

CAREER LADDER

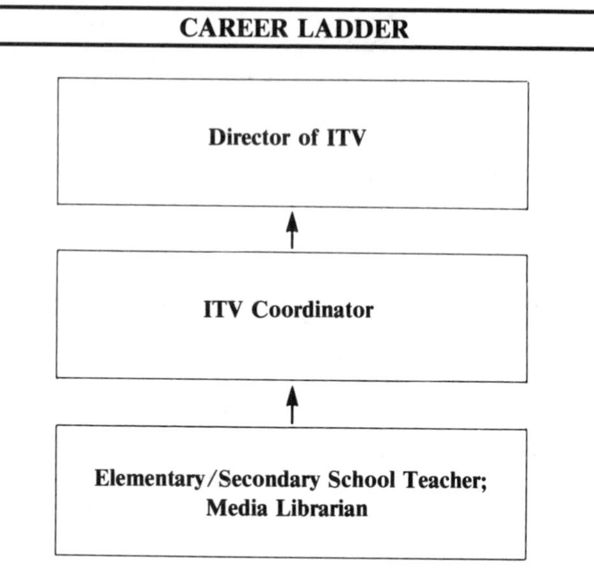

Position Description

An ITV (instructional television) Coordinator is responsible for facilitating the most effective use of televised programs in a school or school district. He or she collaborates with classroom teachers to devise and help implement methods to improve the learning process with the use of instructional television programming.

Usually ITV Coordinators are employed directly by a school district or system, where they work mainly with elementary and middle (junior high) school teachers, or, sometimes, are assigned to work with a public TV station. Some are full-time members of that station's staff. Nearly one-half of the public stations employed such people as Directors of Instructional Services in 1980, according to the Corporation for Public Broadcasting (CPB). ITV Coordinators are not usually employed in institutions of higher education.

ITV Coordinators usually work out of a school district office or administrative headquarters and spend most of their time in schools and classrooms, helping teachers to get full use of television programs in an educational setting. They often conduct in-service training programs on the use of instructional television for groups of teachers, either via TV or in person.

An ITV Coordinator tries to develop a team-teaching environment in which an instructor appearing in an ITV program supplements the classroom teacher with informative and highly visualized television presentations. The ITV Coordinator also prepares and distributes lesson guides, manuals, and ancillary materials for use by the classroom teacher. The teacher then prepares the students for the TV lesson, supervises the actual viewing, answers questions, leads discussions, makes assignments, and gives tests on the material. The ITV Coordinator encourages this learning approach and assists the teacher in making maximum use of television instructional materials.

Some large school districts employ as many as four ITV Coordinators, each assigned to a particular geographic or subject area. In small systems, the duties are often combined with those of other ITV or media employees, such as the Media Librarian.

As a rule, an ITV Coordinator reports to a Director of Instructional Television or a Director of Media Services. The position is not usually a supervisory job.

Additionally, the ITV Coordinator:

- helps classroom teachers and curriculum specialists to acquire, preview, and select ITV programs;

• develops ITV transmission schedules with classroom teachers and administrators;
• conducts research and evaluates the extent and effectiveness of ITV use in classrooms.

Salaries

1980 earnings ranged from $11,000 for relatively inexperienced individuals, to more than $18,000 for those with extensive background and seniority, according to professional estimates. As in other educational occupations, salaries are based on the level of academic achievement and seniority. Individuals with graduate degrees and substantial experience in a school system often earn more, particularly in large urban systems.

At public TV stations, ITV Coordinators usually earn more than their peers in school systems. For the related position of Director of Instructional Services, 1980 salaries ranged from $9,500 to $40,000, according to the CPB. The average was $21,300, which was an increase of 9.5 percent over the previous year.

Employment Prospects

Large school system TV operations often employ full-time ITV specialists. Sometimes a Media Librarian is assigned to or assumes the role of ITV Coordinator. The majority of public television stations usually employ more than one individual to develop and encourage instructional television and video utilization. There is, however, little turnover in the field and the growth of new ITV systems in today's educational environment is slow. Chances of getting a job as an ITV Coordinator are, therefore, only fair.

Advancement Prospects

The possibilities for career advancement are poor, according to professional observers. Most ITV Coordinators are professional teachers who seek or are assigned to the position, and their career progress often depends upon increasing their subject area competence or administrative abilities in the system.

There is little turnover at the next level, the more responsible position of Director of ITV, and the competition is heavy. Those who wish to advance their careers are usually required to seek other media and Director of ITV positions in larger school systems or at universities or community colleges. A few find opportunities in national educational organizations, which occasionally employ specialists in the use of ITV.

Education

A bachelor's degree in education is required, and a master's degree in media is preferable. Courses in television and media production, curriculum design, and educational psychology are necessary in obtaining the position, as well as some in-service training or seminars in instructional television use.

Experience/Skills

One or two years of classroom teaching and some work in television, video, and other media production or use are preferable. The experience may be gained at an organization producing ITV programs, or in a classroom where extensive use is made of instructional television and video. Work as a Media Librarian, with ITV responsibilities, for at least one or two years may also be considered sufficient background for the position.

ITV Coordinators must be enthusiastic proponents of televised instruction. They should be well-organized, have leadership qualities, and excellent interpersonal and language skills.

Minority/Women's Opportunities

The majority of ITV Coordinators in 1980 were white females, according to professional estimates. Some estimates put the figure at more than 65 percent. Minority-group employment is considerably lower, and is mainly in urban centers and large school systems and districts. Most estimates indicate that employment of members of minorities is around 15 percent nationwide.

Unions/Associations

There are no unions that serve as representatives or bargaining agents for ITV Coordinators. As teachers, however, some are represented by the National Education Association (NEA) or the American Federation of Teachers (AFT).

Many individuals belong to the Association for Educational Communications and Technology (AECT) and its Division of Telecommunications (DOT), or to the Information Film Producers of America (IFPA) to share mutual concerns and advance their careers. In addition, some are members of the Educational Film Library Association. (EFLA).

COMMUNICATIONS INSTRUCTOR

CAREER PROFILE

Duties: Teaching college-level courses in advertising, journalism, film, and television

Alternate Title(s): Assistant Professor; Associate Professor; Professor

Salary Range: $13,000 to $35,000+

Employment Prospects: Fair

Advancement Prospects: Fair

Prerequisites:

Education—Doctorate degree preferable

Experience—Some college-level teaching; professional experience helpful

Special Skills—Teaching aptitude; verbal and writing abilities; leadership qualities; ability to motivate others

CAREER LADDER

Department Chairperson; Dean

↑

Communications Instructor

↑

Graduate Assistant; Graduate School

Position Description

Communications Instructor is a general title for someone who teaches courses in any of the communications disciplines. Only courses that have a relationship to television, however, will be discussed here, including advertising, journalism, film, and TV. They may be taught at a community college, 4-year college, or university. Such an Instructor can hold any rank in the teaching hierarchy, from lecturer to full professor, and may teach in an undergraduate, graduate, or continuing education program.

Obviously, no two colleges are alike. Communications courses are taught in various departments and schools at different institutions. Some have full curricula and others offer only a few courses in a particular field. There are universities that have separate schools of communications or journalism. Many smaller colleges teach only a handful of TV-related courses and offer them through the theater, English, speech, or other department. There is, therefore, no clear-cut pattern in the way college-level communications classes are presented.

A Communications Instructor specializing in advertising is frequently a member of a business or journalism school or department. In addition to introductory advertising courses, he or she may teach classes in specific areas such as television and radio commercials, copywriting, or market research.

A journalism Instructor is often a faculty member of a separate school or department of journalism or of a department that is part of a communications school. Because of the importance of television news, there are frequently specialized courses offered. An Instructor may teach classes in news writing, news gathering techniques, and news management as related to broadcasting.

Since film and television are so closely connected, many schools teach them in a single department. If there is a large TV-film curriculum, it is often taught in a separate communications school. If only a few courses are available, they may be taught as part of a theater department. A Communications Instructor can teach in a number of areas, including mass communications theory, the history of broadcasting, television production, film-making, and broadcast management.

Additionally, the Communications Instructor:

- advises students regarding registration, special projects, internships, and graduate theses;
- serves on department and university committees for administrative or curriculum policy making;
- develops proposals for research grants in his or her specialty and administers grant projects.

Salaries

Salaries for Communications Instructors in 1980 ranged from $13,000 for beginners to more than $35,000 for full professors with considerable experience, reputation, and seniority. The median salary for jobs listed with the placement service of the Speech Communication Association (SCA) was $15,000 in 1979-80. Communications Instructors at community colleges are generally paid less at the upper teaching levels than their peers at 4-year institutions.

Employment Prospects

Chances of getting a position as a Communications Instructor are only fair. This appointment requires considerable academic study and apprenticeship. Some professional experience in the field being taught is extremely helpful.

The Association of Communication Administration (ACA) of the SCA reports that there were 449 job listings in the SCA Placement Service in 1979-80. Half were for new employees, mostly at the assistant professor level, in broadcasting, mass communications, and film departments, predominantly in the midwest and the south.

Advancement Prospects

Faculty at institutions of higher education usually begin as instructors and are promoted through the successive ranks of assistant professor, associate professor, and finally, full professor. Occasionally, a Communications Instructor can become chairperson of the department and, in some instances, dean of the school. Although there is relatively little turnover, some Communications Instructors get more rewarding positions and career advancement at more prestigious or larger institutions.

Promotion is usually based on publishing in professional media and a good teaching record. Most colleges and universities emphasize research and writing. Sometimes recognition is given for professional activity in the field, such as producing a documentary. Overall advancement opportunities are fair.

Education

A beginning Communications Instructor should have a doctorate from a recognized and accredited university, although some departments or schools will employ an experienced individual with particular expertise in a discipline and a master's degree. Many community colleges prefer to employ Ph.D.s but will accept master's degree holders. A few accept individuals with bachelor's degrees, but extensive course work and, perhaps, a growing reputation in a specific field are required. Most Communications Instructors have their graduate degrees in speech communications, with an emphasis on a particular discipline.

Experience/Skills

Most colleges and universities expect beginning instructors to have had some experience as graduate teaching assistants. It is extremely helpful if the candidate has also had some professional experience in his or her field.

For assistant, associate, and full professors, successful teaching background is usually a requirement, but the emphasis is on the individual's record of research, published works, projects in the field, and, perhaps, standing in the professional world. Community colleges usually place more emphasis on teaching experience and less on research and writing.

Communications Instructors must have good verbal and writing skills, good leadership qualities, the ability to motivate, and excellent interpersonal skills. Good Instructors are creative, imaginative, and often innovative. Above all, they like to teach.

Minority/Women's Opportunities

According to most academic observers, about 35 percent of Communications Instructors in 1980 were white females. The percentage of minority-group members was lower, except at major universities in urban centers or at schools with a significant minority-group enrollment. All public institutions of higher education are extremely active in affirmative action and equal employment opportunity (EEO) programs, and aggressively recruit minority members and female employees.

Unions/Associations

Some faculty members belong to the American Association of University Professors (AAUP), the American Federation of Teachers (AFT), or the National Education Association (NEA), which bargain for salary increases and benefits on their behalf. Some are members of a state educational association. Many are not represented by any organization.

Many Communications Instructors who teach television studies belong to the Broadcast Education Association (BEA) and the SCA to share common professional concerns. Advertising Instructors often belong to the Business/Professional Advertising Association (BPAA). Journalism Instructors may belong to the Association for Education in Journalism (AEJ). Many female Communications Instructors belong to Women in Communications Inc. (WIC).

MEDIA INSTRUCTOR

CAREER PROFILE

Duties: Teaching college-level courses in educational media

Alternate Title(s): Assistant Professor; Associate Professor; Professor

Salary Range: $13,000 to $30,000+

Employment Prospects: Fair

Advancement Prospects: Fair

Prerequisites:

Education—Minimum of a master's degree in education; doctorate degree preferable

Experience—Minimum of one to two years teaching and media laboratory supervision

Special Skills—Teaching aptitude; knowledge of media; verbal and writing abilities; leadership qualities; ability to motivate others

CAREER LADDER

Department Chairperson; Dean

↑

Media Instructor

↑

Graduate Assistant; Graduate School

Position Description

Media Instructor is a general title for someone who teaches courses that pertain to the educational uses of a wide variety of media, including television and video. The courses are offered for undergraduate, graduate, or in-service teacher education credit at a community college, 4-year college, or university. He or she instructs people who are, or are going to be, teachers as to when, how, and why to use media as part of the classroom instructional process.

Theoretical courses taught by the Media Instructor explain how media work in the learning process to spark interest, to enhance classroom instruction, and to supplement printed materials. In practical courses and laboratories, the Media Instructor discusses specific media, their advantages, availabilities, limitations, and costs. In addition, he or she explains what kinds are appropriate for various learning situations, how to operate equipment, what programs can be used, and how to create media projects inexpensively.

In the past, Media Instructors dealt with the traditional supplementary materials, such as film strips, slides, and 16mm films. While they continue to use these formats, most are also teaching about new resources, including computer-assisted instruction, programmed learning, electronic learning laboratories, television, and video. It is the Media Instructor's responsibility to educate and train his or her students in the theory, use, and techniques of all such learning aids.

A Media Instructor not only teaches classes, but also may be involved with his or her institution's media center, which is where media are produced or obtained for other faculty members teaching any subject. The Media Instructor also conducts individual laboratory sessions detailing how to run and operate specific equipment, holds seminars on new techniques, and lectures in other education classes to highlight media use. He or she advises other educators about media applications in their subjects. As a result, Media Instructors are not only members of the faculty, but also serve as administrative and support employees. They, therefore, often teach only one or two classes each semester.

As a rule, a Media Instructor is a member of the faculty of the school or a department of education, and can hold any rank from lecturer to full professor.

Additionally, the Media Instructor:

- evaluates and selects instructional materials and equipment for classroom use and lab training;
- supervises student workers at media instructional or learning resource centers;
- prepares grant proposals for additional media support.

Salaries

Media Instructors' salaries range from $13,000 for beginners at community colleges to more than $30,000 for full professors with excellent reputations, lengthy service, and administrative duties. Some who have administrative duties are employed on a 12-month basis. All enjoy the standard institutional fringe benefits, including group health and life insurance and retirement plans.

Employment Prospects

During the 1960s and the 1970s, Media Instructors were in great demand. The federal, state, and local emphasis on education, and particularly on new techniques of instruction, created a demand for qualified teachers in the new media. This is less true today, and employment prospects are only fair.

Candidates are usually screened by a search committee. Selection is made on the basis of academic credentials, references, professional writing, research reputation, and on whether the particular academic strengths meet an instructional need at the institution. Many Media Instructors are chosen from the ranks of graduate school students or graduate assistants.

Advancement Prospects

Opportunities for advancement are fair. Teachers usually move up through the ranks of assistant professor, associate professor, and finally to full professor. Promotion is competitive, and is based largely on an individual's research, publication of papers, and teaching abilities. A Media Instructor may become chairperson of the media or education department or may, in some cases, become dean. Some leave the educational profession for higher-paying positions in private industry. Others use their video, television, and training skills to obtain more financially rewarding jobs in business or health media operations.

Education

A Media Instructor should have at least a master's degree in education, although a doctorate is preferred. Courses and training in educational psychology, communications, learning theory, and curriculum design are also required.

Experience/Skills

Media Instructors must have experience operating television cameras, switchers, audio equipment, and videotape and videodisc players and accessories. At least one or two years of experience in teaching and media laboratory supervision is usually required. Media Instructors are usually excellent teachers with knowledge of the traditional instructional support equipment, as well as the new media, their operation, and use in education.

Good verbal and writing skills are essential, as are creativity, an innovative approach to problem-solving, and leadership qualities. Individuals should enjoy teaching, and be able to motivate students.

Minority/Women's Opportunities

In 1980, approximately 30 percent of the Media Instructors were white females, according to academic observers. Minority-group member employment was not as high, except at institutions that had significatnt student enrollments of minority members. All public institutions are actively involved in affirmative action and equal employment opportunity programs, and actively seek women and minority-group employees.

Unions/Associations

Some Media Instructors belong to the American Federation of Teachers (AFT), the American Association of University Professors (AAUP), or the National Education Association (NEA), which act as bargaining agents for salary increases and benefits. In addition, most are active in the Association for Educational Communications and Technology (AECT) and in their particular state associations of that organization, to share mutual concerns and advance their careers. Some belong to the University Film Producers Association (UFPA) and others to the Educational Film Library Association (EFLA).

MANAGER OF AUDIO-VISUAL SERVICES

CAREER PROFILE

Duties: Responsibility for the development and use of audio-visual media in private industry

Alternate Title(s): Manager of Media Services

Salary Range: $19,000 to $44,000

Employment Prospects: Good

Advancement Prospects: Fair

Prerequisites:

Education—Undergraduate degree in communications, educational media, or radio-TV; master's degree preferable

Experience—Minimum of three to four years in broadcast TV or educational media

Special Skills—Organizational ability; knowledge of media; administrative talent

CAREER LADDER

Training Director; Director of Corporate Communications

↑

Manager of Audio-Visual Services

↑

Supervisor of Media Services

Position Description

A Manager of Audio-Visual Services is in charge of a corporate media center within any industry. He or she is responsible for the planning, development, and use of audio-visual and media services to satisfy the company's training and communication needs. He or she develops ideas for media projects that use traditional audio-visual materials as well as film, television, and video to support corporate policies and objectives for employee training, public relations, marketing, sales, and internal communications.

According to audio-visual analyst Tom Hope ("AV Media Outlook," *Photomethods*, June 1979), business and industry account for 52 percent of all audio-visual purchases. To meet the requirements of corporate communications, a Manager of Audio-Visual Services supervises the selection and use of a wide range of media formats. He or she creates videocassettes for training, and to introduce manufacturing procedures or new company policies. Projects may also include the production of periodic company video newsletters, video product displays, slide, film, or video presentations, annual video stockholder reports, and public and community relations videotapes.

For a large company with many employees, the Manager of Audio-Visual Services might establish video playback stations in a number of geographic locations, and distribute TV training and employee communications videocassettes to them. In smaller firms, he or she may be responsible for contracting with outside TV or media production companies to produce videocassettes, films, slides, or multi-media shows, for internal company use.

In a major company, the Manager of Audio-Visual Services may supervise from five to 20 people, and oversees the Supervisor of Media Services, who produces the projects. In a small firm, the Manager of Audio-Visual Services may be the only media person in the company. He or she usually reports to the training director of the company, or, increasingly in larger firms, to the director of corporate communications.

Additionally, the Manager of Audio-Visual Services:

- determines overall policies and budgets for audio-visual and media use in the corporation;
- establishes specific budgets and schedules for in-house or outside production services;
- develops research and evaluation systems to measure the effectiveness of audio-visual use.

Salaries

Managers of Audio-Visual Services earned from $19,000 to $44,000, according to a 1981 survey by Tom Richter for the International Television

Association (ITVA). The average was $29,900. Most companies provide liberal fringe benefits, including health and life insurance, and retirement plans. Some individuals participate in stock option programs, and a few receive annual bonuses based on performance goals.

Employment Prospects

Although this is not usually an entry-level job, except in small companies, the opportunities for employment are good because of increasing growth. In larger companies, the Supervisor of Media Services is frequently promoted to the position when openings occur.

Most industry observers expect that the use of media in corporate training and communications will increase at the rate of 30 to 40 percent annually in the next few years. Another indication of expected growth, is the number of college graduates in media who are entering the field. According to Darryl Sink (in *Educational Media Yearbook 1980*, Brown and Brown, Libraries Unlimited, Littleton, CO), 20 percent of recent media graduates were employed in business or industry.

Advancement Prospects

Opportunities for advancement are only fair because of the heavy competition. Some Managers of Audio-Visual Services are promoted to training directors within their own companies on the strength of their experience and media background. At larger organizations, where external communications is emphasized, an individual may be promoted to director of corporate communications. Some Managers of Audio-Visual Services move to the same position in larger companies to advance their careers, and others take positions as training directors or directors of corporate communications in smaller companies in other industries.

Education

An undergraduate degree in communications, educational media, or radio-TV is the minimum requirement, and a master's degree is strongly recommended. In a study of 326 corporate non-broadcast video professionals, Mark Johnson (*Educational and Industrial Television*, January 1980) found that 41.9 percent had a degree in communications and radio-TV, and 11.8 in education. In a study of 265 business training and media personnel, Les Streid (*Instructional Innovator*, January 1981) found that 45 percent had a bachelor's degree and 36 percent had a master's degree.

Courses in media, education, educational technology, and instructional development are important. Studies in mass communications or radio-TV, and course work in television production and videotape editing, along with some training in traditional audio-visual materials are also helpful.

Experience/Skills

Three to four years of experience in broadcast television or educational media is necessary. Experience in a training environment is vital, as is familiarity with the use of a wide variety of media and equipment in support of instructional or communications objectives.

A Manager of Audio-Visual Services must be a highly organized media generalist with substantial administrative skills. He or she must be capable of analyzing the economic results of media use in various applications, and of designing and implementing instructional programs. Most are bright, creative, and aggressive individuals with excellent interpersonal skills.

Minority/Women's Opportunities

Opportunities for women are excellent. In 1980, nearly 35 percent of the Managers of Audio-Visual Services were white females, according to industry observers. The opportunities for minority-group members are not as good, but minority employment in the position has risen during the past five years, particularly in urban areas and in large corporations.

Unions/Associations

There are no unions that serve as bargaining agents for Managers of Audio-Visual Services. Individuals often belong to a number of professional organizations, however, to share mutual concerns and advance their careers. Among them are the International Television Association (ITVA), the Audio-Visual Management Association (AVMA), the National Audio-Visual Association (NAVA), the International Association of Business Communicators (IABC), the American Society for Training and Development (ASTD), and the Industrial Training and Education Division (ITED) of the Association for Educational Communications and Technology (AECT).

SUPERVISOR OF MEDIA SERVICES

CAREER PROFILE

Duties: Producing audio-visual projects and managing a media center in business and industry

Alternate Title(s): Media Center Manager

Salary Range: $16,000 to $32,000

Employment Prospects: Good

Advancement Prospects: Good

Prerequisites:

Education—Undergraduate degree in mass communications, educational media, or radio-TV; graduate degree preferable

Experience—Minimum of two to three years in media production

Special Skills—Technical aptitude; creativity; organizational ability

CAREER LADDER

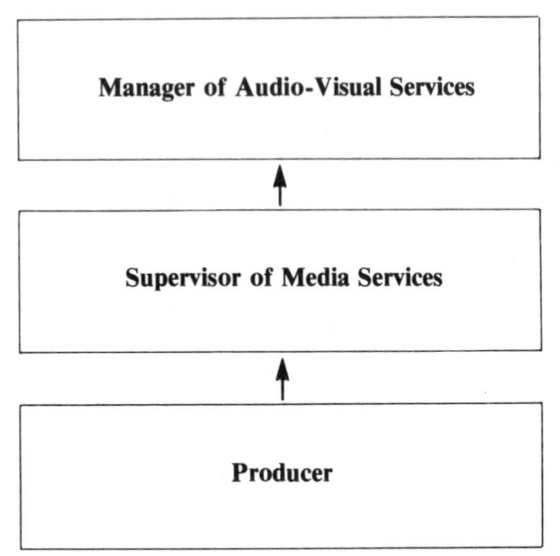

Position Description

A Supervisor of Media Services is usually the second in command at a corporate media center, reporting directly to a Manager of Audio-Visual Services. In the absence of the Manager, he or she acts as the top administrator of the center. The main responsibility of the Supervisor of Media Services is to oversee the production of media projects and manage the day-to-day operation of the center. He or she creates films, slides, audio, and television and videocassette programs for the center and for other units within the company.

Often a Supervisor of Media Services coordinates a number of simultaneous media projects, oversees all pre-production stages, ensures timely and efficient completion, and supervises post-production editing. He or she works with Instructional Designers, heads of manufacturing departments, public relations directors, marketing personnel, and training directors to produce multi-media slide presentations, instructional television training programs, corporate-image films, and other media materials.

At most media centers in business and industry, the production staff is quite small, with a few individuals who double in many capacities for routine in-house productions. *Private Television Communications: Into the Eighties* (Brush and Brush, International Television Association, Berkeley Heights, NJ, 1981) indicates that the median staff size of corporate members of the ITVA in 1980 was 3.5 people.

At smaller audio-visual operations, the Supervisor of Media Services often writes and produces videotape training programs and other media projects. He or she sometimes hires and oversees independent commercial production firms or freelancers for writing and actual production. At larger companies he or she often acts as Production Manager and supervises a full-time in-house staff, that may include Producers, Directors, Instructional Designers, and a number of production and engineering people.

Additionally, the Supervisor of Media Services:

- assists in developing overall operating and capital budgets for the department;
- administers budgets for specific media productions;
- implements research and feedback mechanisms to evaluate the effectiveness of media projects.

Salaries

Salaries ranged from $16,000 for people at smaller companies to more than $32,000 for individuals with a good deal of supervisory responsibility, according to the 1981 Annual Salary Survey of the ITVA conducted by Tom Richter. The average salary was $24,000. The majority of Supervisors participate in company health, life insurance, and pension plans.

Employment Prospects

Opportunities in private industry are good. In a survey conducted by Mark Johnson ("Do Colleges Prepare Students for Non-Broadcast Media?" *Educational and Industrial Television*, January 1980), the respondents indicated that they expected a 47 percent increase in hiring in the field over the next five years. The areas with higher-than-average employment growth were the oil, gas, and retailing industries. The use of audio-visual and media techniques for increased productivity and intra-company communications as well as for teleconferencing and the new information technologies of the office-of-the-future, offer good opportunities for alert and diligent individuals.

Advancement Prospects

A well-trained, ambitious communications specialist should find that the opportunities for advancement are good.

With the increasing need for efficiency and productivity, newer technologies, such as two-way cable TV, word processing equipment, videodiscs, and teleconferencing networks are being harnessed by business to improve communications and training. A bright individual with some experience in media, who can adapt to other methods of originating, processing, and transmitting information, should find opportunities for career growth.

Some Supervisors of Media Services assume broader responsibilities as Managers of Audio-Visual Services within their own companies, but there is relatively little turnover in that position. Some move to smaller firms in other industries, or to government, health, or education operations to assume more general media responsibility, often in a public relations capacity. A few retrain to become information processors, working with and for data processing and telemetrics companies to develop new techniques of transmitting and sharing information.

Education

A bachelor's degree in mass communications, educational media, or radio-TV is usually the minimum requirement, and a graduate degree in one of these fields is often preferable. Undergraduate or graduate level courses in instructional design, television and film production, script writing, photography, and industrial psychology are also helpful.

Experience/Skills

Two to three years as a Producer of media projects and programs is usually necessary for the job, as is experience in all media production techniques, including audio, video, television, slides, film, and still photography. Some instructional design experience is often mandatory.

A candidate should have initiative, enthusiasm, good technical skills, and be a creative self-starter. A Supervisor of Media Services must be organized and have excellent verbal and script writing abilities as well as strong interpersonal skills.

Minority/Women's Opportunities

The employment of women in this position has increased rapidly in the past five years, according to industry observers. The opportunities for members of minorities, however, are not as good, except in the major urban centers and in large corporations.

Unions/Associations

There are no unions that serve as bargaining agents for Supervisors of Media Services. Some individuals, however, belong to the organizations cited in the career entry for Manager of Audio-Visual Services. In addition, some are members of the Association for Multi-Image (AMI), Information Film Producers of America (IFPA), and the Division of Educational Media Management (DEMM), the Division of Telecommunications (DOT), and the Media Design and Production Division (MDPD) of the Association for Educational Communications and Technology (AECT).

CHIEF, AUDIO-VISUAL SERVICES

CAREER PROFILE

Duties: Administering media services for a government agency

Alternate Title(s): Director of Audio-Visual Services; Media Resources Director

Salary Range: Under $25,000 to $40,000

Employment Prospects: Poor

Advancement Prospects: Poor

Prerequisites:

Education—Minimum of an undergraduate degree in radio-TV, communications, or educational media; graduate degree preferable

Experience—Three to five years in educational media or communications, and media administration

Special Skills—Organizational abilities; leadership qualities; administrative and interpersonal skills

CAREER LADDER

Director of Public Affairs; Director of Communications; Training Director

↑

Chief, Audio-Visual Services

↑

Audio-Visual Program Officer

Position Description

A Chief, Audio-Visual Services is responsible for the administration of media services at a local, county, state, or federal government agency for training, internal communications, and public information purposes. It is his or her responsibility to help identify the agency's training and information needs that can best be satisfied with media support. Government agencies use media primarily for specialized and distinct internal staff training needs, particularly to inform and instruct employees who are geographically separated from the organization's headquarters, and to promote and explain the department's service to the general public.

After consulting with administrators, agency heads, and public information officers, the Chief, Audio-Visual Services plans media support materials. He or she supervises the production of original charts, graphs, slides, film, video, and other media programs, or obtains them from outside sources. He or she notifies agency heads and administrators about the availability of video and other formats for agency use.

The federal government is the largest employer of audio-visual specialists. About 44 executive departments and agencies were actively involved in the production or acquisition of films, slides, video programming, and other media. The largest is the Department of Defense, but agencies ranging from the Department of Agriculture to the Veterans Administration use and produce audio-visual materials. Some state government executive and legislative branches also operate media departments. At the county and city level, police, fire, or safety departments occasionally maintain media centers.

A Chief, Audio-Visual Services may supervise from two to more than ten workers, depending on the size of the agency and the extent that it uses video and other media. In small operations, he or she may supervise a Media Specialist, Media Librarian, and, occasionally, a Media Technician. In larger centers that produce video and 16mm film programs, he or she may supervise Audio-Visual Program Officers and their small staffs. At most levels of government, the Chief, Audio-Visual Services usually reports to a director of public affairs or information, or to a director of training, of communications, or of personnel.

Additionally, the Chief, Audio-Visual Services:

- evaluates, previews, rents, or purchases video programs and films for internal training use;
- supervises the operation of audio-visual and video equipment;
- develops and administers capital and operating budgets for the media center.

Salaries

Salaries for government employees are determined by the grade of the position within a local, state, or federal civil service classification system. The federal government classifies its employees according to a GS (government service) rating that assigns a salary range for specific job occupations, from GS-1 to GS-18. Most local, county, and state governments operate under a similar system.

Most Chiefs, Audio-Visual Services in the federal civil service system are ranked at GS-12, GS-13, or a higher level. Starting salaries in 1980 were $27,000 and $32,000 respectively. Most have worked for some time in the position and earn up to $40,000 depending on their length of service and on their automatic annual increases. Those in similar positions at non-federal government agencies, however, usually receive lower salaries, starting at under $25,000, depending on the size of the agency and the classification of the position. All government employees receive health insurance and participate in a pension plan.

Employment Prospects

Almost every government agency maintains some audio-visual equipment and materials, and some use television and video extensively. The federal government, however, has recently placed a moratorium on the expenditure of funds, particularly for the production of films and video programs. In addition, the general tightening of budgets in local, county, and state agencies has created cutbacks in audio-visual production. Nevertheless, some Audio-Visual Program Officers with management skills and seniority are occasionally promoted. Overall opportunities for the position of Chief, Audio-Visual Services at this time, however, are poor.

Advancement Prospects

The general lack of growth in audio-visual and other media expenditures at all levels of government has made opportunities at newly established centers extremely limited. A move to a larger agency in a similar position within the same governmental structure often means only a lateral transfer of responsibility, with little or no increase in grade or salary.

The position is viewed by many as the climax to a successful career, but some take similar positions in business for advancement. A few further their education and obtain higher GS ratings as directors of public affairs, directors of communications, or as training directors within other governmental systems.

Education

An examination or other preclassification evaluation is usually necessary for civil service positions and the candidate must meet at least the minimum established requirements for the job. Most systems are extremely broad in nature but this job usually requires a minimum of a bachelor's degree in radio-TV, communications, or educational media. In a few cases, a master's degree in one of the fields is required. At some agencies that use media primarily for external public information, degrees and course work in communications and radio-TV production are helpful. At agencies where the emphasis is on media production for internal training, course work in educational media and instructional design is useful.

Experience/Skills

Federal civil service usually requires three to five years of experience in some area of educational media or communications, and a background in media administration is usually expected.

Candidates must be well-organized and capable of leadership and administration. They should possess excellent interpersonal skills to work with a diversity of people, and be enthusiastic promoters of governmental use of video and other audio-visual techniques.

Minority/Women's Opportunities

Despite strong affirmative action programs in recent years and the requirements of the Equal Employment Opportunities Act, the majority of Chiefs, Audio-Visual Services in 1980 were white males. Female and minority-group member employment has increased greatly in the job, however, particularly in the federal service in Washington, DC.

Unions/Associations

There are no unions that specifically represent Chiefs, Audio-Visual Services as bargaining agents. Such governmental employees are sometimes represented by the American Federation of State, County, and Municipal Employees (AFSCME). Some in the federal government belong to the Federal Educational Technology Association (FETA) to share common concerns and advance their careers. Other people belong to specific divisions or affiliates of the Association for Educational Communications and Technology (AECT).

AUDIO-VISUAL PROGRAM OFFICER

CAREER PROFILE

Duties: Operating a specific media section in a government agency

Alternate Title(s): Audio-Visual Branch Head; Media Specialist

Salary Range: $15,000 to $30,000

Employment Prospects: Poor

Advancement Prospects: Poor

Prerequisites:

Education—Undergraduate degree in radio-TV, communications, or educational media

Experience—One to two years of college production courses or professional media work in education and training

Special Skills—Creativity; writing ability; interpersonal skills; supervisory talent

CAREER LADDER

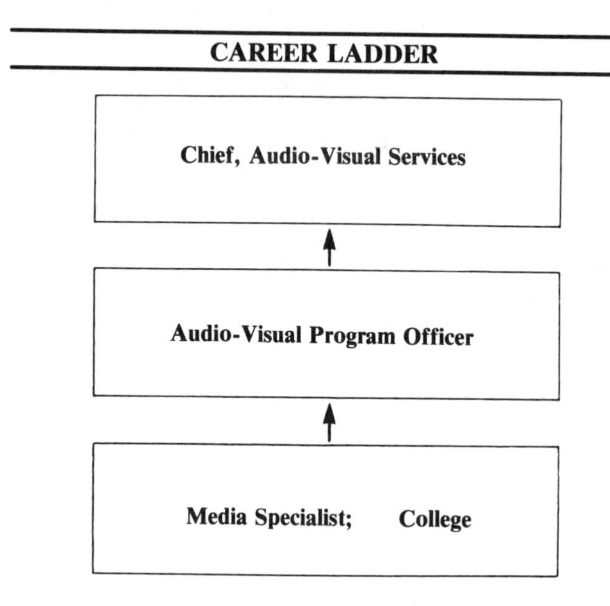

Position Description

An Audio-Visual Program Officer is in charge of the daily operation of a department within a larger media center in a federal, state, city, or local government agency. The specific department may concentrate on exhibits, photography, graphics, motion pictures, radio-TV, or video. It is his or her responsibility to develop and produce training and public service programs and media support materials in response to requests from various divisions or departments within the agency.

At the federal level, such individuals are sometimes referred to as Audio-Visual Branch Heads. At smaller media centers of the federal government and at state agencies, he or she may be classified as a Media Specialist, working in and occasionally supervising employees who are assigned to specific media areas, including radio-TV and video. At the very small media centers of local and county governments, Media Specialists often perform general, more diverse audio-visual duties.

In most settings, however, an Audio-Visual Program Officer specializes in a particular media area and provides operational and support activities in that area as part of the overall service of the media center to the entire agency. He or she usually works on assignment, and conceives and develops productions and supporting materials to meet defined agency goals and objectives. He or she consults regularly with department heads and administrators to identify media projects and to inform them of progress.

When film and video production is undertaken within the government facility, the Audio-Visual Program Officer serves in a capacity similar to a broadcast Production Manager, and may supervise a Producer, Director, Floor Manager, Camera Operator, and, on occasion, Technicians. Often, he or she writes and coordinates a particular project, contracting the actual production to a commercial production firm.

An Audio-Visual Program Officer usually supervises a team of from one to four employees who develop and produce programs and video materials for specific projects. He or she usually reports to the Chief, Audio-Visual Services.

Additionally, the Audio-Visual Program Officer:

- develops the cost analysis and budgeting for particular programs or services;
- prepares the capital and operating budget for his or her unit;
- assists in the development of research and utilization studies that evaluate the effectiveness of specific training programs;
- supervises outside production services on specific projects.

Salaries

Most governmental salaries are determined by the grade level assigned to a particular position. The federal government grades most Audio-Visual Program Officers at GS-10 or GS-11, with a beginning salary range in 1980 of from $20,500 to $22,500. Most earn more, however, since many agencies rank the job at a higher grade and since salaries are affected by seniority. An Audio-Visual Program Officer with several years of experience could earn about $30,000.

At the state, county, or local governmental level, salaries are generally lower, and depend on the size of the agency, its media involvement, grade level of the job, and seniority. Earnings at a smaller agency range from $15,000 for a beginning employee, to over $25,000 for an experienced person with seniority.

Virtually all government employees receive health insurance and pension plans.

Employment Prospects

The general tightening of funds at all levels of government has meant fewer opportunities for employment in media in general, and as an Audio-Visual Program Officer specifically. Fewer new audiovisual services are expected to be developed. Many existing media center budgets have been lowered, and a few centers may be eliminated. Opportunities for employment at this time are poor.

Advancement Prospects

Opportunities for advancement are also poor at all levels of government at present. Some individuals with management skills and seniority, however, are promoted to the position of Chief, Audio-Visual Services when openings occur. Others seek similar positions in the private sector in health, business, or industry to advance their careers.

Education

For most government positions, an examination or preclassification is required and the results compared with the minimum requirements for the job. When hiring an Audio-Visual Program Officer, most state and federal agencies require a bachelor's degree, or equivalent experience, in radio-TV, communications, or educational media. Specific course work in television and video production is necessary. At agencies that emphasize the development of training programs, course work in instructional design is helpful.

Experience/Skills

One to two years of college course work in production techniques, or professional experience as a Media Specialist in the health care field or similar position in another education and training setting is usually required.

Minority/Women's Opportunities

Many women and minority-group members hold the job, particularly at the federal level, but the majority of Audio-Visual Program Officers in 1980 were white males, according to government sources.

There are more opportunities for women and members of minorities in government than in similar positions in non-governmental organizations. Affirmative action programs and the Equal Employment Opportunity Act have opened up more opportunities for minorities and women in the past ten years.

Unions/Associations

There are no specific unions that act as bargaining agents for Audio-Visual Program Officers, although some are represented by local or state government employee unions. Others belong to the Division of Telecommunications (DOT) or other sections of the Association for Educational Communications and Technology (AECT) to share mutual concerns and advance their careers. Still others belong to the Federal Educational Technology Association (FETA) for the same purposes.

MEDIA RESOURCE MANAGER

CAREER PROFILE

Duties: Administration of audio-visual services in a health care facility

Alternate Title(s): Director of Media Services; Director of Audio-Visual Services; Director of Biomedical Communications

Salary Range: $18,000 to $35,200+

Employment Prospects: Fair

Advancement Prospects: Poor

Prerequisites:

Education—Minimum of an undergraduate degree in radio-TV, communications, or educational media; graduate degree preferable

Experience—Minimum of three to four years of multi-media production for a health care or teaching facility

Special Skills—Organizational ability; knowledge of media technology; leadership qualities; administrative capability; interpersonal skills

CAREER LADDER

Director of Health Education

↑

Media Resource Manager

↑

Media Specialist

Position Description

A Media Resource Manager is the chief administrator of the audio-visual center serving a hospital or health care school. He or she is an expert in the cost-efficient development and application of a variety of media to health care education and training. He or she works at hospitals, and at medical, nursing, veterinary, dental, allied health care, and other health-related schools.

In whatever facility, a Media Resource Manager develops programs for human resource development, training, continuing and patient education, internal communications, and public relations that meet the specific objectives of professional departments of the institution. He or she consults regularly with training directors, physicians, nurses, and faculty regarding their use of media support programs. It is his or her responsibility to identify the training and communications needs to be satisfied and to determine the best media approach. The Media Resource Manager supervises the production of health-oriented multi-image slide presentations, instructional films and video programs, as well as other audio or visual projects. He or she informs the health administrative staff of available media programs and equipment.

A Media Resource Manager may supervise from one to four employees, including a Media Specialist, Media Technician, Instructional Designer, Media Librarian, and Cinematographer. When TV and video training is an integral part of the media services, he or she may oversee a unit of up to seven employees, including production and engineering professionals. A staff of this size, however, is employed only in the largest hospitals and video facilities.

In a relatively small number of health care schools, an individual with the related title, Director of Biomedical Communications, fills the role of the Media Resource Manager, but often has a more extensive educational background and greater responsibilities. At the larger health-related schools, a Director of Biomedical Communications supervises audio-visual services, medical television, biomedical photography, medical illustration, educational development, printing services, and other units. He or she was in charge of an average of 23 people in the 1980-81 academic year, according to a survey by the Association of Biomedical Communications Directors (ABCD).

At most hospitals, the Media Resource Manager reports to a director of health education. In some small institutions, he or she reports directly to the hospital administrator.

Additionally, the Media Resource Manager:

• selects, evaluates, and purchases media equipment and programs;

• provides for the operation of media equipment for health care workers or patients;

• develops and administers capital, operating, and staffing budgets for the media center.

Salaries

In 1980, salaries for Media Resource Managers ranged from $18,000 for people with degrees and some background in small hospitals to $25,000, for experienced individuals with seniority at major hospitals, according to industry observers.

Salaries for Directors of Biomedical Communications were considerably higher, averaging $35,200 in 1980-81, according to the ABCD.

Salaries for individuals in either a hospital or health care school vary according to the size of the operation and its geographic location.

Employment Prospects

Opportunities are only fair, with the best chances for employment in large medical schools and hospitals. The extensive experience and education required for the position, and the need to persuade hospital administrators that qualified media experts are necessary, have tended to limit employment opportunities. In general, however, the use of video and other media at health education facilities has increased considerably in the past five years.

Advancement Prospects

There is little turnover in the field, promotion opportunities are generally poor, and competition is strong. Advancement is generally limited to the extent of the individual's further academic study and experience. Candidates should also be willing to move to a new location to get a better position.

To many individuals, becoming a Media Resource Manager is the climax of a successful career in communications and media in the health field. Others get further education in patient education, management, and instructional and organizational development, to become directors of health education at smaller health facilities. Still others move to similar but better-paid positions in media and audio-visual services at larger organizations.

Education

A bachelor's degree in radio-TV, communications, or, preferably, educational media is usually required for a Media Resource Manager's job, and a master's degree is preferred. Courses in instructional development, manpower training, and medical education are helpful. To become a Director of Biomedical Communications, a master's degree is virtually mandatory, and a doctorate is preferable.

Experience/Skills

A minimum of three to four years of experience in multi-media production at a health care or teaching facility is usually required. Many Media Resource Managers have been Media Specialists, and have had experience in producing media training programs. Directors of Biomedical Communications usually have had some years in educational and instructional development and training.

A Media Resource Manager or Biomedical Communications Director must be well-organized, knowledgeable about a variety of media, and have excellent interpersonal skills to deal with a diversity of professional colleagues. He or she must be a good administrator and leader with an enthusiastic desire to improve the use of media in the health sciences.

Minority/Women's Opportunities

In 1980, the majority of Media Resource Managers in hospitals were white males, according to industry observers. An increasing percentage of white females and members of minorities have, however, held the position in the past ten years, particularly in urban areas. Nearly five percent of the Directors of Biomedical Communications in 1980-81 were women, according to the ABCD.

Unions/Associations

There are no unions that serve as bargaining agents for Media Resource Managers or Directors of Biomedical Communications. Some individuals, however, belong to the American Society for Health Manpower Education and Training (ASHET) of the American Hospital Association (AHA) to share mutual ideas and concerns in health and training. For the same reasons, others belong to the Health Education Media Association (HEMA), the Association for Educational Communications and Technology (AECT), the Association of Medical Illustrators (AMI), or the Biological Photographic Association (BPA). Nearly all Directors of Biomedical Communications belong to the Health Sciences Communications Association (HeSCA), and most also belong to the ABCD.

MEDIA SPECIALIST

CAREER PROFILE

Duties: Responsibility for the production and utilization of media in a hospital or health care school

Alternate Title(s): Audio-Visual Specialist; Media Resource Specialist; Media Coordinator

Salary Range: $9,500 to $19,000+

Employment Prospects: Good

Advancement Prospects: Fair

Prerequisites:

Education—Minimum of undergraduate degree in educational media, radio-TV, or communications; master's degree preferable

Experience—One or two years in an education or training media center, or equivalent part-time work in a college

Special Skills—Multi-media production capability; technical aptitude; creativity; writing abilities; interpersonal skills

CAREER LADDER

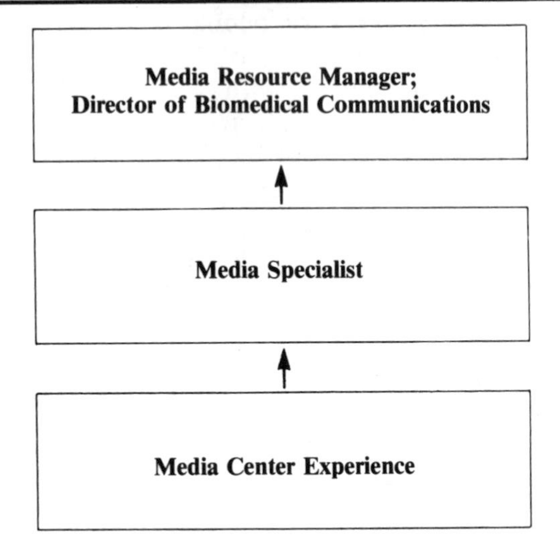

Position Description

A Media Specialist develops and plans for the use of a wide variety of audio-visual support materials to satisfy the instructional, training, continuing education, and community relations needs within a health care or training facility. He or she either produces the necessary materials or selects video and other media programs from outside sources that will best assist in meeting defined educational goals and objectives.

Usually, a Media Specialist works on assignment and conceives and develops film and video programs and slide presentations. He or she may work in a hospital, or a school of medicine, dentistry, nursing, veterinary medicine, allied health care, or other health specialty. The Media Specialist consults with physicians, nurses, and the training staff to determine their media needs and also conducts demonstrations and trains staff in the use of equipment.

In the larger media centers at medical schools or hospitals, a Media Specialist may be a Producer or Director of short demonstration or training programs, or may serve in another capacity in video production. In medium size centers, he or she usually serves as an audio-visual generalist, adept at and experienced with a wide range of equipment and media, including video.

In the smaller health care audio-visual centers, Media Specialists are often responsible for producing many of the traditional educational materials used in patient education or staff training. These may include a series of slides, short films, overhead transparencies, or audio tapes. In many facilities, the Media Specialist is usually the only individual on staff with audio-visual experience.

It is often the responsibility of a Media Specialist to plan and execute simple television instruction and demonstration programs, using non-broadcast video equipment. Frequently, these programs are shot and recorded on location in various parts of the facility, and are played back later. For these simple productions, a Media Specialist acts as a one-person production team.

A Media Specialist normally reports to a Media Resource Manager or to the related position of Director of Biomedical Communications. It is not usually a supervisory job.

Additionally, the Media Specialist:

- assists in the development of studies on the effectiveness of media instruction;
- schedules the playback of media programs within the health facility;
- helps to prepare and administer the capital and operating budgets for the media center.

Salaries

Salaries in 1980 ranged from $9,500 for beginners at small hospitals to more than $15,000 for experienced individuals with seniority at larger hospitals, according to industry sources. Media Specialists at medical schools earned considerably more, according to a survey by the Association of Biomedical Communications Directors (ABCD). Salaries for the related title of Media Coordinator averaged $16,200 in the 1980-81 academic year, while those for people who were primarily Producers averaged $19,000, also according to the ABCD. Media Specialists in hospitals and medical schools participate in all staff benefits, including health insurance and pension programs.

Employment Prospects

Chances of getting a job as a Media Specialist for a trained, diligent individual are generally good. The increased use of audio-visual media, and particularly video, in the health education field has created a number of opportunities. In a survey of the health field reported in (*Video in the 80s*, Dranov, Moore, and Hickey, Knowledge Industry Publications, White Plains, NY, 1980), it was learned that 90 percent of the respondents employed full-time video staffs and that more than one-half were comprised of from one to three employees. Staffs of seven or more were reported by 25 percent of the medical users of video. The opportunities for employment were usually better in 1980 for Media Specialists in the health sciences than for their peers in education and government.

Advancement Prospects

A Media Specialist who has the technical expertise, personal skills, and the necessary education sometimes is promoted to Media Resource Manager in a hospital, and, on occasion, to Director of Biomedical Communications in a medical or other school of health care. Competition for these positions is heavy and opportunities for such growth is only fair.

Personal interests or talents lead others to seek career opportunities at larger health care facilities by concentrating on video as a Producer or Director, on film as a Cinematographer, or, occasionally on curriculum or training development as an Instructional Designer. Usually, however, their ultimate goal is still to become a Media Resource Manager or a Director of Biomedical Communications. Some Media Specialists use their skills to obtain more specialized positions in business, education, or government media centers.

Education

A minimum of a bachelor's degree in educational media, radio-TV, or communications is usually required. Many larger health training facilities and hospitals require a master's degree. Courses in the production of all types of audio-visual materials and in television and film production are necessary. Some courses in the health sciences, health science education, and instructional design are useful.

Experience/Skills

One or two years of general audio-visual experience in an education or training media center, or part-time employment (while a student) in a college's media or television production unit, is usually required.

Media Specialists should have strong backgrounds in audio-visual and video production, and know how to operate a variety of technical equipment. Most good Media Specialists are creative, and have good writing abilities and excellent interpersonal skills.

Minority/Women's Opportunities

Opportunities in the health education field are increasing for both women and minority-group members. Nearly 35 percent of the Media Specialists in hospitals were white females, according to industry observers. In training facilities at medical schools, 29 percent of the Media Specialists were females in the 1980-81 academic year, according to the ABCD. While the percentage of minority-group employment is not as high, opportunities have increased for both males and females during the past five years, particularly in urban areas.

Unions/Associations

There are no unions that specifically represent or bargain for Media Specialists, but some individuals are represented by unions as a part of their staff membership in hospitals or health care facilities. Many belong to the American Society for Health Manpower Education and Training (ASHET) of the American Hospital Association (AHA), the Health Education Media Association (HEMA), the Association of Medical Illustrators (AMI), or the Biological Photographic Association (BPA) to share common concerns and advance their careers. In addition, some Media Specialists belong to the Association for Multi-Image (AMI) and the Media Design and Production Division of the Association for Educational Communications and Technology (AECT).

APPENDIX I—DEGREE AND NON-DEGREE PROGRAMS

A Colleges and Universities

The preferred minimum educational requirement for the majority of positions discussed in this book is a bachelor's degree. Of the 3,200 colleges and universities in the United States, more than 2,000 have classes in the general area of communications, according to the Speech Communication Association (SCA). Some 1,067 present course work in TV and film, according to the American Film Institute (AFI), and offer more than 3,300 undergraduate and 970 graduate courses specifically in television.

Television courses are often taught in several academic departments, and some schools offer interdisciplinary or interdepartmental programs leading to a degree. Although it is sometimes possible to major in another field and minor in communications, educational media, film, journalism, or radio-TV, many institutions offer degree programs in these communications areas.

This section contains the names and addresses of approximately 235 four-year colleges and universities that offer programs leading to a bachelor's degree in speech, journalism, communications, or education. Most of these institutions offer a major in, or emphasis on, radio-television or media. Information about these and many other colleges, universities, and community colleges offering courses in radio-TV and film can be obtained from the SCA (5105 Backlick Road, Annandale, VA 22003) or the AFI (The John F. Kennedy Center for the Performing Arts, Washington, DC 20566). Additional information about college courses in television is available from the Broadcast Education Association (c/o The National Association of Broadcasters, 1771 N Street, NW, Washington, DC 20036).

ALABAMA

Auburn University
Speech Communication Department
Haley Center
Auburn, AL 36830

University of Alabama
Department of Broadcast and Film Communication
PO Box D
University, AL 35486

ARIZONA

Arizona State University
Department of Mass Communications
Tempe, AZ 85281

University of Arizona
Department of Radio-Television
College of Fine Arts
Tucson, AZ 85721

ARKANSAS

Arkansas State University
Division of Radio-Television
State University, AR 72467

University of Arkansas, Fayetteville
Communications Department
Fayetteville, AR 72701

University of Arkansas, Little Rock
Department of Radio, TV, and Film
33rd and University
Little Rock, AR 72204

CALIFORNIA

California State University, Chico
Center for Information and Communication Studies
Chico, CA 95926

California State University, Fresno
Department of Radio-TV-Cinema
Cedar and Shaw Avenues
Fresno, CA 93740

California State University, Fullerton
Communications Department
Fullerton, CA 92634

California State University, Long Beach
Radio-Television Department
1250 Bellflower Blvd
Long Beach, CA 90840

California State University, Los Angeles
Department of Speech Communication and Drama
5151 State University Drive
Los Angeles, CA 90034

California State University, Northridge
Radio-TV-Film Department
18111 Nordhoff Street
Northridge, CA 91330

San Diego State University
Telecommunications and Film Department
College Avenue
San Diego, CA 92182

San Francisco State University
Broadcast Communication Arts Department
1600 Holloway Avenue
San Francisco, CA 94132

San Jose State University
Theatre Arts Department
School of Humanities and the Arts
Seventh and San Fernando
San Jose, CA 95192

Stanford University
Communication Department
Redwood Hall
Stanford, CA 94305

University of California, Los Angeles
Theater Arts Department
405 Hilgard Avenue
Los Angeles, CA 90024

University of Southern California
Annenberg School of Communications
University Park
Los Angeles, CA 90007

COLORADO

University of Colorado, Boulder
Communications Department
School of Journalism
Boulder, CO 80302

University of Denver
Mass Communications Department
2490 South Gaylord
Denver, CO 80210

CONNECTICUT

University of Bridgeport
Journalism/Communication Department
84 Iranistan Avenue
Bridgeport, CT 06602

Western Connecticut State College
Audiovisual and Television Center
181 White Street
Danbury, CT 06810

DISTRICT OF COLUMBIA

American University
School of Communication
Washington, DC 20016

George Washington University
Speech and Drama Department
2115 G Street NW
Washington, DC 20056

Howard University
Television and Film Department
School of Communications
2600 Fourth Street NW
Washington, DC 20059

FLORIDA

Florida Atlantic University
Communication Department
Boca Raton, FL 33431

Florida State University
Communications Department
329 PSA Building
Tallahassee, FL 32306

Florida Technological University
Communication Department
PO Box 25000
Orlando, FL 32816

University of Florida
Broadcasting Department
315 Stadium Building
Gainesville, FL 32611

University of Miami
Communications Department
PO Box 248127
Coral Gables, FL 33124

University of South Florida
Mass Communications Department
College of Arts and Letters
Tampa, FL 33620

GEORGIA

Georgia Southern College
Speech-Drama Department
Landrum Box 8091
Statesboro, GA 30458

University of Georgia
Radio-TV-Film Department
School of Journalism
Athens, GA 30602

HAWAII

University of Hawaii
Educational Communication and Technology Department
1776 University Avenue
Wist Hall 105
Honolulu, HI 96822

IDAHO

University of Idaho
Radio-TV Department
School of Communication
Moscow, ID 83843

ILLINOIS

Columbia College
Television Department
600 South Michigan Avenue
Chicago, IL 60605

Illinois State University
Information Sciences Department
Normal, IL 61761

Loyola University of Chicago
Communication Arts Department
820 North Michigan Avenue
Chicago, IL 60611

Northern Illinois University
Speech Communications Department
De Kalb, IL 60115

Northwestern University
Radio, Television, and Film Department
School of Speech
Evanston, IL 60201

Southern Illinois University, Carbondale
Radio-Television Department
Communications Building
Carbondale, IL 62901

Southern Illinois University, Edwardsville
Mass Communications Department
Box 73
Edwardsville, IL 62025

University of Illinois, Urbana-Champaign
Radio-Television Department
School of Humanities
2090 Foreign Languages Building
Urbana, IL 61801

Western Illinois University
Communication Arts and Sciences Department
Macomb, IL 61455

INDIANA

Ball State University
Center for Radio and Television
Muncie, IN 47306

Butler University
Radio-Television Department
46th and Sunset
Indianapolis, IN 46208

Indiana University
Telecommunications Department
Radio and Television Center
College of Arts and Sciences
Bloomington, IN 47401

Indiana University-Purdue University
Speech and Theater Department
925 West Michigan Street
Indianapolis, IN 46202

Purdue University
Department of Communication
Heavilon Hall
West Lafayette, IN 47907

Purdue University, Calumet Campus
Communication and Creative Arts Department
Hammond, IN 46323

University of Evansville
Center for the Study of Communications
PO Box 329
Evansville, IN 47702

IOWA

Buena Vista College
Mass Communications Department
Fourth and College
Storm Lake, IA 50588

Clarke College
Journalism-Communication Department
1500 Clarke Drive
Dubuque, IA 52001

Drake University
Radio-Television Department
25th and University
Des Moines, IA 50311

Iowa State University
Speech-Telecommunications Arts Department
College of Sciences and Humanities
21 Exhibit Hall
Ames, IA 50011

University of Iowa
Division of Broadcasting and Film
Department of Speech and Dramatic Art
Iowa City, IA 52242

University of Northern Iowa
Educational Media Department
Cedar Falls, IA 50613

KANSAS

Fort Hays State University
Radio-Television Department
Hays, KS 67601

Kansas State University
Department of Journalism and Mass Communications
College of Arts and Sciences
Kedzie Hall
Manhattan, KS 66502

University of Kansas
Radio-Television-Film Department
217 Flint Hall
Lawrence, KS 66045

Wichita State University
Speech Communication Department
Wichita, KS 67208

KENTUCKY

Eastern Kentucky University
Department of Mass Communications
Richmond, KY 40475

Murray State University
Journalism and Radio-TV Department
Murray, KY 42071

Northern Kentucky University
Communication Department
Highland Heights, KY 41076

University of Kentucky
Communications Department
Lexington, KY 40506

Western Kentucky University
Communication and Theater Department
IWCFA 190
Bowling Green, KY 42101

LOUISIANA

Grambling State University
Speech and Drama Department
Main Street
Grambling, LA 71245

Louisiana State University
Speech Department
Baton Rouge, LA 70803

Loyola University
Communications Department
New Orleans, LA 70118

McNeese State University
Speech Department
Lake Charles, LA 70609

University of New Orleans
Drama and Communications Department
Lakefront Street
New Orleans, LA 70122

University of Southwestern Louisiana
Broadcast Services
Speech Department
Box 4-2091
Lafayette, LA 70504

MAINE

University of Maine at Orono
Broadcasting/Film Division
School of Performing Arts
Orono, ME 04473

MARYLAND

College of Notre Dame of Maryland
Communication Arts Department
4701 North Charles Street
Baltimore, MD 21210

Loyola College
Communication Arts Department
4501 North Charles Street
Baltimore, MD 21210

Towson State University
Speech and Mass Communications Department
Baltimore, MD 21204

University of Maryland
Radio-Television-Film Division
College Park, MD 20742

MASSACHUSETTS

Boston College
Speech, Communication, and Theater Department
College of Arts and Sciences
McGuinn Hall 501
Chestnut Hill, MA 02167

Boston University
Department of Broadcasting and Film
School of Public Communication
640 Commonwealth Avenue
Boston, MA 02215

Emerson College
Mass Communication Department
148 Beacon Street
Boston, MA 02116

Simmons College
Department of Communications
300 The Fenway
Boston, MA 02115

University of Massachusetts
Communication Studies Department
Amherst, MA 01002

Worcester State College
Department of Media
486 Chandler Street
Worcester, MA 01602

MICHIGAN

Central Michigan University
Broadcast and Cinematic Arts
340 Moore Hall
Mount Pleasant, MI 48858

Eastern Michigan University
Speech and Dramatic Arts Department
129 Quirk
Ypsilanti, MI 48197

Michigan State University
Telecommunication Department
East Lansing, MI 48824

Northern Michigan University
Speech Department
Marquette, MI 49855

University of Michigan
Speech Communication and Theatre Department
2020 Frieze Building
Ann Arbor, MI 48109

Wayne State University
Speech Department
Detroit, MI 48202

Western Michigan University
Communication Arts and Sciences Department
Kalamazoo, MI 49008

MINNESOTA

Bemidji State University
Community Media Department
Bemidji, MN 56601

Moorhead State University
Speech Department
Moorhead, MN 56560

St. Cloud State University
Mass Communications Department
St. Cloud, MN 56301

University of Minnesota, Duluth
Speech-Communication Department
Duluth, MN 55812

University of Minnesota, Minneapolis/St. Paul
Speech-Communications Department
Folwell Hall
Minneapolis, MN 55455

MISSISSIPPI

Jackson State University
Mass Communications Department
PO Box 17112
Jackson, MS 39217

Mississippi State University
Communication Department
PO Drawer NJ
Mississippi State, MS 39762

University of Mississippi
Radio and Television Program
Department of Speech and Theatre
College of Liberal Arts
University, MS 38677

University of Southern Mississippi
Radio, Television, and Film Department
Hardy Street and US 49
Hattiesburg, MS 39401

MISSOURI

Central Missouri State University
Department of Mass Communication
Warrensburg, MO 64093

Northwest Missouri State University
Speech and Theatre Department
Maryville, MO 64468

Saint Louis University
Department of Communication
3733 West Pine
St. Louis, MO 63108

Stephens College
Television, Radio, and Film Department
Columbia, MO 65201

University of Missouri-Columbia
Division of Radio-Television-Film
Speech and Dramatic Art Department
200 Swallow Hall
Columbia, MO 65201

University of Missouri, Kansas City
Department of Communication Studies
College of Arts and Sciences
5100 Rockhill Road
Kansas City, MO 64110

MONTANA

Montana State University
Film, Television, and Photography Department
Bozeman, MT 59715

University of Montana
Radio-Television Department
School of Journalism
Missoula, MT 59801

NEBRASKA

Creighton University
Journalism/Mass Communication Department
2500 California
Omaha, NE 68178

Kearney State College
Radio-Television Department
25th Street and Ninth Avenue
Kearney, NE 68847

University of Nebraska
Television-Radio-Film Department
Lincoln, NE 68508

University of Nebraska, Omaha
Communication Department
PO Box 688
Omaha, NE 68101

Wayne State College
Division of Humanities
Wayne, NE 68787

NEVADA

University of Nevada, Las Vegas
Department of Communication Studies
4505 Maryland Parkway
Las Vegas, NV 89154

NEW HAMPSHIRE

Franklin Pierce College
Mass Communications Department
Rindge, NH 03461

University of New Hampshire
Theater and Communication Department
Paul Arts Center
Durham, NH 03824

NEW JERSEY

Fairleigh Dickinson University, Teaneck
Department of Communications, Speech, and Theatre
College of Liberal Arts
1000 River Road
Teaneck, NJ 07666

Jersey City State College
Media Arts Department
2039 Kennedy Boulevard
Jersey City, NJ 07305

Montclair State College
Speech and Theater Department
Valley Road and Normal Avenue
Upper Montclair, NJ 07043

Seton Hall University
Communications Department
South Orange, NJ 07042

William Paterson College of New Jersey
Communications Department
300 Pompton Road
Wayne, NJ 07470

NEW MEXICO

New Mexico State University
Journalism and Mass Communication Department
PO Box 3J
Las Cruces, NM 88003

University of New Mexico
Theatre Arts Department
Albuquerque, NJ 87131

NEW YORK

Adelphi University
Communications Department
South Avenue
Garden City, NY 11530

Canisius College
Communication Department
2001 Main Street
Buffalo, NY 14208

City University of New York, Brooklyn College
Television/Radio Department
School of Humanities
Bedford Avenue and Avenue H
Brooklyn, NY 11210

City University of New York, Hunter College
Communications Department
695 Park Avenue
New York, NY 10021

City University of New York, Lehman College
Speech and Theatre Department
Mass Communication Division
Bedford Park Boulevard W
Bronx, NY 10468

City University of New York, Queens College
Department of Communication Arts and Sciences
Flushing, NY 11367

College of Mount Saint Vincent
Communication Arts Department
Riverdale, NY 10471

C.W. Post Center of Long Island University
Department of Theatre and Film
Northern Boulevard
Greenvale, NY 11548

Fordham University
Communications Department
Bronx, NY 10458

Hofstra University
Communication Arts Department
Hempstead Turnpike
Hempstead, NY 11550

Ithaca College
Department of Television-Radio
School of Communications
Danby Road
Ithaca, NY 14850

Marist College
Communication Arts Department
North Road
Poughkeepsie, NY 12601

Medaille College
Media Communications Department
18 Agazziz Circle
Buffalo, NY 14214

New York Institute of Technology, Old Westbury
Communication Arts Department
Wheatley Road
Old Westbury, NY 11568

New York University
Undergraduate Institute of Film and Television
School of the Arts
51 West Fourth Street
New York, NY 10003

State University of New York at Buffalo
Center for Media Study
310 Hochstetter Avenue
Buffalo, NY 14222

State University of New York, College at Buffalo
Speech and Theatre Arts Department
1300 Elmwood Avenue
Buffalo, NY 14222

State University of New York, College at New Paltz
Speech Department
New Paltz, NY 12561

State University of New York, College at Oswego
Department of Communication Studies
College of Arts and Sciences
Sheldon Hall
Oswego, NY 13126

Syracuse University
Telecommunications/Film Department
S.I. Newhouse School of Public Communications
215 University Place
Syracuse, NY 13210

NORTH CAROLINA

Appalachian State University
Communication Arts Department
Boone, NC 28608

North Carolina State University
Speech Communication and Theatre Arts Department
Crosby Hall 302
Greensboro, NC 27411

Shaw University
Radio-Television-Film Department
118 East South Street
Raleigh, NC 27602

University of North Carolina at Chapel Hill
Department of Radio, Television, and Motion Pictures
College of Arts and Sciences
Swain Hall 044A
Chapel Hill, NC 27514

Wake Forest University
Speech Communication and Theatre Arts Department
PO Box 7347
Winston-Salem, NC 27109

Western Carolina University
Speech-Theater Arts Department
Cullowhee, NC 28723

NORTH DAKOTA

University of North Dakota
Speech Department
College of Arts and Sciences
Grand Forks, ND 58202

OHIO

Antioch College
Communication-Video Department
Yellow Springs, OH 45387

Ashland College
Speech Department
College Avenue
Ashland, OH 44805

Bowling Green State University
Radio/Television/Film Area
School of Speech Communication
413 South Hall
Bowling Green, OH 43403

Kent State University
Telecommunications Department
School of Speech
511 Wright Hall
Kent, OH 44242

Miami University
Communication and Theater Department
Oxford, OH 45056

Ohio State University
Communications Department
Columbus, OH 43210

Ohio University
Radio-Television Department

College of Fine Arts
Athens, OH 45701

University of Akron
Mass Media Communication Department
Akron, OH 44325

University of Toledo
Communication Department
2901 West Bancroft Street
Toledo, OH 43606

Xavier University
Communication Arts Department
Victory and Dana Streets
Cincinnati, OH 45207

OKLAHOMA

Oklahoma State University
School of Journalism and Broadcasting
Stillwater, OK 74078

Oral Roberts University
Communication Arts Department
7777 South Lewis
Tulsa, OK 74171

University of Oklahoma
H.H. Herbert School of Journalism
and Mass Communications
Norman, OK 73019

University of Tulsa
Communications Department
600 South College
Tulsa, OK 74104

OREGON

Lewis and Clark College
Communications Department
Portland, OR 97219

Oregon State University
Speech Communication Department
College of Liberal Arts
Corvallis, OR 97331

Southern Oregon State College
Speech Communications Department
1250 Siskiyou Boulevard
Ashland, OR 97520

University of Oregon
Telecommunications Area
Department of Speech
Villard Hall
Eugene, OR 97403

University of Portland
Creative and Communicative
Arts Department
5000 North Willamette Boulevard
Portland, OR 97203

PENNSYLVANIA

California State College
Speech Communication Department
California, PA 15419

East Stroudsburg State College
Speech Communication and
Theatre Arts Department
East Stroudsburg, PA 18301

Edinboro State College
Speech Communication Department/
Radio-TV Center
128 MRLC
Edinboro, PA 16412

Geneva College
Speech Communication Department
Beaver Falls, PA 15010

Indiana University of Pennsylvania
Learning Resources and
Mass Media Department
Stouffer G-16
Indiana, PA 15701

Kutztown State College
Department of Telecommunications
Kutztown, PA 19530

La Salle College
English Department
20th and Olney Avenue
Philadelphia, PA 19141

Pennsylvania State University
Speech Communication Department
912 Sparks Building
University Park, PA 16802

Shippensburg State College
Communication-Journalism Department
Shippensburg, PA 17257

Slippery Rock State College
Communication Department
Slippery Rock, PA 16057

Temple University
Department of Radio-Television-Film
School of Communications and Theater
Broad and Montgomery Streets
Philadelphia, PA 19122

University of Pennsylvania
Annenburg School of Communications
3620 Walnut Street
Philadelphia, PA 19104

University of Scranton
English-Communications Department
Linden Street
Scranton, PA 18510

Westminster College
Speech and Theatre Department
New Wilmington, PA 16142

York College of Pennsylvania
English and Speech Department
York, PA 17405

SOUTH CAROLINA

Bob Jones University
Division of Cinema
Wade Hampton Boulevard
Greenville, SC 29614

University of South Carolina
Media Arts Department
Columbia, SC 29208

SOUTH DAKOTA

South Dakota State University
Speech Department
Brookings, SD 57006

University of South Dakota
Mass Communication Department
Vermillion, SD 57069

TENNESSEE

Memphis State University
Department of Speech and Drama
Division of Communication Studies
Central Avenue
Memphis, TN 38152

Middle Tennessee State University
Mass Communications Department
PO Box 51
Murfreesboro, TN 37130

Southwestern at Memphis
Communication Arts Department
2000 North Parkway
Memphis, TN 38112

University of Tennessee, Knoxville
Department of Broadcasting
College of Communications
295 Communication and University
Extension Building
Knoxville, TN 37916

University of Tennessee, Martin
English/Communications Department
Martin, TN 38238

TEXAS

Abilene Christian University
Mass Communications Department
PO Box 7568
Abilene, TX 79601

Baylor University
Division of Radio-Television-Film
Waco, TX 76706

Corpus Christi State University
Communications Department
PO Box 6010
Corpus Christi, TX 78411

East Texas State University
Department of Journalism
and Graphic Arts
PO Box D, ET Station
Commerce, TX 75428

North Texas State University
Division of Radio/Television/Film
Department of Speech Communication
Denton, TX 76203

Prairie View A&M University
Mass Communication Department
PO Box 2546
Prairie View, TX 77445

Sam Houston State University
Speech and Drama Department
Huntsville, TX 77340

Southern Methodist University
Division of Broadcast-Film Art
Meadows School of the Arts
Dallas, TX 75275

Stephen F. Austin State University
Communication Department
PO Box 3048, SFA Station
Nacogdoches, TX 75961

Sul Ross State University
Speech Communication and
Theatre Department
PO Box 43
Alpine, TX 79830

Texas Christian University
Speech Communication/Radio-
Television-Film Department
Fort Worth, TX 76129

Texas Southern University
Telecommunications Department
School of Communication
Wheeler Avenue
Houston, TX 77004

Texas Tech University
Mass Communications Department
Lubbock, TX 79409

University of Houston
Communications Department
Houston, TX 77004

University of Texas at Arlington
Communication Department
Arlington, TX 76019

University of Texas at Austin
Department of Radio-Television-Film
School of Communication
Austin, TX 78712

UTAH

Brigham Young University
Communications Department
117D EWLC
Provo, UT 84602

Southern Utah State College
Communication Department
Cedar City, UT 84720

University of Utah
Communication Department
Salt Lake City, UT 84112

Weber State College
Communication Department
Harrison Boulevard
Ogden, UT 84408

VERMONT

Lyndon State College
Media-Communications Department
Lyndonville, VT 05851

VIRGINIA

James Madison University
Communication Arts Department
Harrisonburg, VA 22801

Norfolk State College
Mass Communications Department
2400 Corprew Avenue
Norfolk, VA 23504

Virginia Commonwealth University
Mass Communications Department
901 West Franklin Street
Richmond, VA 23284

**Virginia Polytechnic Institute
and State University**
Communications Program
Blacksburg, VA 24061

WASHINGTON

Central Washington University
Mass Media Program
Communications Department
Ellensburg, WA 98926

Eastern Washington State College
Radio-Television Department
Cheney, WA 99004

Pacific Lutheran University
Communication Arts Department
Tacoma, WA 98447

University of Washington
School of Communications
College of Arts and Sciences
DS-40
Seattle, WA 98195

Washington State University
Communications Department
Pullman, WA 99163

Western Washington State College
Speech Department
Bellingham, WA 98225

WEST VIRGINIA

Marshall University
Speech Department
Huntington, WV 25701

West Virginia State College
Department of Communications
PO Box 8
Institute, WV 25112

WISCONSIN

Marquette University
Broadcast Communication Department
Milwaukee, WI 53233

University of Wisconsin, Green Bay
Communication Department
Green Bay, WI 54302

University of Wisconsin, Madison
Department of Communication Arts
Vilas Communication Hall
821 University Avenue
Madison, WI 53706

University of Wisconsin, Milwaukee
Department of Mass Communications
Milwaukee, WI 53201

University of Wisconsin, Oshkosh
Radio-Television Division
Speech Department
Arts and Communication Center
Oshkosh, WI 54901

University of Wisconsin, Platteville
Communications Department
Platteville, WI 53818

University of Wisconsin, Stevens Point
Communications Department
Stevens Point, WI 54481

University of Wisconsin, Stout
Media Technology Department
Menomonie, WI 54751

University of Wisconsin, Superior
Communication Arts Department
1800 Grand Avenue
Superior, WI 54880

University of Wisconsin, Whitewater
Speech Communications Department
900 West Main Street
Whitewater, WI 53190

WYOMING

University of Wyoming
Communication and Broadcasting Department
PO Box 3341
University Station
Laramie, WY 82071

B Workshops and Seminars

The following is a list of selected workshops and seminars with their sponsors' names and addresses and the general subjects they cover. Most last for two or three days, and a modest fee is charged for attending. In some cases, scholarships are available. Others are more commercial in nature and run for several days at higher fees.

American Film Institute
Center for Advanced Film Studies
Directing Workshop for Women
2021 North Western Avenue
Beverly Hills, CA 90027
women's workshop in television and film directing

Boston Film/Video Foundation
39 Brighton Avenue
Allston, MA 02134
workshops in video writing, directing, acting

Columbia Pictures
711 Fifth Avenue
New York, NY 10022
workshops in television acting and writing

Community Antenna Television Association (CATA)
1825 K Street NW
Washington, DC 20006
technical seminars in cable TV

Development Communications Associates
111 South Fairfax Street
Alexandria, VA 22314
workshops and seminars in video production, editing, script writing

Electronic Industries Association (EIA)
2001 Eye Street NW, Suite 405
Washington, DC 20006
seminars in home video retailing management

Focus on Media Productions
1888 Century Park East, Suite 10
Los Angeles, CA 90067
seminars in women's opportunities and career planning in television

International Radio & Television Society
College Conference
420 Lexington Avenue
New York, NY 10171
seminars (for 25 students) in cable TV, consumer electronics, and related fields

The Media School
90 Park Avenue
New York, NY 10016
seminars in media and advertising

Media Study/Buffalo
207 Delaware Avenue
Buffalo, NY 14202
general video workshop

National Academy of Television Arts and Sciences
110 West 57th Street, Suite 301
New York, NY 10019
workshop in television script development

North American Television Institute
Knowledge Industry Publications, Inc.
701 Westchester Avenue
White Plains, NY 10604
seminars in video management, training, lighting, sound, editing, script writing, directing

School of Television Arts in Production, Performing and Editing (TAPE)
Lincoln Center
18 West 61st Street
New York, NY 10023
workshops and seminars in television, pay cable TV, performance careers, programming, home video, writing, auditioning, editing

Script Development Workshop
420 West 46th Street
New York, NY 10036
workshop in television script writing, acting, directing

Smith-Mattingly Productions Ltd.
2560 Huntington Avenue
Alexandria, VA 22303
workshops in video production and electronic editing

Sony Corporation of America
700 West Artesia Boulevard
Compton, CA 90220
workshops in video equipment and utilization

Yellow Ball Workshop
62 Tarbell Avenue
Lexington, MA 02173
workshop in television animation

Young Filmmakers/Video Arts
Directors Project
4 Rivington Street
New York, NY 10002
workshops in television directing

C Internships, Apprenticeships, and Training Programs

The selected companies and organizations listed below offer special programs in various television-related disciplines. Some pay a modest stipend and some offer no money but may grant college credit. (Contact the individual sponsor for specific information.)

There are other programs throughout the country and the interested applicant should check with local stations (especially public TV), advertising agencies, and cable companies in his or her area for further leads.

In addition to the name of the sponsoring group and its address, each listing provides a brief description of the type of program offered.

Alternate Media Center
Apprenticeship Program
144 Bleecker Street
New York, NY 10012
apprenticeships in cable television

American Advertising Federation
1225 Connecticut Avenue NW
Washington, DC 20036
local internships in television advertising

American Film Institute
2021 North Western Avenue
Beverly Hills, CA 90027
internships in television news directing and film production

American Guild of Authors and Composers
40 West 57th Street
Suite 410
New York, NY 10019
training programs in music lyric writing (including TV)

American Television & Communications Corp.
20 Inverness Place East
Englewood, CO 80112
technical training programs in cable TV

Association for Educational Communications and Technology
Convention Internship Program
1126 Sixteenth Street NW
Washington, DC 20036
internship in convention and association activities related to educational telecommunications

Association of Motion Picture and Television Producers
8480 Beverly Boulevard
Los Angeles, CA 90048
training programs in TV assistant directing, camera operation, makeup, publicity, and script supervision

Boston Film/Video Foundation
39 Brighton Avenue
Allston, MA 02134
internships in video writing, directing, and acting

Broadway/Hollywood Video Productions
PO Box 1314
Englewood Cliffs, NJ 07632
internships in video production

California State University
Sports Information Bureau
1250 Bellflower Boulevard
Long Beach, CA 90840
internships in broadcast and cable television sports

Canadian Broadcasting Corp.
CBC Institute of Scenography for TV and Film
PO Box 500, Station A
Toronto, Ont. M5W 1E6
training program in television scenic design

Channel 12
Grassroots Television
Box 2006
Aspen, CO 81611
internships in television production

Connecticut Public Television
24 Summit Street
Hartford, CT 06106
internships in television production, cinematography, graphics, and set construction

Contract Services Administration Trust Fund
8480 Beverly Boulevard
Hollywood, CA 90048
training program in television cinematography

Educational Broadcasting Corp.
WNET-TV
356 West 58th Street
New York, NY 10019
internships in television production, management, and music

Edward Shaw Productions
9465 Wilshire Boulevard
Beverly Hills, CA 90212
internships in television production

Electronic Industries Association
Service Technician Development Program
2001 Eye Street NW
Washington, DC 20006
training programs for minorities in basic television repair

International Alliance of Theatrical Stage Employees (IATSE)
1515 Broadway, Suite 601
New York, NY 10036
apprenticeships in television art direction, makeup, costuming, and properties

International Radio & Television Society
420 Lexington Avenue
New York, NY 10170
general internships in television, communications, and related fields

Kapp Television
PO Box 1747
Yakima, WA 98907
internships in television production and news writing, editing, and reporting

KATU Television
Fisher Broadcasting, Inc.
2153 NE Sandy Boulevard
PO Box 8799
Portland, OR 97208
internships in television operations, sales, public affairs, administration, news, programming, and production

KBLE-TV 3
Colonial Cablevision Corporation

334 Broadway
Revere, MA 02151
internships in cable TV production, news, and sales

KDUB-TV
One Dubuque Plaza
Dubuque, IA 52001
internships in television news, production, and engineering

KPLR-TV
4935 Lindell Boulevard
St. Louis, MO 63108
internships in television production, news, promotion, and management

Media Study/Buffalo
207 Delaware Avenue
Buffalo, NY 14202
high school internships in video

National Association of Broadcasters
1771 N Street NW
Washington, DC 20036
training programs in television management

National Audio-Visual Association
3150 Spring Street
Fairfax, VA 22031
workshops in audio-visual techniques in business, education, and health

New Jersey Public TV
1573 Parkside Avenue
Trenton, NJ 08938
general internships in public television

RKO General, Inc.
WNAC-TV
RKO General Building
Government Center
Boston, MA 02114
internships in television news production, promotion, graphics, and music

Rocky Mountain Lab
Center for Cinema & Television Arts
Box 234
Aspen, CO 81612
apprenticeships in video art

Teleprompter Corp.
888 Seventh Avenue
New York, NY 10019
internships for minorities in cable TV

Warner Amex Cable Communications, Inc.
QUBE
1201 Olentangy River Road
Columbus, OH 43212
internships in cable TV production, public affairs, special projects, and master control, videotape library, and traffic procedures

WAVE-TV
PO Box 23970
Louisville, KY 40232
internships for minorities in television production and news

WBAL-TV
3800 Hooper Avenue
Baltimore, MD 21211
internships in television production

WCHS-TV
1111 Virginia Street East
Charleston, WV 25301
internships in television news

WCVB-TV
5 TV Place
Needham, MA 02192
internships in television news, community service, and public relations

WEAR-TV
Box 12278
Pensacola, FL 32581
internships in television production and news

WFRV-TV
1181 East Mason Street
Gree Bay, WI 54301
internships in television production and news

WNEW-TV
205 East 67th Street
New York, NY 10021
internships in general television production

WTLV News
PO Box TV-12
Jacksonville, FL 32231
internships in television news production and reporting

WTOL-TV
Cosmos Broadcasting Corporation
PO Box 715
Toledo, OH 43695
internships in television news

D Fellowships, Scholarships, and Grants

The organizations listed below award fellowships, scholarships, and grants for training in television, video, and telecommunications. Some are specifically for members of minority groups and/or women.

The names of additional sources can be gotten through the appropriate offices at most colleges and universities.

American Association of Advertising Agencies
200 Park Avenue
New York, NY 10166
minority student fellowship program for internship in an advertising agency

American Association of University Women
Education Foundation/Applications
2401 Virginia Avenue, NW
Washington, DC 20037
one-year research fellowships for women studying any subject

Broadcast Education Association
Harold E. Fellows Memorial Scholarship
217 Flint Hall
University of Kansas
Lawrence, KS 66045
scholarships for students with a parent who has been employed by the National Association of Broadcasters

Corporation for Public Broadcasting
1111 16th Street, NW
Washington, DC 20036
grants to public TV stations for use in training women and members of minority groups

grants to public TV stations for training Graphic Artists (co-sponsored by the National Endowment for the Arts)

funding for minorities for year-long master's degree program in television and radio management at Ohio University

Directors Guild Benevolent and Educational Foundation College Scholarships
7950 Sunset Boulevard
Hollywood, CA 90046
financial assistance for completing a film project while in college

Information Film Producers of America
Scholarship Competition
3518 Cahuenga Boulevard, Suite 313
Hollywood, CA 90068
scholarship for an outstanding cinema student

National Academy of Television Arts and Sciences
110 West 57th Street
New York, NY 10019
scholarship to ten schools to be awarded to outstanding students in communications

National Audio-Visual Association
Past Presidents' Memorial Scholarship
Audio-Visual Center
Indiana University
Bloomington, IN 47401
scholarship for graduate audio-visual studies

National Endowment for the Arts
Visual Arts Program
Mail Stop 501
Washington, DC 20506
fellowships for video artists

Society of Motion Picture and Television Engineers
Scholarships and Grants
862 Scarsdale Avenue
Scarsdale, NY 10583
scholarships for the undergraduate study of film and television technology and science

University Film Association
Scholarship Awards Program
Department of Radio/TV/Film
University of Texas, Austin
Austin, TX 78712
scholarships for film/video production and research

APPENDIX II— UNIONS AND ASSOCIATIONS

The unions and associations listed here are all closely related to the careers discussed in this book. Those that are mentioned in the job descriptions are included as well as others that are of importance in the television, video, and telecommunications fields.

The name and acronym (if one is commonly used) of each group is included in addition to its address and telephone number.

A Unions

Actors Equity Association (Equity)
165 West 46th Street
New York, NY 10036
212-869-8530

Affiliated Property Craftsmen
7429 Sunset Boulevard
Hollywood, CA 90046
213-876-2320

American Association of University Professors (AAUP)
One Dupont Circle NW
Washington, DC 20036
202-466-8050

American Communications Association (ACA)
Communications Trade Division
111 Broadway
New York, NY 10006
212-267-2374

American Federation of Musicians (AFM)
1500 Broadway
New York, NY 10036
212-869-1330

1777 North Vine Street, Suite 410
Hollywood, CA 90028
213-461-3441

American Federation of State, County and Municipal Employees (AFSCME)
1625 L Street NW
Washington, DC 20036
202-452-4800

American Federation of Teachers (AFT)
11 Dupont Circle NW
Washington, DC 20036
202-797-4400

American Federation of Television and Radio Artists (AFTRA)
1350 Avenue of the Americas
New York, NY 10019
212-265-7700

American Guild of Authors and Composers (AGAC)
40 West 57th Street
New York, NY 10019
212-757-8833

American Guild of Musical Artists (AGMA)
1841 Broadway
New York, NY 10023
212-265-3687

American Guild of Variety Artists (AGVA)
1540 Broadway
New York, NY 10036
212-765-0800

Authors Guild
234 West 44th Street
New York, NY 10036
212-398-0838

Broadcast-Television Recording Engineers (BTRE)
3518 Cahuenga Boulevard West, Suite 307
Hollywood, CA 90068
213-851-5515

Communications Workers of America (CWA)
1925 K Street NW
Washington, DC 20006
202-785-6700

Composers and Lyricists Guild of America
10999 Riverside Drive, Suite 100
North Hollywood, CA 91602
213-985-4102

Directors Guild of America (DGA)
7950 Sunset Boulevard
Hollywood, CA 90046
213-656-1220

Dramatists Guild
234 West 44th Street
New York, NY 10036
212-398-9366

Illustrators and Matte Artists of Motion Picture, Television, and Amusement Industries
7715 Sunset Boulevard
Hollywood, CA 90046
213-876-2010

International Alliance of Theatrical Stage Employees (IATSE)
1515 Broadway
New York, NY 10036
212-730-1770

Local 1 (Carpenters, Property Masters)
1775 Broadway
New York, NY 10019
212-489-7710

Local 33 (Carpenters, Property Masters)
4605 Lankershim Boulevard, Suite 833
North Hollywood, CA 91602
213-985-0633

Local 52 (Carpenters, Stagehand/ Grips, Property Masters)
221 West 57th Street, 11th Floor
New York, NY 10019
212-765-0741

Local 644 (Camera Operators)
250 West 57th Street, Suite 1723
New York, NY 10019
212-247-3860

Local 706 (Makeup Artists, Hair Stylists)
11519 Chandler Boulevard
Hollywood, CA 91601
213-877-2776

Local 798 (Makeup Artists, Hair Stylists)
1790 Broadway
New York, NY 10019
212-757-9120

Local 816 (Scenic Artists)
7429 Sunset Boulevard
Hollywood, CA 90046
213-876-1440

Local 847 (Scenic Designers)
7715 Sunset Boulevard, Suite 210
Hollywood, CA 90046
213-876-2010

Local 876 (Art Directors)
7715 Sunset Boulevard
Hollywood, CA 90046
213-876-4330

Local 892 (Costume Designers)
11286 Westminster
Los Angeles, CA 90019
213-397-3162

International Brotherhood of Electrical Workers (IBEW)
1125 15th Street NW
Washington, DC 20005
202-833-7000

International Sound Technicians
15840 Ventura Boulevard, Suite 303
Encino, CA 91436
213-981-0452

National Association of Broadcast Employees and Technicians AFL-CIO (NABET)
7101 Wisconsin Avenue, Suite 1303
Bethesda, MD 20814
301-657-8420

80 East Jackson Boulevard
Chicago, IL 60604
312-922-2462

1776 Broadway, Suite 1900
New York, NY 10019
212-265-3500

1800 Argyle
Los Angeles, CA 90028
213-462-7485

National Education Association (NEA)
1201 16th Street NW
Washington, DC 20036
202-833-4000

Producers Guild of America (PGA)
9201 Beverly Boulevard
Los Angeles, CA 90048
213-651-0084

Screen Actors Guild (SAG)
7750 Sunset Boulevard
Hollywood, CA 90046
213-876-3030

1700 Broadway
New York, NY 10019
212-957-5370

Screen Extras Guild (SEG)
3629 Cahuenga Boulevard West
Los Angeles, CA 90068
213-851-4301

United Brotherhood of Carpenters and Joiners of America (UBC)
101 Constitution Avenue
Washington, DC 20001
202-546-6206

United Scenic Artists (USA)
1540 Broadway
New York, NY 10036
212-575-5120

Writers Guild of America East (WGA)
555 West 57th Street
New York, NY 10019
212-245-6180

Writers Guild of America West (WGA)
8955 Beverly Boulevard
Los Angeles, CA 90048
213-550-1000

B Associations

Academy of Television Arts and Sciences
4605 Lankershim Boulevard, Suite 800
North Hollywood, CA 91602
213-506-7880

Advertising Council
825 Third Avenue
New York, NY 10022
212-758-0400

Alpha Epsilon Rho
College of Journalism
University of South Carolina
Columbia, SC 29208
803-777-6783

American Advertising Federation (AAF)
1225 Connecticut Avenue NW
Washington, DC 20036
202-659-1800

50 California Street
San Francisco, CA 94111
415-421-6867

American Association of Advertising Agencies (4As)
200 Park Avenue
New York, NY 10017
212-682-2500

American Association of Community and Junior Colleges (AACJC)
One Dupont Circle
Washington, DC 20036
202-293-7050

American Film Institute (AFI)
John F. Kennedy Center for the Performing Arts
Washington, DC 20566
202-828-4040

American Institute of Graphic Arts (AIGA)
1059 Third Avenue
New York, NY 10021
212-752-0813

American Library Association (ALA)
50 East Huron Street
Chicago, IL 60611
312-944-6780

American Management Association (AMA)
135 West 50th Street
New York, NY 10020
212-486-8100

American Marketing Association (AMA)
222 Riverside Plaza
Chicago, IL 60606
312-648-0536

American Meteorological Society (AMS)
45 Beacon Street
Boston, MA 02108
617-227-2425

American Society for Health Manpower Education and Training (ASHET)
c/o The American Hospital Association
840 North Lake Shore Drive
Chicago, IL 60611
312-280-6111

American Society for Training and Development (ASTD)
PO Box 5307
Madison, WI 53705
608-274-3440

American Society of TV Cameramen (ASTVC)
Washington Street
PO Box 296
Sparkill, NY 10976
914-359-5985

American Sportscasters Association (ASA)
150 Nassau Street
New York, NY 10038
212-227-8080

American Women in Radio and Television (AWRT)
1321 Connecticut Avenue NW
Washington, DC 20036
202-296-0009

Art Directors Club of New York (ADC)
488 Madison Avenue
New York, NY 10022
212-838-8140

Association for Communication Administration (ACA)
5105 Backlick Road
Annandale, VA 22003
703-750-0534

Association for Education in Journalism (AEJ)
c/o Del Brinkman
University of Kansas
Lawrence, KS 66045
913-864-4735

Association for Educational Communications and Technology (AECT)
1126 16th Street NW
Washington, DC 20036
202-833-4180

Divisions (selected):
- Division of Educational Media Management (DEMM)
- Division of Instructional Development (DID)
- Division of School Media Specialists (DSMS)
- Division of Telecommunications (DOT)
- Industrial Training and Education Division (ITED)
- Media Design and Production Division (MDPD)

Affiliates (selected):
- American Student Media Association (ASMA)
- Association for Multi-Image (AMI) (separate listing)
- Federal Educational Technology Association (FETA) (separate listing)
- Health Education Media Association (HEMA) (separate listing)
- Information Film Producers of America (IFPA) (separate listing)
- Minorities in Media (MIMS)
- Women in Instructional Technology (WIT)

Association for Multi-Image (AMI)
Los Angeles Community College District
855 North Vermont Avenue
Los Angeles, CA 90029
213-660-4825

Association of Audio-Visual Technicians (AAVT)
PO Box 9716
Denver, CO 80209
303-733-3137

Association of Biomedical Communications Directors (ABCD)
c/o Joseph Day
Director of Educational Media
Georgetown University Medical Center
3900 Reservoir Road NW
Washington, DC 20007
202-625-2211

Association of Independent Commercial Producers (AICP)
100 East 42nd Street
New York, NY 10017
212-867-5720

Association of Independent Video and Filmmakers (AIVF)
625 Broadway
New York, NY 10012
212-473-3400

Association of Maximum Service Telecasters (MST)
1735 DeSales Street NW
Washington, DC 20036
202-347-5412

Association of Media Producers (AMP)
1707 L Street NW
Washington, DC 20036
202-296-4710

Association of Medical Illustrators (AMI)
5820 Wilshire Boulevard
Los Angeles, CA 90036
213-937-5514

Association of Motion Picture and Television Producers (AMPTP)
8480 Beverly Boulevard
Los Angeles, CA 90048
213-653-2200

Association of Radio-Television News Analysts (ARTNA)
190 Riverside Drive, Suite 6-B
New York, NY 10024
212-799-2528

Audio-Visual Management Association (AVMA)
Box 821
Royal Oak, MI 48068
313-549-6582

Audio-Visual Technicians Association (AVTA)
See Association of Audio-Visual Technicians

Biological Photographic Association (BPI)
Box 2603
West Durham Station
Durham, NC 27705
919-493-4854

Broadcast Advertising Producers Society of America
Box 2230
Grand Central Station
New York, NY 10017

Broadcast Education Association (BEA)
1771 N Street NW
Washington, DC 20036
202-293-3518

Broadcast Financial Management Association (BFMA)
360 North Michigan Avenue, Suite 910
Chicago, IL 60601
312-332-1295

Broadcast Music, Inc. (BMI)
320 West 57th Street
New York, NY 10019
212-586-2000

Broadcasters' Promotion Association (BPA)
PO Box 5102
Lancaster, PA 17601
717-626-4524

Business/Professional Advertising Association (BPAA)
205 East 42nd Street
New York, NY 10017
212-661-0222

Cable Advertising Bureau (CAB)
767 Third Avenue
New York, NY 10017
212-751-7770

Cable Television Administration and Marketing Society (CTAM)
c/o Tom Johnson
Daniels & Associates Inc.
2930 East Third Avenue
Denver, CO 80206
303-321-7550

Cable Television Information Center (CTIC)
1800 North Kent Street
Arlington, VA 22209
703-528-6836

Community Antenna Television Association (CATA)
1825 K Street NW
Washington, DC 20006
202-659-2612

Corporation for Public Broadcasting (CPB)
1111 16th Street NW
Washington, DC 20036
202-293-6160

Cosmetic Career Women (CCW)
M-160 Taft Hotel
777 Seventh Avenue
New York, NY 10019
212-765-8406

Economics News Broadcasters Association (ENBA)
1629 K Street NW, Suite 554
Washington, DC 20006
202-296-5689

Educational Communication Association (ECA)
National Press Building, Room 822
Washington, DC 20004
202-393-6267

Educational Film Library Association (EFLA)
43 West 61st Street
New York, NY 10023
212-246-4533

Electronic Industries Association (EIA)
The Consumer Electronics Group
2001 Eye Street NW
Washington, DC 20006
202-457-4919

Electronic Representatives Association (ERA)
233 East Erie Street
Chicago, IL 60611
312-649-1333

Federal Communications Bar Association (FCBA)
PO Box 57109
Washington, DC 20037

Federal Communications Commission (FCC)
1919 M Street NW
Washington, DC 20554
202-655-4000

Federal Educational Technology Association (FETA)
c/o Chief, Instructional Systems Development Branch
U.S. Coast Guard Headquarters
PTE-4, TP-41
2100 Second Street SW
Washington, DC 20590
202-426-2746

Health Education Media Association (HEMA)
Box 771
Riverdale, GA 30274
404-997-0449

Health Sciences Communications Association (HeSCA)
Office of the Manager
2343 North 115th Street
Wauwatosa, WI 53226
414-258-2528

Hollywood Radio and Television Society (HRTS)
5315 Laurel Canyon Boulevard
No. Hollywood, CA 91607
213-769-4313

Information Film Producers of America (IFPA)
Film and Video Communicators
750 East Colorado Boulevard
Pasadena, CA 91101
213-795-7866

Information Industry Association (IIA)
316 Pennsylvania Avenue SW
Washington, DC 20003
202-544-2969

Institute of Electrical and Electronics Engineers (IEEE)
345 East 47th Street
New York, NY 10017
212-644-7908

International Advertising Association (IAA)
475 Fifth Avenue
New York, NY 10017
212-684-1583

International Association of Business Communicators (IABC)
870 Market Street, Suite 928
San Francisco, CA 94102
415-433-3400

International Association of Independent Producers (IAIP)
Box 1933
Washington, DC 20013
202-638-5595

International Radio and Television Society (IRTS)
420 Lexington Avenue
New York, NY 10017
212-867-6650

International Society of Certified Electronics Technicians (ISCET)
2708 West Berry Street
Fort Worth, TX 76109
817-921-9101

International Tape/Disc Association (ITA)
10 Columbus Circle, Suite 2270
New York, NY 10019
212-956-7110

International Television Association (ITVA)
136 Sherman Avenue
Berkeley Heights, NJ 07922
201-464-6747

Joint Council on Educational Telecommunications (JCET)
1126 16th Street NW
Washington, DC 20036
202-659-9740

Kappa Tau Alpha
Box 838
University of Missouri
Columbia, MO 65205
314-882-4852

Motion Picture Association of America (MPAA)
522 Fifth Avenue
New York, NY 10036
212-840-6161

Music Teachers National Association (MTNA)
2113 Carew Tower
Cincinnati, OH 45202
513-421-1420

National Academy of Television Arts and Sciences
110 West 57th Street
New York, NY 10019
212-586-8424

National Academy of Video Arts and Sciences
c/o Independent Producers Registry
146 East 49th Street
New York, NY 10017
212-486-0471

National Association of Broadcasters (NAB)
1771 N Street, NW
Washington, DC 20036
202-293-3500

National Association of Educational Broadcasters (NAEB)
1346 Connecticut Avenue NW
Washington, DC 20036

National Association of Farm Broadcasters (NAFB)
PO Box 119
Topeka, KS 66601
913-272-3456

National Association of Government Communicators (NAGC)
PO Box 7127
Alexandria, VA 22307
703-768-4546

National Association of Independent TV Producers and Distributors (NAITPD)
375 Park Avenue
New York, NY 10022
212-751-0600

National Association of MDS Service Companies (NAMSCO)
1629 K Street NW, Suite 520
Washington, DC 20006
202-296-5775

National Association of Public Television Stations (NAPTS)
475 L'Enfant Plaza SW
Washington, DC 20024
202-488-5010

National Association of Recording Merchandisers (NARM)
1060 Kings Highway North, Suite 200
Cherry Hill, NJ 08034
609-795-5555

National Association of Retail Dealers of America (NARDA)
2 North Riverside Plaza
Chicago, IL 60606
312-454-0944

National Association of State Educational Media Professionals
Minnesota Department of Education
Capitol Square Building, Room 606
550 Cedar Street
St. Paul, MN 55101
612-296-6114

National Association of Television and Electronic Servicers of America (NATESA)
5908 South Troy Street
Chicago, IL 60629
312-476-6363

National Association of Television Program Executives (NATPE)
Box 5275
Lancaster, PA 17601
717-626-4424

National Audio-Visual Association (NAVA)
3150 Spring Street
Fairfax, VA 22031
703-273-7200

National AudioVisual Center (NAC)
Washington, DC 20409
202-763-1896

National Cable Television Association (NCTA)
1724 Massachusetts Avenue NW
Washington, DC 20036
202-775-3550

National Cable Television Institute (NCTI)
Box 27277
Denver, CO 80227
303-697-4967

National Electronic Distributors Association (NEDA)
1480 Renaissance Drive, Suite 214
Park Ridge, IL 60068
312-298-9747

National Electronic Service Dealers Association (NESDA)
2708 West Berry Street
Fort Worth, TX 76109
817-921-9061

National Federation of Local Cable Programmers
3700 Far Hills Avenue, Suite 109
Kettering, OH 45429

National Hairdressers and Cosmetologists Association (NHCA)
3510 Olive Street
St. Louis, MO 63103
314-534-7980

National Society for Performance and Instruction (NSPI)
1126 16th Street NW, Suite 315
Washington, DC 20036
202-861-0777

National Writers Club
1450 South Havana, Suite 620
Aurora, CO 80012
303-751-7844

P.E.N.
American Center
47 Fifth Avenue
New York, NY 10003
212-255-1977

Phi Delta Kappa
Eighth Street and Union Avenue
Bloomington, IN 47401
812-339-1156

Public Broadcasting News Producers Association
Box 14301
Hartford, CT 06106

Public Relations Society of America (PRSA)
845 Third Avenue
New York, NY 10022
212-826-1750

Public Telecommunications Financial Management Association (PTFMA)
c/o Boris Frank, WHA-TV
University of Wisconsin
Madison, WI 53706
608-263-2121

Radio and Television Correspondents Association (RTCA)
Senate Radio-Television Gallery
U.S. Capitol, Room S-312
Washington, DC 20510
202-224-6421

Radio and Television Research Council (RTRC)
c/o Bernard Lipsky
Foote, Cone & Belding
200 Park Avenue
New York, NY 10017
212-880-9000

Radio-Television News Directors Association (RTNDA)
1735 DeSales Street NW
Washington, DC 20036
202-737-8657

Sales and Marketing Executives International
330 West 42nd Street
New York, NY 10017
212-239-1919

Society for Applied Learning Technology (SALT)
50 Culpeper Street
Warrenton, VA 22186
703-347-0055

Society for Technical Communication (STC)
815 15th Street NW, Suite 506
Washington, DC 20005
202-737-0035

Society of Broadcast Engineers (SBE)
Box 50344
Indianapolis, IN 46250
317-842-0836

Society of Cable Television Engineers (SCTE)
1900 L Street NW, Suite 614
Washington, DC 20036
202-293-7841

Society of Motion Picture and Television Engineers (SMPTE)
862 Scarsdale Avenue
Scarsdale, NY 10583
914-472-6606

Society of Professional Journalists, Sigma Delta Chi
35 East Wacker Drive
Chicago, IL 60601
312-236-6577

Society of Women Engineers (SWE)
345 East 47th Street
New York, NY 10017
212-644-7855

Special Libraries Association (SLA)
235 Park Avenue South
New York, NY 10003
212-477-9250

Speech Communication Association (SCA)
5105 Backlick Road, Suite E
Annandale, VA 22003
703-750-0533

Subscription Television Association (STA)
1425 21st Street NW
Washington, DC 20036
202-833-2744

Television Bureau of Advertising (TBA)
1345 Avenue of the Americas
New York, NY 10019
212-397-3456

Television Information Office (TIO)
745 Fifth Avenue
New York, NY 10022
212-759-6800

University Film Association (UFA)
(Formerly University Film Producers Association)
c/o Dr. Charles H. Harpole
Department of Cinema
Southern Illinois University
Carbondale, IL 62901
618-453-2365

Videotape Production Association (VPA)
63 West 83rd Street
New York, NY 10024
212-986-0289

Western Educational Society for Telecommunications (WEST)
c/o Broadcast Services
Ricks College
Rexburg, ID 83440
208-356-2204

Western States Advertising Agencies Association (WSAAA)
5900 Wilshire Boulevard
Los Angeles, CA 90036
213-933-7337

Women in Cable (WIC)
1629 K Street NW, Suite 800
Washington, DC 20006
202-296-4218

Women in Communications (WIC)
Box 9561
Austin, TX 78766
512-345-8922

Writers' Institute
112 West Boston Post Road
Mamaroneck, NY 10543

APPENDIX III—BIBLIOGRAPHY

A Books

The following books and pamphlets should be useful in providing further information on the areas discussed in this book. The publications are grouped according to the sections in Parts I and II, with an additional listing relating to career planning and guidance. In addition to the name and address of the publisher, the year of publication, number of pages, and price are given, when available. Contact the publisher before ordering any title to make sure that it is still in print and to learn its current price.

This bibliography was selected from hundreds of titles in the fields of television, video, and communications, and represents those publications that are most likely to be of interest to the serious reader.

TELEVISION BROADCASTING

Management and Administration

Broadcast Management: Radio, Television
by Ward L. Quaal and James A. Brown
Hastings House
10 East 40th Street
New York, NY 10016
1975, 464 pp, $8.95

Broadcasting Yearbook
Broadcasting Publications
1735 DeSales Street NW
Washington, DC 20036
1982, 1250 pp, $60.00

Case Studies in Broadcast Management
by Howard W. Coleman
Hastings House
10 East 40th Street
New York, NY 10016
1980, 154 pp, $4.95

Introduction to Mass Communications, 5th edition
by Warren K. Agee, Phillip H. Ault, and Edwin Emery
Harper & Row
10 East 53rd Street
New York, NY 10022
1976, 469 pp, $20.95

Mass Communication Law: Cases and Comment, 3rd edition
by Donald M. Gillmor and Jerome A. Barron
West Publishing
170 Old Country Road
Mineola, NY 11501
1979, 1008 pp, $22.95

The New York Times Encyclopedia of Television
by Les Brown
Harper & Row
10 East 53rd Street
New York, NY 10022
1977, 490 pp, $20.00

Television Factbook
Television Digest
1836 Jefferson Place NW
Washington, DC 20006
1982, 1252 and 1420 pp (2 vols), $169.00

TV Facts
by Cobbett Steinberg
Facts on File, Inc.
460 Park Avenue South
New York, NY 10016
1980, 565 pp, $19.95

Programming

The Cool Fire: How to Make it in Television
by Bob Shanks
Random House
201 East 50th Street
New York, NY 10022
1976, 318 pp, $4.95

Television and Radio Announcing, 3rd edition
by Stuart W. Hyde
Houghton Mifflin Co.
Wayside Road
Burlington, MA 01803
1979, 467 pp, $15.95

TV Feature Film Source Book
Broadcast Information Bureau
100 Lafayette Drive
Syosset, NY 11791
1982, 900 pp, $219.00

Production

Artist's Market
edited by Betsy Wones
Writer's Digest Books
9933 Alliance Road
Cincinnati, OH 45242
1978, 583 pp, $10.95

Directing the Television Commercial
by Ben Gradus
Hastings House
10 East 40th Street
New York, NY 10016
1981, 236 pp, $16.95

Making Films Your Business
by Mollie Gregory
Schocken Books
200 Madison Avenue
New York, NY 10016
1979, 256 pp, $6.95

Opportunities in Graphic Communications
by George Reinfeld
VGM Career Horizons
8259 Niles Center Road
Skokie, IL 60077
1981, 160 pp, $4.46

The Technique of Television Production
by Gerald Millerson
Hastings House
10 East 40th Street
New York, NY 10016
1979, 365 pp, $17.50

Television Production
by Alan Wurtzel

McGraw-Hill Book Company
1221 Avenue of the Americas
New York, NY 10020
1979, 624 pp, $23.95

Television Production: Disciplines and Techniques
by Thomas D. Burrows and Donald N. Wood
William C. Brown Co.
2460 Kerper Boulevard
Dubuque, IA 52001
1978, 354 pp, $11.95

Television Production Workbook
by Herbert Zettl
Wadsworth Publishing Co.
10 Davis Drive
Belmont, CA 94002
1977, 200 pp, $5.95

The Television Program: Its Direction and Production
by Edward Stasheff, Rudy Bretz, John Gartley, and Lynn Gartley
Hill and Wang
19 Union Square West
New York, NY 10003
1976, 243 pp, $5.95

Television: The Director's Viewpoint
by John W. Ravage
Westview Press
5500 Central Avenue
Boulder, CO 80301
1978, 184 pp, $16.50

News

The Broadcast News Process
by Frederick Shook and Dan Lattimore
Morton Publishing Company
2700 East Bates Avenue
Denver, CO 80210
1979, 369 pp, $10.95

Newsgathering
by Ken Metzler
Prentice-Hall
Englewood Cliffs, NJ 07632
1979, 375 pp, $12.95

Women in Television News
by Judith S. Gelfman
Columbia University Press
562 West 113th Street
New York, NY 10025
1976, 186 pp, $10.00

Engineering

Engineering Handbook
edited by George Bartlett
National Association of Broadcasters
1771 N Street NW
Washington, DC 20036
1975, 540 pp, $45.00

TV Camera Operation
by Gerald Millerson
Hastings House
10 East 40th Street
New York, NY 10016
1973, 160 pp, $5.95

Advertising

Advertising in the Broadcast Media
by Elizabeth J. Heighton and Don R. Cunningham
Wadsworth Publishing Co.
10 Davis Drive
Belmont, CA 94002
1976, 349 pp, $12.95

Breaking in: A Guide to Creating a Winning Advertising Portfolio for Beginning Art Directors and Copywriters
by Ken Musto
Marie Press
Box 887
Plandome, NY 11030
1981, 64 pp, $6.95

Broadcast Advertising: A Comprehensive Working Textbook
by Sherilyn K. Zeigler and Herbert H. Howard
GRID
4666 Indianola Avenue
Columbus, OH 43214
1978, 341 pp, $14.95

Dictionary of Advertising Terms
Crain Books
740 Rush Street
Chicago, IL 60611
1980, 209 pp, $14.95

Opportunities in Advertising
by Perry C. Groome, Jr.
VGM Career Horizons
8259 Niles Center Road
Skokie, Il 60077
1976, 125 pp, $4.46

The Radio and Television Commercial
by Albert C. Book and Norman D. Cary
Crain Books
740 Rush Street
Chicago, IL 60611
1978, 122 pp, $6.95

Technical/Craft

The Actor's Guide to Breaking into TV Commercials
by Cordland Jessup and S. Lee Alpert
Pilot Books
347 Fifth Avenue
New York, NY 10016
1980, 63 pp, $3.50

The Business of Show Business
by Judith A. Katz
Harper & Row
10 East 53rd Street
New York, NY 10022
1981, 254 pp, $5.25

Exploring Theatre and Media Careers: A Student Guidebook
Superintendent of Documents
U.S. Government Printing Office
Washington, DC 20402
1976, 138 pp, $2.50

How to Break into Motion Pictures, Television Commercials, and Modeling
by Nina Blanchard
Doubleday & Co.
501 Franklin Avenue
Garden City, NY 11530
1978, 240 pp, $8.95

In Front of the Camera
by Bernard Sandler
E. P. Dutton
2 Park Avenue
New York, NY 10016
1981, 180 pp, $13.50

Music Scoring for TV and Motion Pictures
by Marlin Skiles
TAB Books
Monterey and Pinola Avenues
Blue Ridge Summit, PA 17214
1975, 266 pp, $12.95

No Acting Please: A Revolutionary Approach to Acting and Living
by Eric Morris and Joan Hotchkis
Quick Fox, Inc.
33 West 60th Street
New York, NY 10023
1979, 256 pp, $6.95

Opportunities in Acting
by Dick Moore
VGM Career Horizons
8259 Niles Center Road
Skokie, IL 60077
1981, 160 pp, $4.46

The Performer in Mass Media
by William Hawes
Hastings House
10 East 40th Street
New York, NY 10016
1978, 350 pp, $8.95

Theatrical Costume: A Guide to Information Sources
by Jackson Kesler
Gale Research Co.
Book Tower
Detroit, MI 48226
1979, 308 pp, $24.00

TV Acting: A Manual for Camera Performance
by Larry Kirkman, James Hindman, and Elizabeth Monk Daley
Hastings House
10 East 40th Street
New York, NY 10016
1979, 191 pp, $7.85

The TV Scriptwriter's Handbook
by Alfred Brenner
Writer's Digest Books
9933 Alliance Road
Cincinnati, OH 45242
1980, 322 pp, $11.95

Writing for Television and Radio, 3rd edition
by Robert L. Hilliard
Hastings House
10 East 40th Street
New York, NY 10016
1976, 461 pp, $16.50

Writing the Script: A Practical Guide for Films and Television
by Wells Root
Holt, Rinehart & Winston
383 Madison Avenue
New York, NY 10017
1980, 210 pp, $5.95

VIDEO AND TELECOMMUNICATIONS

Consumer Electronics and Home Video

The Home Video Yearbook, 1981-82
edited by Richard Beardsley
Knowledge Industry Publications
701 Westchester Avenue
White Plains, NY 10604
1981, 250 pp, $75.00

The Video Source Book, 3rd edition
The National Video Clearinghouse
100 Lafayette Drive
Syosset, NY 11791
1982, 1400 pp, $95.00

The Video Tape & Disc Guide to Home Entertainment
The National Video Clearinghouse
100 Lafayette Drive
Syosset, NY 11791
1981, 408 pp, $9.95

Cable TV and Related Industries

Arbitron Cable Dictionary
Arbitron Company, Inc.
1350 Avenue of the Americas
New York, NY 10019
1981, 40 pp, $5.00

The Cable/Broadband Communications Book, Volume 2, 1980-1981
edited by Mary Louise Hollowell
Communications Press
1346 Connecticut Avenue NW
Washington, DC 20036
1981, 230 pp, $19.50

The Cable Primer
National Cable Television Association
1724 Massachusetts Avenue NW
Washington, DC 20036
1981, 54 pp, $15.00

Careers in Cable
National Cable Television Association
1724 Massachusetts Avenue NW
Washington, DC 20036
1981, 25 pp, $3.00

CATV Program Origination and Production
by Donald Schiller
Broadcasting Publications, Inc.
1735 DeSales Street NW
Washington, DC 20036
252 pp, $14.95

Designing and Maintaining the CATV and Small TV Studio
by Kenneth Knecht
TAB Books
Monterey and Pinola Avenues
Blue Ridge Summit, PA 17214
1976, 281 pp, $12.95

Future Developments in Telecommunications, 2nd edition
by James Martin
Prentice-Hall
Englewood Cliffs, NJ 07632
1978, 668 pp, $37.50

Education

Educational Media Yearbook 1980
edited by James W. Brown
Libraries Unlimited
PO Box 263
Littleton, CO 80160
1980, 445 pp

Educational Telecommunications
By Donald N. Wood and Donald G. Wylie
Wadsworth Publishing Co.
10 David Drive
Belmont, CA 94002
1977, 370 pp, $9.95

Handbook for Producing Educational and Public-Access Programs for Cable Television
by Rudy Bretz
Educational Technology Publications
140 Sylvan Avenue
Englewood Cliffs, NJ 07632
1976, 160 pp, $15.95

Instructional Media and the Individual Learner
by Erhard U. Heidt
Nichols Publishing Co.
PO Box 96
New York, NY 10024
1978, 180 pp, $17.50

Instructional Television: Status and Directions
edited by Jerrold Ackerman and Lawrence Lipsitz
Educational Technology Publications
140 Sylvan Avenue
Englewood Cliffs, NJ 07632
240 pp, $17.95

Public School, and Academic Media Centers: A Guide to Information Sources
edited by Esther R. Dyer and Pam Berger
Gale Research Co.
Book Tower
Detroit, MI 48226
1981, 237 pp, $38.00

Video in the Classroom: A Guide to Creative Television
by Don Kaplan
Knowledge Industry Publications
701 Westchester Avenue
White Plains, NY 10604
1980, 161 pp, $17.50

Private Industry

How to Write, Direct & Produce Effective Business Films and Documentaries
by Jerry McGuire
Broadcasting Publications, Inc.
1735 DeSales Street NW
Washington, DC 20036
280 pp, $14.95

Managing the Corporate Media Center
by Eugene Marlow
Knowledge Industry Publications
701 Westchester Avenue
White Plains, NY 10604
1981, $24.95

Private Television Communications: Into the Eighties
by Judith M. and Douglas P. Brush
International Television Association
136 Sherman Avenue
Berkeley Heights, NJ 07922
1981, 204 pp, $39.95

Video in the 80s
by Paula Dranov, Louise Moore, and Adrienne Hickey
Knowledge Industry Publications
701 Westchester Avenue
White Plains, NY 10604
1980, 186 pp, $34.95

Health

A Television Primer for Health and Welfare Public Relations
by Roy Danish
Television Information Office
745 Fifth Avenue
New York, NY 10022
1964, 7 pp, $.25

Video in Health
edited by George Van Son
Knowledge Industry Publications
701 Westchester Avenue
White Plains, NY 10604
1981, 200 pp, $29.95

Career Planning and Guidance

Careers in Communication
Association for Communication Administration
5105 Backlick Road
Annandale, VA 22003
1980, 16 pp, $.50

Careers in Film and Television (Factfile #2)
American Film Institute
The John F. Kennedy Center
Washington, DC 20566
1977, $3.00

Careers in Television
(List of publications and articles)
Television Information Office
745 Fifth Avenue
New York, NY 10022
1975, 3 pp, $.25

Communications-Related Occupations
Superintendent of Documents
U.S. Government Printing Office
Washington, DC 20402
1979, 20 pp, $.35

Guide to College Courses in Film and Television
American Film Institute
c/o Peterson's Guides
PO Box 2123
Princeton, NJ 08540
1982, 430 pp, $12.75

The Job Hunt
by Robert B. Nelson
Pragmatic Publications
1233 Ashland Avenue
St. Paul, MN 55104
1981, 64 pp, $2.95

1981 Internships
edited by Kirk Polking and Colleen Cannon
Writer's Digest Books
9933 Alliance Road
Cincinnati, OH 45242
1981, 302 pp, $7.95

Opportunities in Broadcasting
by Elmo Ellis
VGM Career Horizons
8259 Niles Center Road
Skokie, IL 60076
1981, 149 pp, $6.50

TV and Radio Careers
by D.X. Fenten
Franklin Watts, Inc.
730 Fifth Avenue
New York, NY 10019
1976, 65 pp, $4.47

What Color is Your Parachute?
by Richard Nelson Bolles
Ten Speed Press
Box 7123
Berkeley, CA 94707
1980, 328 pp, $6.95

Women in Communications
by Alice Fins
VGM Career Horizons
8259 Niles Center Road
Skokie, IL 60077
1979, 160 pp, $4.46

Women on the Job: Careers in Broadcasting
American Women in Radio and Television
1321 Connecticut Avenue NW
Washington, DC 20036
26 pp, free

Your Future in Television Careers
by David W. Berlyn
Richards Rosen Press
29 East 21st Street
New York, NY 10010
1978, 114 pp, $4.98

B Periodicals

The following list of important periodicals includes magazines, newsletters, newspapers, and membership reports of unions and associations. They provide information and news about various aspects of television, video, and telecommunications. Although the list conforms to the major occupational sections used in this book, many of the publications cover more than one field.

The name, publisher, and address of each periodical is included below as well as the type and frequency of publication, price, and a brief description of its focus.

Although not complete, this list represents a core group of professional periodicals. There are a number of consumer-oriented publications (especially pertaining to home video) that can be found on most newsstands.

TELEVISION BROADCASTING

Management and Administration

BM/E (Broadcast Management/ Engineering)
Broadband Information Services
295 Madison Avenue
New York, NY 10017
magazine, monthly, $18/year
articles on television and radio management and engineering

Broadcasting
Broadcasting Publications, Inc.
1735 DeSales Street NW
Washington, DC 20036
magazine, weekly, $35/year
news for television and radio management with extensive job listings for all TV positions

Journal of Broadcasting
Broadcast Education Association
1771 N Street NW
Washington, DC 20036
magazine, quarterly, $17.50/year
television trends and research for professionals and management

Television Digest
1836 Jefferson Place NW
Washington, DC 20036
newsletter, weekly, $323/year
news of interest to management and programmers in broadcast television

Television/Radio Age
1270 Avenue of the Americas
New York, NY 10019
magazine, bi-weekly, $40/year
articles for TV, radio, and cable TV management

Programming

AIVF Newsletter
Association of Independent Video and Filmmakers
625 Broadway
New York, NY 10012
newsletter, monthly, $25/year
for independent film and video artists; includes information on jobs and internships

American Film
The American Film Institute
The John F. Kennedy Center for the Performing Arts
Washington, DC 20566
magazine, 10 times/year, $20/year
general articles on commercial motion pictures and home video

Facts, Figures & Film
Broadcast Information Bureau
100 Lafayette Drive
Syosset, NY 11791
magazine, monthly, $52/year
for television programming and management executives with an emphasis on syndicated programming

The Hollywood Reporter
PO Box 1431
Hollywood, CA 90028
magazine, daily, $48/year
trade news of the film and entertainment industry; includes job listings

Public Relations Journal
Public Relations Society of America
845 Third Avenue
New York, NY 10022
magazine, monthly, $24/year
articles of interest to the public relations professional

Television Quarterly
National Academy of Television Arts and Sciences
110 West 57th Street
New York, NY 10019
magazine, quarterly, $10/year
articles on all facets of production

Production

The Animator
Northwest Film Study Center
Portland Art Museum
1219 SW Park Avenue
Portland, OR 97205
magazine, quarterly, $3/year
articles on film and video activities in the northwestern United States; includes job listings

Art Direction Magazine
10 East 39th Street
New York, NY 10016
magazine, monthly, $16/year
articles for the graphics professional

Graphics, USA
120 East 56th Street
New York, NY 10022
magazine, monthly, $15/year
articles of interest to art directors and graphic artists

Industrial Photography
United Business Publications
475 Park Avenue South
New York, NY 10016
magazine, monthly, $13.50/year
covers commercial film production and photography

Lighting Dimensions
31706 South Coast Highway
South Laguna, CA 92677
magazine, 7 times/year, $18/year
forum for lighting directors; occasional job listings

News

RTNDA Communicator
Radio-Television News Directors Association
1735 DeSales Street NW
Washington, DC 20036
magazine, monthly, $15/year

articles about the broadcast news profession

Engineering

SMPTE Journal
Society of Motion Picture and Television Engineers
862 Scarsdale Avenue
Scarsdale, NY 10583
magazine, monthly, $35/year
articles reviewing technology and developments in broadcast engineering

Advertising

Advertising Age
Crain Communications
740 Rush Street
Chicago, IL 60611
magazine, weekly, $40/year
trade news about all types of advertising; extensive job listings

Journal of Advertising Research
Advertising Research Foundation
3 East 54th Street
New York, NY 10022
magazine, bi-monthly, $50/year
discusses research for the advertising community with insights into new ideas and developments

Marketing and Media Decisions
342 Madison Avenue
New York, NY 10017
magazine, monthly, $36/year
emphasizes advertising and marketing pertaining to media

Sight & Sound Marketing
Drorbaugh Publications, Inc.
51 East 42nd Street
New York, NY 10017
magazine, monthly, free
discusses various television and video products; for professionals in advertising and marketing

Technical/Craft

Backstage
Backstage Publications
330 West 42nd Street
New York, NY 10036
newspaper, weekly, $32/year
theater news, casting calls, listings of specialty schools in performing, crafts, and commercial and video production

Billboard
Billboard Publications
1515 Broadway
New York, NY 10036
newspaper, weekly, $110/year
emphasis on the music industry with a section on home video

Daily Variety
1400 North Cahuenga Boulevard
Hollywood, CA 90028
newspaper, daily, $50/year
news of the entertainment industry; some job listings

IATSE Official Bulletin
International Alliance of Theatrical Stage Employees
1515 Broadway
New York, NY 10036
magazine, quarterly, $2/year
union news and reports; production articles; agent listings

Newsletter
Writers Guild of America West
8955 Beverly Boulevard
Los Angeles, CA 90048
newsletter, 10 times/year, $18/year to non-members
opinion pieces by writers and reports on Guild activities

Screen Actor
Screen Actors Guild
7750 Sunset Boulevard
Los Angeles, CA 90046
newsletter bi-monthly/magazine bi-annual, $4/year
reports to Guild members; agent listings

Show Business News
134 West 44th Street
New York, NY 10036
newspaper, weekly, $30/year
casting news and production information; agent listings

Theatre Crafts
Rodale Press
250 West 57th Street
New York, NY 10107
magazine, 9 times/year, $15.75/year
articles on theater and television design, staging, and production

Variety
154 West 46th Street
New York, NY 10036
newspaper, weekly, $50/year
news of television, home video, films, and theater; some job listings

VIDEO AND TELECOMMUNICATIONS

Consumer Electronics and Home Video

Dealerscope
115 Second Avenue
Waltham, MA 02154
magazine, monthly, free
news on consumer electronics and home video retailing

Home Video Report
Knowledge Industry Publications
701 Westchester Avenue
White Plains, NY 10604
newsletter, monthly, $225/year
news about home video, cable TV, STV, and MDS

ITA News Digest
International Tape/Disc Association
10 Columbus Circle
New York, NY 10019
magazine, monthly, free
news and articles on the home video and audio industries

Video Marketing Newsletter
1680 Vine Street
Hollywood, CA 90028
newsletter, semi-monthly, $188/year
news of interest to the home video retailer

The Video Programs Retailer
National Video Clearinghouse
100 Lafayette Drive
Syosset, NY 11791
magazine, monthly, free
trade news for the video retail manager

Video Trade News
C.S. Tepfer Publishing Co.
51 Sugar Hollow Road
Danbury, CT 06810
newspaper, monthly, $8/year
news of retailing practices and equipment; some job listings

Video Week
c/o Television Digest
510 Madison Avenue
New York, NY 10022
newsletter, weekly, $327/year
trade news for the home video professional

Videonews
Phillips Publishing
7315 Wisconsin Avenue

Bethesda, MD 20814
newsletter, bi-weekly, $167/year
news on home video and new technology

The Videoplay Report
C.S. Tepfer Publishing Co.
51 Sugar Hollow Road
Danbury, CT 06810
newsletter, bi-weekly, $75/year
trade news of home video, education and business technology, products and programs

Cable TV, STV, and MDS

Cable News
Phillips Publishing
7315 Wisconsin Avenue
Bethesda, MD 20814
newsletter, bi-weekly, $167/year
news of cable TV and related industries

Cablevision
Titsch Publishing, Inc.
2500 Curtis Street
Denver, CO 80205
magazine, weekly, $54/year
articles on the cable industries; job listings

CTIC CableReports
Cable Television Information Center
1800 North Kent Street
Arlington, VA 22209
newsletter, monthly, $165/year
reports on new technology and public policy issues

The Videocassette & CATV Newsletter
Martin Roberts & Associates
PO Box 5254N
Beverly Hills, CA 90201
newsletter, monthly, $54/year
trade news on cable TV, new technology, and home video retail distribution

View
150 East 58th Street
New York, NY 10155
magazine, monthly, $19.95/year
articles and trade news about cable television programming

Education

AFI Education Newsletter
National Education Services
The American Film Institute
The John F. Kennedy Center for the Performing Arts
Washington, DC 20566
newsletter, bi-monthly, free
information for media teachers

Educational Technology
140 Sylvan Avenue
Englewood Cliffs, NJ 07632
magazine, monthly, $49/year
articles on media, instructional design, and A-V methods for educators

ETV Newsletter
C.S. Tepfer Publishing Co.
51 Sugar Hollow Road
Danbury, CT 06810
newsletter, bi-weekly, $75/year
news for educational communicators; job listings

Instructional Innovator
Association for Educational Communications and Technology
1126 16th Street NW
Washington, DC 20036
magazine, 9 times/year, $18/year to non-members
news and articles on educational media

Journal of Instructional Development
Association for Educational Communications and Technology
1126 16th St NW
Washington, DC 20036
journal, quarterly, $24/year to non-members
research studies and reports for professionals in instructional development

Media and Methods
North American Publishing Co.
401 North Broad Street
Philadelphia, PA 19108
magazine, 9 times/year, $18/year
articles on educational media; job listings

Sightlines
Educational Film Library Association
43 West 61st Street
New York, NY 10023
magazine, quarterly, $15/year to non-members
information for educational, library, and community film and video users

Private Industry

Audio-Visual Communications
United Business Publications
475 Park Avenue South
New York, NY 10016
magazine, monthly, $13.50/year
information on applications of media for business and education; new product news

Audio Visual Directions
Montage Publishing Co.
5173 Overland Avenue
Culver City, CA 90230
magazine, 7 times/year, $15/year
news and articles on training and equipment for education and business

Educational and Industrial Television
C.S. Tepfer Publishing Co.
51 Sugar Hollow Road
Danbury, CT 06810
magazine, monthly, $15/year
articles on instructional and corporate telecommunications

Training
Lakewood Publications
731 Hennepin Avenue
Minneapolis, MN 55403
magazine, monthly, $24/year
information for training directors and communicators in private industry

Training and Development Journal
American Society for Training and Development
PO Box 5307, 6414 Adana Road
Madison, WI 53705
journal, monthly, $30/year to non-members
information and reports for communications personnel in industry

Videography
United Business Publications
475 Park Avenue South
New York, NY 10016
magazine, monthly, $12/year
articles about media production in business and education

VU Marketplace
Knowledge Industry Publications
701 Westchester Avenue
White Plains, NY 10604
newspaper, bi-monthly, $20/year
information on institutional and business use of video hardware and software; some job listings

Health

Biomedical Communications
United Business Publications
475 Park Avenue South
New York, NY 10016
magazine, bi-monthly, $10/year
news and articles for communicators and educators in hospitals and health care schools

INDEX

This index includes all of the job titles discussed in the book. It also lists the alternate titles mentioned in the Career Profiles and cross references them to their appropriate positions. When a career and alternate title are the same, the section of the book in which each appears is noted in parentheses.